Essentials Version

Anatomy and Physiology
Laboratory Textbook

Essentials Version

Anatomy and Physiology
Laboratory Textbook

Stanley E. Gunstream
Pasadena City College

Harold J. Benson
Pasadena City College

Arthur Talaro
Pasadena City College

Kathleen P. Talaro
Pasadena City College

Wm. C. Brown Publishers
Dubuque, Iowa

Book Team

Editor *Edward G. Jaffe*
Developmental Editor *Lynne M. Meyers*
Designer *K. Wayne Harms*
Art Editor *Barbara J. Grantham*
Production Editor *Kay J. Brimeyer*
Visuals Processor *Joseph P. O'Connell*

wcb

Chairman of the Board *Wm. C. Brown*
President and Chief Executive Officer *Mark C. Falb*

wcb

Wm. C. Brown Publishers, College Division

President *G. Franklin Lewis*
Vice President, Editor-in-Chief *George Wm. Bergquist*
Vice President, Director of Production *Beverly Kolz*
Vice President, National Sales Manager *Bob McLaughlin*
Director of Marketing *Thomas E. Doran*
Marketing Communications Manager *Edward Bartell*
Marketing Manager *Craig S. Marty*
Executive Editor *Edward G. Jaffe*
Manager of Visuals and Design *Faye M. Schilling*
Production Editorial Manager *Colleen A. Yonda*
Production Editorial Manager *Julie A. Kennedy*
Publishing Services Manager *Karen J. Slaght*

Anatomy and Physiology Laboratory Textbooks

Benson, Gunstream, Talaro, Tararo	Complete Version—Cat, 4th edition Intermediate Version—Cat, 2nd edition Intermediate Version—Fetal Pig, 1st edition Short Version, 4th edition
Gunstream, Benson, Talaro, Talaro	Essentials Version

Library of Congress Catalog Card Number: 88-70779

ISBN 0-697-03153-5

Printed in the United States of America by Wm. C. Brown Publishers
2460 Kerper Boulevard, Dubuque, IA 52001

10 9 8 7 6 5 4 3

Contents

v

Preface

The primary objective of this *Essentials Version* of the *Anatomy and Physiology Laboratory Textbook* is to present the fundamentals of human anatomy and physiology in a manner that is appropriate for students in allied health programs such as practical nursing, radiologic technology, medical assisting and dental assisting. These students usually take a one semester course in human anatomy and physiology, and need a laboratory text that provides coverage of the fundamentals without the clutter of excessive details and unneeded terminology. While maintaining many of the strengths of the *Short Version,* this *Essentials Version* is shorter, more concise and less rigorous to better serve these students. This brief laboratory text may be accompanied by a textbook that provides more in-depth coverage of topics if that is desired by the instructor.

Each of the thirty-seven exercises begins with a short list of objectives that outline the minimal learning responsibilities for the students. The exercises are basically self-directing, which minimizes the need for lengthy introductions by the instructor.

Each exercise topic is discussed in a simple, concise style that facilitates student learning. The major topics of human anatomy and physiology are covered with a minimum of detail. The necessary key terms are in bold print to aid students in building a vocabulary of anatomical and physiological terms. There are numerous illustrations that are correlated with the text to aid student understanding. Although some clinical relationships are woven into the text, color-coded boxes are inserted to call attention to selected clinical aspects.

Laboratory procedures are distinct from and follow the discussion of the exercise topic. They are presented in a concise, step-wise manner to guide the student. Activities consist of: (1) labeling illustrations, (2) dissections, (3) study of specimens and models, (4) physiological experiments, and (5) microscopic studies. A major dissection specimen, such as the cat or fetal pig, is not included. Instead, freshly killed rats, and beef or sheep parts are used in some exercises, and the study of anatomical models is stressed.

The pedagogical design calls for students to develop an understanding of each exercise topic by (1) labeling the illustrations using information presented in the text and (2) completing corresponding portions of the laboratory report. Where appropriate, students are asked to color code anatomical parts to reinforce learning. Usually, students should complete these activities *before* proceeding with the other laboratory studies.

The *Laboratory Reports* are designed to guide and reinforce student learning and provide a convenient site for recording data and making diagrams of observations. They are located at the back of the book to better control the pagination and to keep the book brief. They are easy to remove and are three-hole punched to fit a standard notebook. The corresponding laboratory report should be removed from the back of the book whenever students *begin* working on an exercise. This prevents page flipping and facilitates completing the laboratory report. Completed laboratory reports should be kept in the student's notebook. Instructors may choose either to collect and grade the laboratory reports or to post the answer keys to allow students to check their own work. The design of the laboratory reports makes them easy to grade.

Students in the target group typically have difficulty in learning the necessary anatomical and physiological terminology, especially the correct spellings. Experience has shown that students learn key terms and their correct spellings more easily if they write the terms. Thus, the laboratory reports are structured so that students are required to write the names of key terms rather than simply provide a matching letter or number.

Microscopic study is another difficult area for these students since they have trouble locating histological structures on slides when using diagrams as a reference. This problem is largely overcome by the inclusion of full-color photomicrographs of common tissues and of selected histological subjects. These are found in the Histology Atlas and in Exercise 36, The Endocrine Glands.

Instructors will find that the listing of required equipment and materials on the first page of each exercise facilitates laboratory preparation. The exercises use standard equipment and materials that are usually available in most biology departments. The only exercise that requires electronic equipment is Exercise 20, The Physiology of Muscle Contraction, which uses the Physiograph.

The accompanying *Instructor's Manual* provides additional aid for the instructor: (1) a composite list of equipment and supplies, (2) operational suggestions, and (3) answer keys for the laboratory reports.

Adopters are encouraged to contact the authors regarding any problems and to submit constructive suggestions that will improve future editions.

We would like to acknowledge the contributions of Bill South and Gary Townsend of Narco Bio-Systems, Houston, Texas. They graciously provided helpful photographs and granted permission for the use of the sample Physiograph® recordings shown in Appendix B.

We also wish to recognize the helpful suggestions provided by the following reviewers: Louise Squitieri, Bronx Community College; Martha Leonard, Midlands Technical College; Ron Hackney, Volunteer State Community College; Robert Thomas, College of the Redwoods; Cynthia A. Bottrell, Scott Community College; Robert E. Nabors, Tarrant County Junior College; Ernest Joe Harber, San Antonio College; Glenn Yoshida, Los Angeles Southwest College.

To the Student

This laboratory textbook has been developed to help you master the fundamentals of human anatomy and physiology, which provide the basis of the health-related professions. An understanding of anatomical structures, physiological processes, terminology, and techniques will give you the basic knowledge that is essential for success in your chosen field.

Each exercise begins with a list of learning objectives to guide your study. The sequence of activities in each exercise is established to facilitate the development of your understanding. Follow the sequence; don't skip around.

The activities consist of: (1) labeling illustrations, (2) dissections, (3) study of specimens and models, (4) physiological experiments, and (5) microscopic study. Follow the directions for each activity with care to enhance your success in the laboratory. Work carefully and thoughtfully. Remember, the objective is to learn the material, not just to complete the exercise.

Each exercise has a *Laboratory Report,* located in the back of the book, that you are to complete as you work through the exercise. Remove it from the book when you *start* the exercise so that you can complete it without page flipping. The laboratory reports are three-hole punched so you can keep the completed laboratory reports in your notebook for future reference.

Labeling Illustrations

Most of the exercises are designed so that you will learn the text of the exercise by labeling the illustrations. All structures to be labeled are described in the text. Correctly labeled illustrations are then used as references when examining specimens and models. Use them for study and review purposes. Usually, the illustrations should be labeled *prior* to coming to the laboratory session.

Dissections

The dissections will include freshly killed rats, and animal parts obtained from a slaughterhouse. The purpose of dissection is to expose anatomical parts for observation and study. Strive to cut as little as possible to achieve this goal. Use the scalpel sparingly. Most dissections can be accomplished with scissors, forceps and probe.

Experiments

Before performing any experiment, read the directions completely through so that you have a good understanding of the experiment. Be sure that you have all of the equipment and materials required to complete the experiment before starting. Then, carefully follow the directions.

Microscopic Study

An examination of prepared slides enables you to visualize and understand the cytological or histological structure of specimens. Correlate your observations with the text, diagrams and photomicrographs. If drawings are required, make them with care and label the structures. In this way, you will understand the microscopic structures more quickly and better prepare yourself for lab practicums.

Laboratory Reports

Complete the laboratory reports independently to maximize your understanding. The laboratory reports are purposely designed so that you must write the key terms involved in the exercise, sometimes more than once. This is done because writing the terms enables you to learn them more easily, especially their correct spellings.

General Operations

Success in the laboratory can be increased by following a few simple guidelines.

1. Be on time to class so that you will hear the comments and directions given by your instructor.
2. Follow your instructor's directions. Take notes on any changes in the equipment, materials, or procedures.
3. Keep your work area free of clutter. Extraneous items should be located elsewhere during the laboratory session.
4. Bring your textbook to the laboratory for reference.
5. Use equipment and materials with care. Report any problems or accidents to your instructor immediately.
6. Work independently, but be cooperative and helpful in team assignments.
7. Do not eat, drink, or smoke in the laboratory.

Essentials Version

Anatomy and Physiology

Laboratory
Textbook

Introduction to Human Anatomy

Objectives

After completing this exercise, you should be able to:

1. Correctly use directional and regional terms in describing the location of anatomical structures.
2. Identify the external body regions and body cavities on charts or a torso model.
3. Contrast the types of body sections used in anatomical studies.
4. Describe the organs and membranes of the body cavities.
5. Define all terms in bold print.

Materials

Human torso model with removable organs

Anatomical description requires the use of specific terminology that communicates precise meaning. In this exercise, you will be introduced to the anatomical terminology that is used to describe: (1) relative positions of structures, (2) body sections or planes, (3) external features and body regions, and (4) body cavities and their membranes.

To communicate effectively about human anatomy, the body must always be considered in a standard position, the **anatomical position.** This position is shown in Figures 1.1 and 1.2. Note that the subject is erect with arms at the sides and with the palms of the hands facing forward.

Oral and written communications among health-care personnel rely on anatomical terminology for precise meaning about body parts and positions.

Directional Terms

The relative position of body structures is communicated through the use of **directional terms.** These terms typically occur as pairs with the members of each pair having

Table 1.1. Directional Terms.

Term	Meaning
Anterior (ventral)	Toward the front or abdominal surface of the body.
Posterior (dorsal)	Toward the back of the body.
Superior (cephalad)	Toward the head.
Inferior (caudad)	Away from the head.
Medial	Nearer the midline of the body.
Lateral	Farther from the midline of the body.
Proximal	Nearer the attachment of an extremity to the trunk.
Distal	Farther from the attachment of an extremity to the trunk.
External (superficial)	Toward or on the body surface.
Internal (deep)	Away from the body surface.
Parietal	Pertaining to the outer boundary of body cavities.
Visceral	Pertaining to the internal organs.

opposite meanings. Most of the terms may apply to an organ as well as to the body as a whole. Commonly used directional terms are shown in Table 1.1.

Body Sections

To observe the relative positions of internal structures, it is necessary to view them in sections that have been cut through the body. These sections or planes are applicable to both the whole body and to individual organs. See Figure 1.1.

Sagittal sections are parallel to the longitudinal (long) axis of the body and divide the body into left and right portions. A **midsagittal section** is made along the **median line** and divides the body into *equal* left and right halves.

Frontal (coronal) sections divide the body into anterior and posterior portions and are perpendicular to sagittal sections but parallel to the longitudinal axis.

Transverse (horizontal) sections divide the body into superior and inferior portions and are perpendicular to the longitudinal axis of the body.

Figure 1.1 Body sections and surfaces.

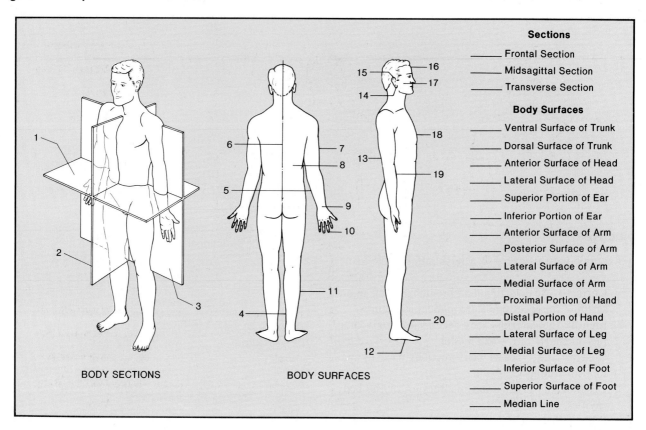

Sections
_____ Frontal Section
_____ Midsagittal Section
_____ Transverse Section

Body Surfaces
_____ Ventral Surface of Trunk
_____ Dorsal Surface of Trunk
_____ Anterior Surface of Head
_____ Lateral Surface of Head
_____ Superior Portion of Ear
_____ Inferior Portion of Ear
_____ Anterior Surface of Arm
_____ Posterior Surface of Arm
_____ Lateral Surface of Arm
_____ Medial Surface of Arm
_____ Proximal Portion of Hand
_____ Distal Portion of Hand
_____ Lateral Surface of Leg
_____ Medial Surface of Leg
_____ Inferior Surface of Foot
_____ Superior Surface of Foot
_____ Median Line

BODY SECTIONS   BODY SURFACES

Assignment

Label Figure 1.1 by placing the correct numbers in front of the labels listed.

Regional Terminology

Various terms have been applied to specific regions of the body to facilitate localization. Refer to Figures 1.2 and 1.3 as you study this section on terminology.

Trunk

The anterior surface of the trunk may be subdivided into two pectoral, two groin and the abdominal regions. The upper chest area is designated as the **pectoral** or **mammary region.** The anterior trunk that is not covered by the ribs is the **abdominal region.** The depressed area where the thigh joins the abdomen is the **groin.**

The **dorsum** or posterior surface of the trunk may be subdivided into the costal, lumbar, and gluteal regions. The **costal region** is that portion over the ribs. The lower back between the ribs and hips is the **lumbar** or **loin region.** The **gluteal region** refers to the **buttocks,** rounded eminences of the rump. The lateral surface of the trunk adjacent to the lumbar region is the **flank.** The armpit area is known as the **axilla.**

Upper Extremity

The **shoulder** is formed by the attachment of the upper extremity to the trunk. The upper arm is called the **brachium** while the forearm is the **antebrachium.** The elbow area on the posterior surface of the arm is the **cubital area.** The surface on the opposite side of the elbow is the **antecubital area.**

Lower Extremity

The upper part of the lower extremity is the **thigh,** and the lower portion between the knee and ankle is called the **leg.** The anterior surface of the knee is the **patellar area** while the posterior surface of the knee is the **popliteal region.** The sole of the foot is the **plantar surface.**

Abdominal Divisions

Physicians and nurses divide the abdominal surface into nine areas to aid in the location of underlying internal organs. Note in Figure 1.3 that these nine areas are formed by the intersection of two vertical and two transverse planes.

The **umbilical region** is the central area that includes the umbilicus. On each side of the umbilical region are the left and right **lumbar regions.** Just above the umbilical region is the **epigastric region** that is between the left and

Figure 1.2 Regional terminology.

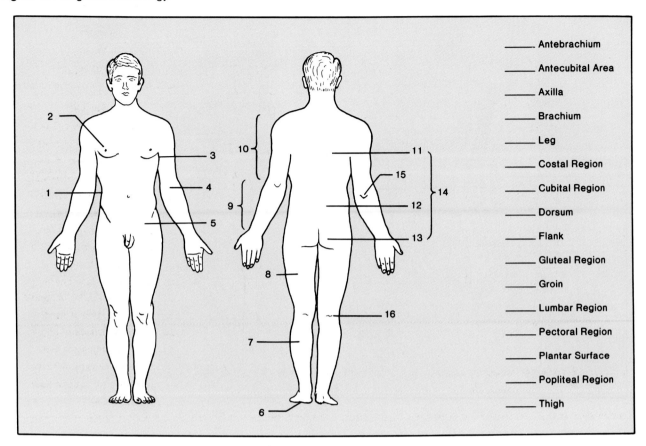

_____	Antebrachium
_____	Antecubital Area
_____	Axilla
_____	Brachium
_____	Leg
_____	Costal Region
_____	Cubital Region
_____	Dorsum
_____	Flank
_____	Gluteal Region
_____	Groin
_____	Lumbar Region
_____	Pectoral Region
_____	Plantar Surface
_____	Popliteal Region
_____	Thigh

Figure 1.3 Abdominal regions.

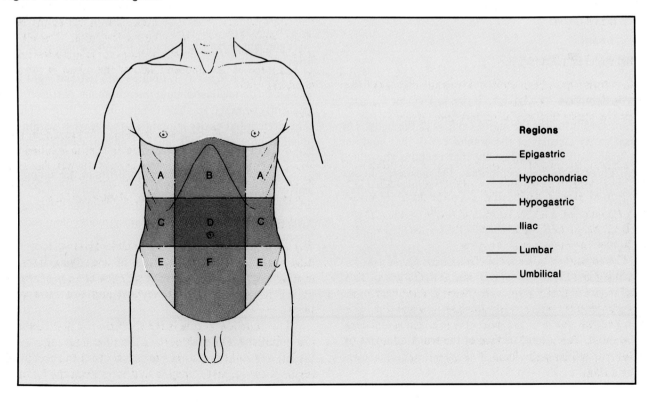

Regions

_____ Epigastric

_____ Hypochondriac

_____ Hypogastric

_____ Iliac

_____ Lumbar

_____ Umbilical

Figure 1.4 Body cavities.

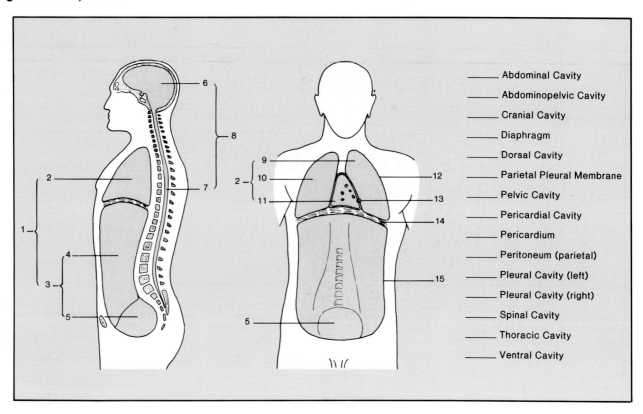

Labels to the right of figure:

— Abdominal Cavity
— Abdominopelvic Cavity
— Cranial Cavity
— Diaphragm
— Dorsal Cavity
— Parietal Pleural Membrane
— Pelvic Cavity
— Pericardial Cavity
— Pericardium
— Peritoneum (parietal)
— Pleural Cavity (left)
— Pleural Cavity (right)
— Spinal Cavity
— Thoracic Cavity
— Ventral Cavity

right **hypochondriac regions.** Below the umbilical region is the **hypogastric,** or **pubic, region** that is between the left and right **iliac regions.**

Assignment

Label Figures 1.2 and 1.3.

Body Cavities

Figure 1.4 illustrates the major cavities of the body. The **dorsal cavity** consists of the **cranial cavity** that contains the brain, and the **spinal cavity** that contains the spinal cord. Three protective membranes, the **meninges,** lie between the brain and spinal cord and the wall of the dorsal cavity.

The **ventral cavity** consists of the **thoracic cavity,** which houses the lungs, heart, and other thoracic organs, and the **abdominopelvic cavity,** which contains most of the internal organs. A thin, dome-shaped sheet of muscle, the **diaphragm,** separates the thoracic and abdominopelvic cavities.

The thoracic cavity is divided into left and right portions by the **mediastinum,** a membranous partition that contains the heart, trachea, esophagus, and thymus gland. The **pericardial cavity** containing the heart is within the mediastinum. The right and left **pleural cavities** contain the lungs.

The abdominopelvic cavity consists of the **abdominal cavity** that contains the stomach, intestines, liver, gall bladder, pancreas, spleen, and kidneys, and the **pelvic cavity** that contains the urinary bladder, sigmoid colon, rectum, uterus, ovaries, and oviducts.

Figure 1.5 Transverse section through thorax.

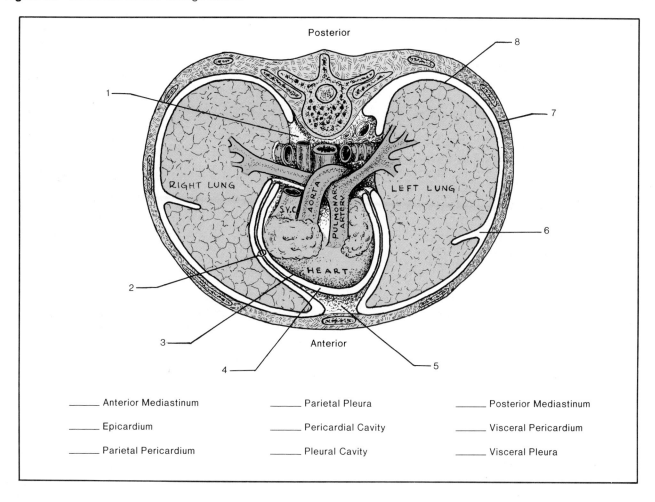

_____ Anterior Mediastinum	_____ Parietal Pleura	_____ Posterior Mediastinum
_____ Epicardium	_____ Pericardial Cavity	_____ Visceral Pericardium
_____ Parietal Pericardium	_____ Pleural Cavity	_____ Visceral Pleura

Membranes of the Ventral Cavity

The ventral cavity is lined with **serous membranes** that provide a smooth, moist, friction-reducing surface for the enclosed internal organs.

Thoracic Cavity Membranes

The walls of the pleural cavities are lined by the **parietal pleurae,** while the lungs are covered with the **visceral (pulmonary) pleurae.** Note in Figure 1.5 that the pleurae are continuous with the mediastinum. The potential space between the visceral and parietal pleurae is the **pleural cavity.**

The heart is tightly enveloped by the **visceral pericardium,** or **epicardium,** a serous membrane, and is surrounded by the double-layered **parietal pericardium.** The inner layer of the parietal pericardium is a serous membrane, but the outer one is fibrous and very strong. The pericardial cavity is the potential space between the visceral and parietal pericardia.

Figure 1.6 Transverse section through the abdominal cavity.

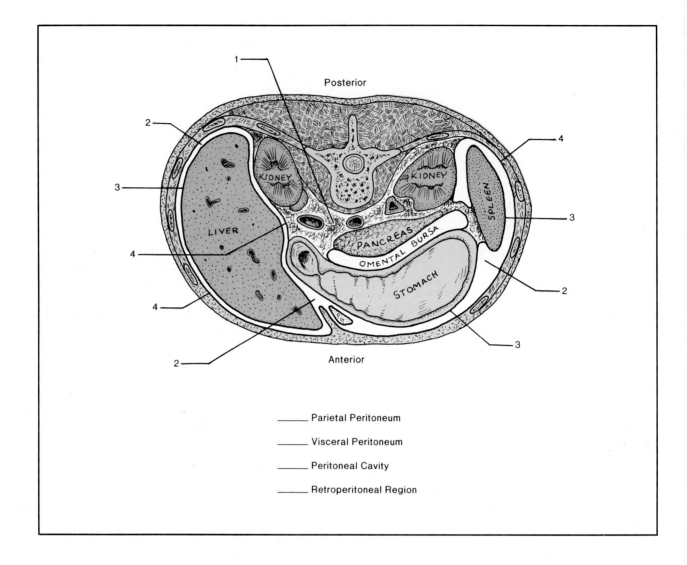

_____ Parietal Peritoneum

_____ Visceral Peritoneum

_____ Peritoneal Cavity

_____ Retroperitoneal Region

Abdominal Cavity Membranes

The **peritoneum** lines the abdominal cavity. It extends inferiorly only to cover the upper portion of the urinary bladder and does not line the pelvic cavity. The double-layered folds of the peritoneum are called **mesenteries,** and they provide support for the abdominal organs.

The **parietal peritoneum** lines the inner abdominal wall, while the **visceral peritoneum** covers the surface of the abdominal organs. The potential space between the visceral and parietal peritoneum is the **peritoneal cavity.**

See Figure 1.6. The kidneys are located in the **retroperitoneal region,** and only part of their surfaces are covered with the parietal peritoneum.

Assignment

1. Label Figures 1.4, 1.5, and 1.6.
2. Complete Section A of the laboratory report.
3. Using a human torso model: (a) locate the surface features and body cavities, (b) identify the major organs of each cavity, and (c) locate the organs within each of the nine abdominal areas.
4. Complete the laboratory report.

Body Organization

Objectives

After completing this exercise, you should be able to:

1. List the organ systems and describe their general functions.
2. List the component organs for each organ system.
3. Locate and identify the organs in the ventral cavity of the rat.
4. Define all terms in bold print.

Materials

Freshly killed rat
Dissecting pan with wax bottom
Dissecting instruments and pins

The functions of the body are performed by specific organs working in a coordinated fashion. An **organ** is a structure that has a definite form and specific function, and is composed of two or more tissues. The heart, for example, is an organ composed of muscle, epithelial and other tissues. An **organ system** is a group of organs functioning together to perform specific functions. The circulatory system is an example of an organ system.

In this exercise, you will study a summary of the components and general functions of the ten organ systems of the body in order to gain an overview of body organization and function. Then, you will dissect a freshly killed rat to observe the mammalian body organization, body cavities and organs of the ventral cavity.

The Integumentary System

The skin, including hair, nails, associated glands and sensory receptors, constitutes the integumentary system. The **skin** consists of two layers, an outer **epidermis** and an inner **dermis,** which protect the underlying tissues from mild abrasions, excessive water loss, microorganisms, and ultraviolet radiation. Perspiration secreted by sweat glands contains water and waste materials similar to dilute urine. The evaporation of perspiration cools the body surface.

The Skeletal System

The skeletal system forms the framework of the body, and provides support and protection for softer organs and tissues. It consists of **bones, cartilages,** and **ligaments.** In conjunction with skeletal muscles, the skeleton forms lever systems that enable movement. In addition, **red bone marrow** produces blood cells.

The Muscular System

The contraction of muscles provides the force that enables movement. **Skeletal muscles** are attached to bones by **tendons** and constitute nearly half of the body weight.

The Nervous System

The nervous system is a complex, highly-organized system consisting of the **brain, spinal cord, cranial** and **spinal nerves,** and **sensory receptors.** These components work together to enable rapid perception and interpretation of the environment, and to coordinate responses to stimuli. The human brain is responsible for the intelligence, will, self-awareness, and emotions characteristic of humans.

The Circulatory System

The circulatory system is often subdivided into the cardiovascular and lymphatic systems.

The Cardiovascular System

The **heart, arteries, veins, capillaries, spleen,** and **blood** constitute the cardiovascular system. These components work together to transport materials such as oxygen, carbon dioxide, nutrients, wastes, and hormones throughout the body. Contractions of the heart circulate the blood through the blood vessels. Blood, consisting of cells and plasma, is the transporting agent and also provides the primary defense against disease organisms. The spleen serves as a blood reservoir and removes worn-out red blood cells from circulation.

The Lymphatic System

The lymphatic system consists of **lymphoid tissue** and a network of **lymphatic vessels** that collect fluid from interstitial spaces and return it to large veins under the collar bones. Tissue or interstitial fluid is called **lymph** after it enters a lymphatic vessel. En route, lymph passes through **lymph nodes**, nodules of lymphoid tissue, that remove cellular debris and microorganisms that may be present. Lymphoid tissue, collectively called the **reticuloendothelial system,** is found in many organs such as the spleen, thymus, tonsils, adenoids, liver, intestines, and bone marrow.

The Respiratory System

The exchange of gases between the atmosphere and the blood is enabled by the respiratory system. It consists of air passageways and gas exchange organs. The passageways are the **nasal cavity, nasopharynx, larynx, trachea,** and **bronchi.** The gas exchange organs are the **lungs.**

The Digestive System

The mechanical and chemical breakdown of food into absorbable nutrients is accomplished by the digestive system. Mechanical breakdown is mainly by the chewing of food. Chemical breakdown primarily occurs in the stomach and small intestine. The major parts of the digestive system are the **mouth, pharynx, esophagus, stomach, small** and **large intestines, rectum, anus, pancreas,** and **liver.**

The Urinary System

The nitrogenous wastes of metabolism, and excess water and minerals are removed from the blood and body by the urinary system. The **kidneys** remove wastes and excess materials from the blood to form urine, which is carried by two **ureters** to the **urinary bladder** for temporary storage. Subsequently, urine is voided via the **urethra.**

The Endocrine System

The endocrine system consists of small masses of glandular tissue that secrete hormones. Hormones, which are chemical messengers, are absorbed by the blood and transported throughout the body, where they bring about the chemical control of body functions. The larger masses of endocrine tissues occur in the **endocrine glands: pituitary, thyroid, parathyroid, thymus, adrenal, pancreas, pineal, ovaries,** and **testes.** Smaller masses of hormone-producing tissues occur in other organs such as the kidneys, heart and placenta, and in the digestive tract.

The Reproductive System

Continuity of the species is the function of the male and female reproductive systems. The male reproductive system consists of a pair of sperm-producing **testes** located in the **scrotum,** two tubes, the **vasa deferentia,** which carry sperm to the **urethra, accessory glands** that secrete fluids for sperm transport, and the **penis,** the male copulatory organ. The female reproductive system consists of a pair of **ovaries** that produce ova, two **uterine tubes** that carry the ova to the **uterus, accessory glands** that secrete lubricating fluids, and the **vagina,** the female copulatory organ and birth canal.

Assignment

Complete the laboratory report for this exercise.

Rat Dissection

You will work in pairs to perform this part of the exercise. Your principal objective in the dissection is to *expose the organs for study, not to simply cut up the animal.* Most cutting will be performed with scissors. The scalpel blade will be used only occasionally, but the flat blunt end of the handle will be used frequently for separating tissues.

Skinning the Ventral Surface

1. Pin the four feet to the bottom of the dissecting pan as illustrated in Figure 2.1. Before making any incision examine the oral cavity. Note the large **incisors** in the front of the mouth, which are used for biting off food particles. Force the mouth open suf-

Figure 2.1 Incision is started on the median line with a pair of scissors.

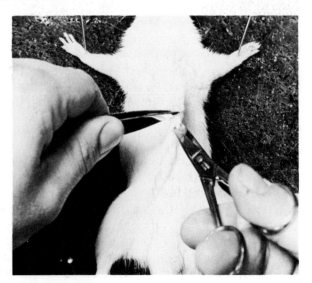

Figure 2.2 First cut is extended from the incision up to the lower jaw.

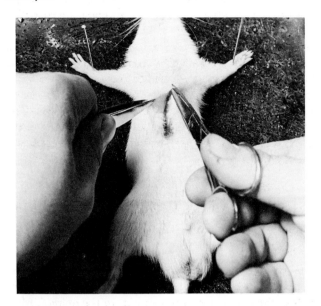

Figure 2.3 Completed incision from the lower jaw to the anus.

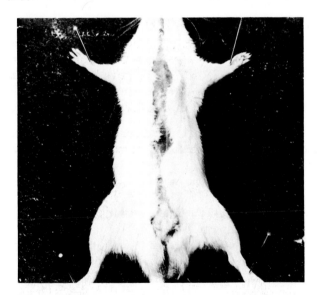

Figure 2.4 Skin is separated from musculature with scalpel handle.

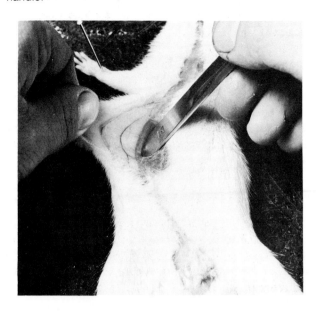

Figure 2.5 Incision of musculature is begun on the median line.

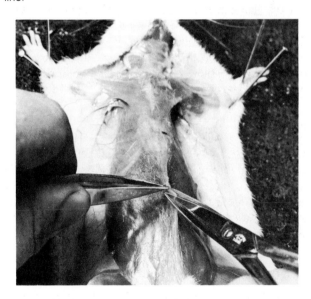

ficiently to examine the flattened **molars** at the back of the mouth. These teeth are used for grinding food into small particles. Note that the **tongue** is attached at its posterior end. Lightly scrape the surface of the tongue with a scalpel to determine its texture. The roof of the mouth consists of an anterior **hard palate** and a posterior **soft palate.** The throat is called the **pharynx,** which is a component of both the digestive and respiratory systems.

2. Lift the skin along the mid-ventral line with your forceps and make a small incision with scissors as shown in Figure 2.1. Cut the skin upward to the lower jaw, turn the pan around, and complete this incision to the anus, cutting around both sides of the genital openings. The completed incision should appear as in Figure 2.3.

3. With the handle of the scalpel, separate the skin from the musculature as shown in Figure 2.4. The fibrous connective tissue that lies between the skin and musculature is the **superficial fascia.**

Figure 2.6 Lateral cuts at base of rib cage are made in both directions.

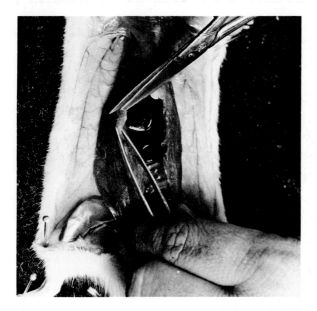

Figure 2.7 Flaps of abdominal wall are pinned back to expose viscera.

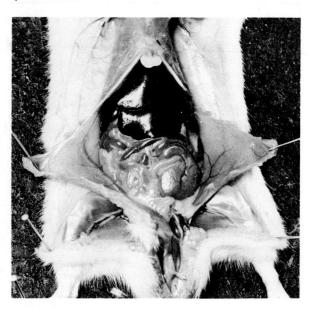

4. Skin the legs down to the "knees" and "elbows" and pin the stretched out skin to the wax. Examine the surfaces of the **muscles** and note that **tendons,** which consist of tough fibrous connective tissue, attach the muscles to the skeleton. Covering the surface of each muscle is another thin, gray felt-like layer, the **deep fascia.** Fibers of the deep fascia are continuous with fibers of the superficial fascia, so that considerable force with the scalpel handle is necessary to separate the two membranes.
5. At this stage your specimen should appear as in Figure 2.5. If your specimen is a female, the mammary glands will probably remain attached to the skin.

Opening the Abdominal Wall

1. As shown in Figure 2.5, make an incision through the abdominal wall with a pair of scissors. To make the cut it is necessary to hold the muscle tissue with a pair of forceps. **Caution:** Avoid damaging the underlying viscera as you cut.
2. Cut upward along the midline to the rib cage and downward along the midline to the genitalia.
3. To completely expose the abdominal organs make two lateral cuts near the base of the rib cage—one to the left and the other to the right. See Figure 2.6. The cuts should extend all the way to the pinned back skin.
4. Fold out the flaps of the body wall and pin them to the wax as shown in Figure 2.7. The abdominal organs are now well exposed.
5. Using Figure 2.8 as a reference, identify all of the labeled viscera without moving the organs out of place. Note in particular the position and structure of the **diaphragm.**

Figure 2.8 Viscera of a female rat.

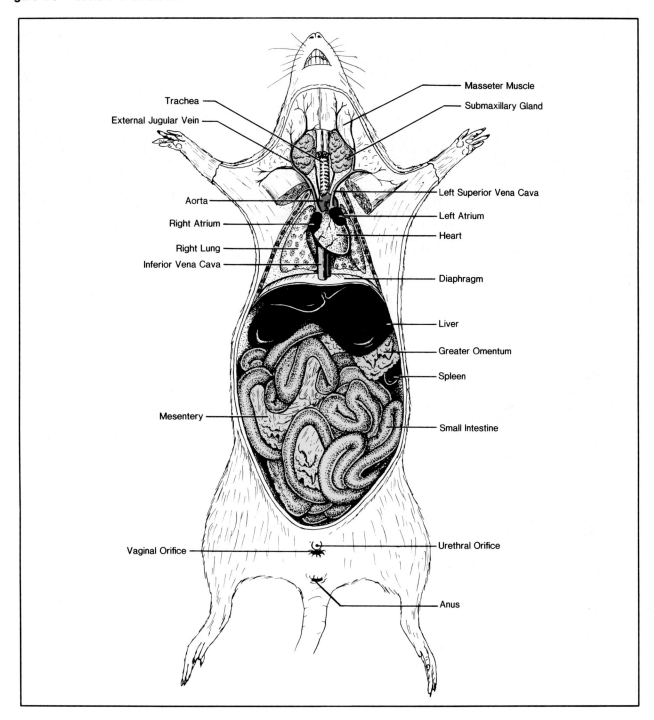

Trachea

External Jugular Vein

Aorta

Right Atrium

Right Lung

Inferior Vena Cava

Mesentery

Vaginal Orifice

Masseter Muscle

Submaxillary Gland

Left Superior Vena Cava

Left Atrium

Heart

Diaphragm

Liver

Greater Omentum

Spleen

Small Intestine

Urethral Orifice

Anus

Figure 2.9 Rib cage is severed on each side with scissors.

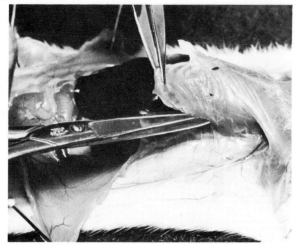

Figure 2.10 Diaphragm is cut free from edge of rib cage.

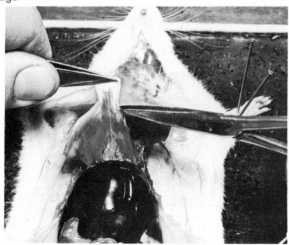

Figure 2.11 Thoracic organs are exposed as rib cage is lifted off.

Figure 2.12 Specimen with heart, lungs, and thymus gland removed.

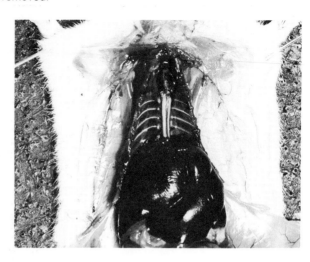

Examination of Thoracic Cavity

1. Using your scissors, cut along the left side of the rib cage as shown in Figure 2.9. Cut through all of the ribs and connective tissue. Then, cut along the right side of the rib cage in a similar manner.
2. Grasp the xiphoid cartilage of the sternum with forceps as shown in Figure 2.10 and cut the diaphragm away from the rib cage with your scissors. Now you can lift up the rib cage and look into the thoracic cavity.
3. With your scissors, complete the removal of the rib cage by cutting any remaining attachment tissue.
4. Now, examine the structures that are exposed in the thoracic cavity. Refer to Figure 2.8 and identify all the structures that are labeled.
5. Note the pale-colored **thymus gland,** which is located just above the heart. Remove this gland.
6. Carefully remove the thin **pericardial membrane** that encloses the **heart.**
7. Remove the heart by cutting through the major blood vessels attached to it. Gently sponge away pools of blood with *Kimwipes* or other soft tissues.
8. Locate the **trachea** in the neck region and the **larynx** (voice box) at the anterior end of the trachea. Trace the trachea posteriorly to where it divides into two **bronchi** that enter the **lungs.** Squeeze the lungs with your fingers, noting how elastic they are. Remove the lungs.
9. Probe under the trachea to locate the soft tubular **esophagus** that runs from the oral cavity to the stomach. Excise a section of the trachea to reveal the esophagus as illustrated in Figure 2.13.

Figure 2.13 Viscera of a male rat (heart, lungs, and thymus removed).

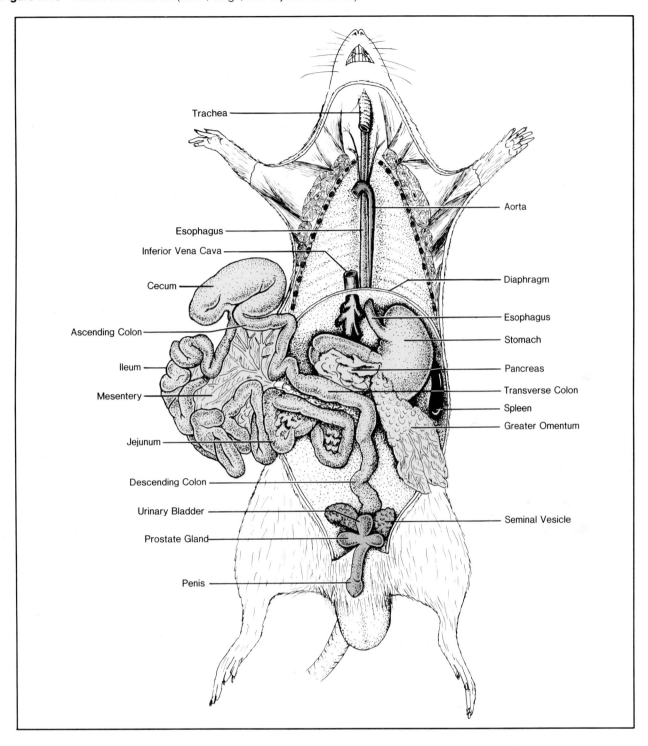

Trachea

Aorta

Esophagus

Inferior Vena Cava

Cecum

Diaphragm

Ascending Colon

Esophagus

Ileum

Stomach

Mesentery

Pancreas

Transverse Colon

Spleen

Jejunum

Greater Omentum

Descending Colon

Urinary Bladder

Prostate Gland

Seminal Vesicle

Penis

Figure 2.14 Abdominal cavity of a female rat (intestines and liver removed).

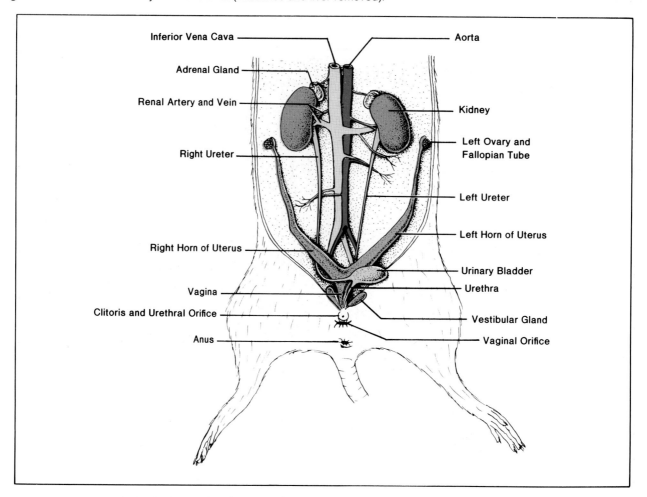

Inferior Vena Cava

Adrenal Gland

Renal Artery and Vein

Right Ureter

Right Horn of Uterus

Vagina

Clitoris and Urethral Orifice

Anus

Aorta

Kidney

Left Ovary and Fallopian Tube

Left Ureter

Left Horn of Uterus

Urinary Bladder

Urethra

Vestibular Gland

Vaginal Orifice

Deeper Examination of Abdominal Organs

1. Lift up the lobes of the reddish-brown liver and examine them. Note that rats lack a **gall bladder.** *Carefully excise the liver* and wash out the abdominal cavity. The stomach and intestines are now clearly visible.

2. Lift out a portion of the intestines and identify the membranous **mesentery,** which holds the intestines in place. It contains blood vessels and nerves that supply the digestive tract. If your specimen is a mature healthy animal, the mesenteries will contain considerable fat.

3. Now, lift the intestines out of the abdominal cavity, cutting the mesenteries, as necessary, for a better view of the organs. Note the great length of the small intestine. Its name refers to its diameter, not its length. The first portion of the small intestine, which is connected to the stomach, is called the **duodenum.** At its distal end the small intestine is connected to a large sac-like structure, the **cecum.**

The *appendix* in humans is a *vestigial* portion of the cecum. The cecum communicates with the **large intestine.** This latter structure consists of the **ascending, transverse, descending,** and **sigmoid** segments. The last of these segments empties into the **rectum.**

4. Try to locate the **pancreas,** which is embedded in the mesentery alongside the duodenum. It is often difficult to see. Pancreatic enzymes enter the duodenum via the **pancreatic duct.** See if you can locate this minute tube.

5. Locate the **spleen,** which is situated on the left side of the abdomen near the stomach. It is reddish brown and held in place with mesentery.

6. Remove the digestive tract by cutting through the esophagus next to the stomach and through the sigmoid colon. You can now see the descending **aorta** and the **inferior vena cava.** The aorta carries blood posteriorly to the body tissues. The **inferior vena cava** vein returns blood from the posterior regions to the heart.

Figure 2.15 Abdominal cavity of male rat (intestines and liver removed).

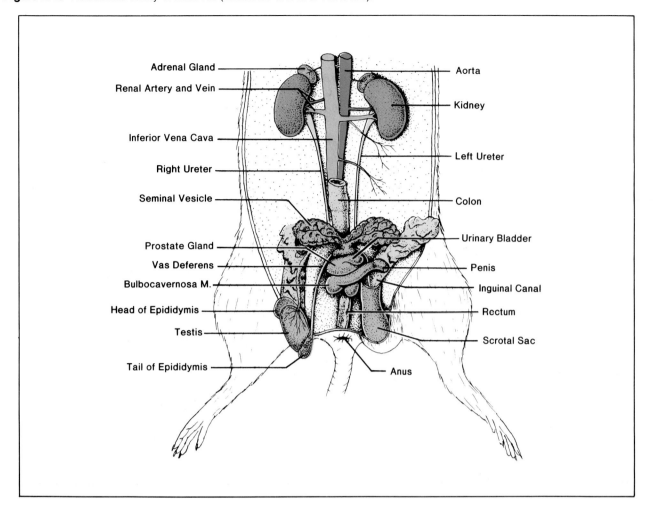

Adrenal Gland

Renal Artery and Vein

Inferior Vena Cava

Right Ureter

Seminal Vesicle

Prostate Gland

Vas Deferens

Bulbocavernosa M.

Head of Epididymis

Testis

Tail of Epididymis

Aorta

Kidney

Left Ureter

Colon

Urinary Bladder

Penis

Inguinal Canal

Rectum

Scrotal Sac

Anus

7. Peel away the peritoneum and fat from the posterior wall of the abdominal cavity. *Removal of the fat will require special care to avoid damaging important structures.* This will make the kidneys, blood vessels, and reproductive structures more visible. Locate the two **kidneys** and **urinary bladder.** Trace the two **ureters,** which extend from the kidneys to the bladder. Examine the anterior surfaces of the kidneys and locate the **adrenal glands,** which are important components of the endocrine gland system.

8. **Female.** If your specimen is a female, compare it with Figure 2.14. Locate the two **ovaries,** which lie lateral to the kidneys. From each ovary a **fallopian tube,** or **oviduct,** leads posteriorly to join the **uterus.** Note that the uterus is a Y-shaped structure joined to the **vagina.** If your specimen appears to be pregnant, open up the uterus and examine the developing embryos. Note how they are attached to the uterine wall.

9. **Male.** If your specimen is a male, compare it with Figure 2.15. The **urethra** is located in the **penis.** Apply pressure to one of the **testes** through the wall of the **scrotum** to see if it can be forced up into the **inguinal canal.** Carefully dissect out a testis, **epididymis,** and **vas deferens** from one side of the scrotum and, if possible, trace the vas deferens over the urinary bladder to where it penetrates the **prostate gland** to join the urethra.

In this cursory dissection you have become acquainted with the respiratory, circulatory, digestive, urinary, and reproductive systems. Portions of the endocrine system have also been observed. If you have done a careful and thoughtful rat dissection, you should have a good general understanding of the basic structural organization of the human body. Much that we see in rat anatomy has its human counterpart.

Clean-up. Dispose of the specimen as directed by your instructor. Scrub your instruments with soap and water, rinse and dry them.

The Microscope

Objectives

After completing this exercise, you should be able to:

1. Identify the parts of a compound microscope and describe the function of each.
2. Describe and demonstrate the correct way to: (a) carry a microscope, (b) clean the lenses, (c) focus with each objective, and (d) calculate the total magnification.
3. Define all terms in bold print.

Materials

Compound microscope
Prepared slide of the letter "e"

The effective use of a **compound microscope** is essential for the study of cells and tissues. This exercise will introduce you to the parts, care, and use of a compound microscope.

Parts of the Microscope

Figure 3.1 shows the major parts of a compound microscope. Refer to it as you study this section. Your microscope may be somewhat different than the one illustrated, but you will be able to relate the illustration and the discussion below to your microscope without difficulty.

The **base** is the bottom portion of the microscope. It contains the **lamp.** Ideally, the lamp should have a voltage control to vary intensity of light. The microscope shown in Figure 3.1 has the voltage control on a separate transformer (not shown). Some microscopes have the voltage control built into the base. The lowest possible voltage that provides a good image should be used since this will extend the life of the lamp.

Many microscopes have a **neutral density filter** that may be moved over the bulb to prevent excessive brightness. The **neutral density filter control lever** is on the base.

The **arm** rises from the base and supports the rest of the microscope. It serves as a convenient "handle" when carrying a microscope. The **body tube** has an **ocular lens** at the upper end and a **revolving nosepiece,** to which the **objective lenses** are attached, at the lower end.

Microscopes may be either binocular as in Figure 3.1 or monocular. The magnification of the ocular is usually 10×. Student microscopes usually have three objectives attached to the revolving nosepiece. The shortest objective is the **low power objective,** and it has a magnification of 10×. The longest objective is the **oil immersion objective,** which has a magnification of 100×. The intermediate length objective is the **high-dry objective** with a magnification of 45×. The magnifications of the oil immersion and high-dry objectives may vary slightly in different models of microscopes.

Figure 3.1 The compound microscope.

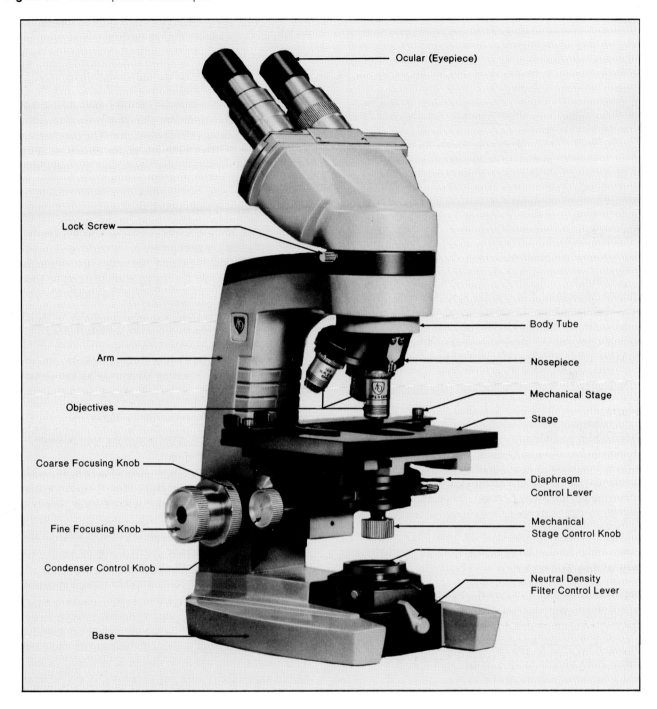

Ocular (Eyepiece)

Lock Screw

Arm

Objectives

Coarse Focusing Knob

Fine Focusing Knob

Condenser Control Knob

Base

Body Tube

Nosepiece

Mechanical Stage

Stage

Diaphragm Control Lever

Mechanical Stage Control Knob

Neutral Density Filter Control Lever

Figure 3.2 The slide must be properly positioned as the retainer lever is moved to the right.

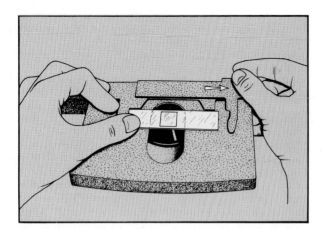

The **stage** is the platform on which a microscope slide is placed for viewing. The opening in the center of the stage is the **stage aperture.** The **mechanical stage** is the device on the stage surface that holds the slide and enables precise movement of the slide. It is operated by the **mechanical stage control knobs.** Figure 3.2 shows how to place a slide in the mechanical stage.

Below the stage is the **condenser** that concentrates the light on the microscope slide. It may be raised or lowered by the **condenser control knob.** Usually, it should be raised to its highest position.

Just below the condenser is the **diaphragm** that regulates the amount of light entering the condenser. It may be opened and closed with the **diaphragm control lever.**

There are two focusing knobs. The **coarse focusing knob** has a larger diameter and is used to bring objects into rough focus. The **fine focusing knob** has a smaller diameter and is used to bring objects that are in rough focus into sharp focus.

Magnification

The **total magnification** is determined by the power of the ocular and objective being used. It is calculated by multiplying the power of the ocular by the power of the objective. For example, a $10\times$ ocular and a $45\times$ objective yield a total magnification of $450\times$.

Resolution

Resolving power is a function of the wavelength of light and the design of the microscope lenses. The shortest wavelengths of visible light provide maximum resolution. This is why microscopes have a blue light filter over the lamp. Use of the oil immersion objective is required for maximum resolution, and on the best light microscopes it will enable the distinction of microscopic objects that are $0.2~\mu m$ apart. If they are closer together, they will be seen as one object due to a fusion of the images.

Focusing

A microscope is focused by changing the distance between the object on the microscope slide and the objective lens. This is accomplished by using the coarse and fine focusing knobs. The focusing knobs raise and lower either the stage *or* the body tube depending on the type of microscope. Both types of focusing procedures are described below. Determine which method is to be used for your microscope.

As a general rule, you should always start focusing with the low power ($10\times$) objective. The coarse focusing knob is used *only* with the low power objective. With the high-dry and oil immersion objectives use *only* the fine focusing knob.

Focusing with a Movable Stage

1. Place the $10\times$ objective into viewing position by rotating the nosepiece.
2. While looking from the side (not through the ocular), raise the stage with the coarse focusing knob until either it stops or the slide is about 3mm from the objective.
3. While looking through the ocular, slowly lower the stage with the coarse focusing knob until the object comes into view.
4. Use the fine focusing knob to bring the object into sharp focus.

Focusing with a Movable Body Tube

1. Place the $10\times$ objective in viewing position by rotating the nosepiece.
2. While looking from the side (not through the ocular), lower the body tube with the coarse focusing knob until either it stops or the slide is about 3mm from the objective.
3. While looking through the ocular, slowly raise the body tube with the coarse focusing knob until the object comes into view.
4. Use the fine focusing knob to bring the object into sharp focus.

Switching Objectives

Modern microscopes are usually **parcentric** and **parfocal.** This means that when an object is centered in the field and in sharp focus with one objective, it will be centered and in focus when another objective is rotated into the viewing position. However, it may be necessary to make slight adjustments to re-center the object with the mechanical stage or bring it into sharper focus with the fine focusing knob.

Start your observations with the low power objective, even though you may want to observe the object with the high-dry or oil immersion objective. Once the object is centered and in focus with the low power objective, it is easy to switch to the high-dry objective simply by rotating the nosepiece. Note that the **working distance,** the distance between the objective and the slide, decreases as the power of the objective increases.

The amount of light entering the objective decreases as the power of the objective increases. Thus, you will need to open the diaphragm a bit to provide a light intensity that yields the sharpest image with the high-dry objective.

The easiest and safest way to bring the oil immersion objective into position is to progress from low power to high-dry to oil immersion. However, you may switch directly from the low power objective to the oil immersion objective. Before rotating the oil immersion objective into the viewing position, place a drop of immersion oil on the slide. Also, open the diaphragm to its maximum aperture to increase the amount of light. A *slight adjustment* of the fine focusing knob is all that will be required to bring the object into focus.

Transport and Placement

Care must be used in transporting the microscope from the storage cabinet to your work station. Figure 3.3 shows the correct way to carry a microscope. Note that it is carried in front of you with one hand supporting the base while the other grasps the arm. *Never* carry it with one hand at your side since a microscope in this position is apt to collide with furniture in the lab.

Most modern microscopes have an inclined body tube and a rotatable head that is secured by a set screw. You may use the microscope in either of the positions shown in Figure 3.4. The conventional position is with the arm facing the student. However, when the stage is facing the student, it is more easily accessible. Use the position recommended by your instructor.

Care of the Microscope

Your instructor will assign a specific microscope to you for your use throughout the course. You will be responsible for its care. Do not use a different microscope without permission. Report any problems or malfunction to your instructor immediately.

Lens Care

Develop the habit of cleaning the lenses with **lens paper** before using the microscope. *Use only lens paper for cleaning the lenses.* If liquid gets on the objectives or stage of the microscope, wipe it off immediately with lens paper. If simply wiping the lenses with lens paper doesn't get them clean, it may be necessary to clean them with lens paper

Figure 3.3 The microscope should be held firmly with both hands while carrying it.

Figure 3.4 The microscope position on the right has the advantage of stage accessibility.

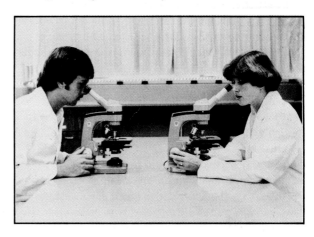

moistened with green soap or xylene. *Use such a procedure only as directed by your instructor. Never try to disassemble any part of the microscope.*

The best way to determine if the ocular is clean is to rotate it between your thumb and forefinger while looking through the microscope. A rotating pattern is evidence of a dirty lens. If cleaning the top lens fails to remove the debris, *inform your instructor. If the ocular is removed to clean the lower lens, it is imperative that a piece of lens paper be placed over the open end of the body tube as shown in Figure 3.5.*

Figure 3.5 After cleaning the lenses, a blast of air from an air syringe removes residual lint.

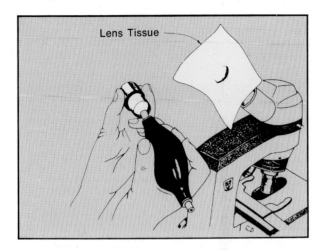

Lens Tissue

An image that appears cloudy or blurred indicates a dirty objective. If cleaning with lens paper moistened with water fails to clear the image, *consult your instructor* about cleaning the lens with xylene.

Routinely wipe off the upper surface of the condenser lens with lens paper to remove any accumulated dust.

Clean Up and Storage

When you have finished using the microscope, complete the following steps before returning it to the cabinet.

1. Remove the slide from the stage.
2. Clean the ocular, objective lenses and the stage.
3. Rotate the nosepiece to place the lowest power objective in the viewing position.
4. Depending on the type of microscope, lower the body tube to its lowest position *or* raise the stage to its highest position.
5. Raise the condenser to its highest position.
6. Adjust the mechanical stage so that it projects minimally from each side of the stage.
7. Place the dust cover over the microscope.

Assignment

Complete Sections A through C on the laboratory report.

Using the Microscope

Obtain the microscope assigned to you and place it on the table in front of you. Compare it with Figure 3.1. Locate on your microscope the parts described above. Manipulate the control knobs to see how they operate. Plug in the light cord and turn on the light. The circle of light that you see when looking through the microscope is called the **microscopic field** or simply the **field.** Clean the lenses. If you are using a binocular microscope, you must adjust the oculars to match the distance between your eyes.

1. Raise the condenser to its highest position.
2. Using the methods described above, place the slide in the mechanical stage, center the "e" over the stage aperture and locate the letter "e" using the low power objective. Practice moving the slide with the mechanical stage until you can center the "e" easily. Note any difference in the orientation of the "e" when viewed with the naked eye and with the microscope.
3. With the "e" centered, vary the amount of light by increasing and decreasing the lamp voltage and/or opening and closing the diaphragm to observe the effect on the sharpness of the image. Note that reduced light improves the contrast of the image. Lower the voltage and adjust the diaphragm to obtain a sharp image.
4. Rotate the high-dry objective into viewing position. Is the "e" centered and in focus? Recenter the "e" and refocus with the fine focusing knob, if necessary. Adjust the amount of light to give a sharp image. Note that a very small part of the "e" (just a few ink spots) is visible and that the **diameter of the field** decreases as the magnification increases.
5. Practice placing the slide in the mechanical stage, locating the "e" with the low power objective, centering and bringing the "e" into sharp focus, switching to the high-dry objective and refocusing and recentering as necessary. When you can do this quickly and easily, you are ready to go on to Exercise 4 Cell Anatomy.
6. Complete the laboratory report.

Cell Anatomy

Objectives

After completing this exercise, you should be able to:

1. Identify the parts of a cell on diagrams or models.
2. Identify the visible parts of a cell when viewed with a microscope.
3. Describe the basic structure of cellular organelles and their functions.
4. Define all terms in bold print.

Materials

Model of a generalized cell
Compound microscope
Microscope slides and cover glasses
Medicine droppers, toothpicks
Methylene blue, 0.01%, in dropping bottles
Culture of *Amoeba*
Prepared slides of human sperm

The **cell** is the structural and functional unit of all organisms including the human body. Cells possess many common anatomical and functional characteristics, although differences in size, shape and internal composition may result from specialization for particular functions. In this exercise, you will (1) study the anatomical features common to human cells and (2) examine the structure of selected cells with a compound microscope.

Electron microscopy has shown the cell to be a complex organization of cellular components called **organelles** (tiny organs). As you study the discussion of the cell that follows, locate the organelles in Figure 4.1, which shows a pancreatic cell, a rather unspecialized cell, as it might appear when observed with an electron microscope.

The Plasma Membrane

Cell contents are separated from the surrounding environment by the **plasma (cell) membrane.** Like all cellular membranes, the plasma membrane consists of two layers of phospholipid molecules, arranged back-to-back, in which protein molecules are embedded. The membrane is flexible, with the consistency of a liquid film. See Figure 4.1.

The **selective permeability** of the plasma membrane allows some molecules to pass through the membrane while preventing the passage of others. In this way, the movement of materials into and out of the cell is controlled by the plasma membrane in both active and passive transport.

The Nucleus

The **nucleus** is a large, spherical organelle surrounded by a double-layered **nuclear envelope.** Unlike other membranes, the nuclear envelope contains pores that facilitate movement of materials between the nucleus and the cytoplasm. Note that a portion of the nuclear envelope is cut away in Figure 4.1.

The nucleus is the control center of the cell because it contains the thread-like **chromosomes** composed of **deoxyribonucleic acid (DNA)** and protein. The genetic information that controls cellular functions is encoded in the DNA molecules. In nondividing cells, the uncoiled and dispersed chromosomes appear as **chromatin granules.** During cell division, they coil tightly to form the familiar rod-like structures characteristic of chromosomes.

A single **nucleolus** is shown in Figure 4.1, although some types of cells may contain two or more nucleoli. Nucleoli are composed of **ribonucleic acid (RNA)** and proteins, and are assembly sites for these molecules that will move into the cytoplasm to form ribosomes.

Figure 4.1 Cellular anatomy (pancreatic cell).

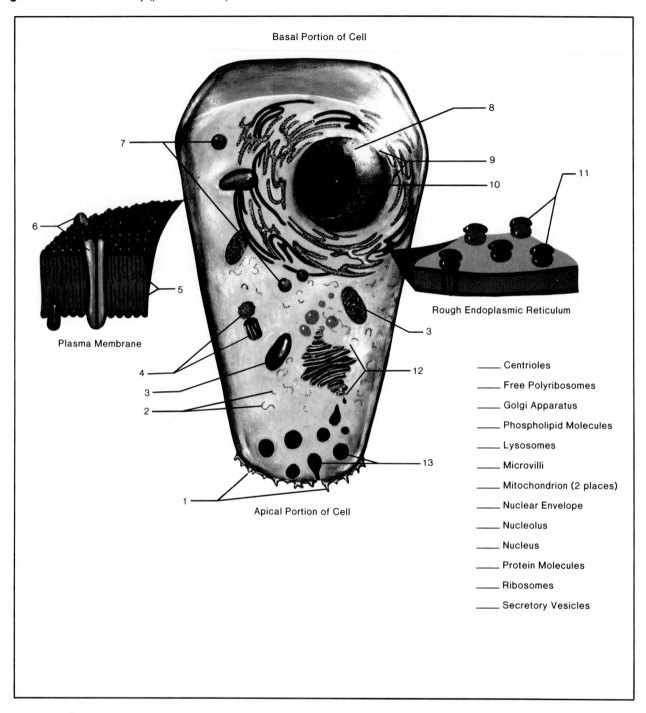

Basal Portion of Cell

Rough Endoplasmic Reticulum

Plasma Membrane

Apical Portion of Cell

_____ Centrioles

_____ Free Polyribosomes

_____ Golgi Apparatus

_____ Phospholipid Molecules

_____ Lysosomes

_____ Microvilli

_____ Mitochondrion (2 places)

_____ Nuclear Envelope

_____ Nucleolus

_____ Nucleus

_____ Protein Molecules

_____ Ribosomes

_____ Secretory Vesicles

The Cytoplasm

The semifluid, heterogeneous **cytoplasm** contains a network of two types of ultra-fine protein fibers that contribute to the support and shape of the cell. **Microtubules** are hollow cylinders that are especially evident in forming the spindle fibers during cell division. **Microfilaments** are solid parallel bundles of contractile proteins that enable cell movement. They are most abundant in muscle cells. The other organelles in the cytoplasm that perform much of the cellular work are noted below.

Endoplasmic Reticulum (ER)

This network of membranes permeates the cytoplasm between the nuclear and plasma membranes and provides channels for the movement of molecules. **Rough endoplasmic reticulum (RER)** has ribosomes on its surface, and provides storage and transport of proteins formed on the ribosomes. **Smooth (agranular) endoplasmic reticulum (AGER)** lacks ribosomes and is the site of a number of chemical reactions.

Golgi Apparatus

This organelle consists of a stack of flattened membranous sacs that are continuous with the ER. It receives substances from the ER and packages them in membranous **secretory vesicles** for export from the cell.

Ribosomes

These tiny organelles are composed of RNA and protein, and serve as sites of protein synthesis. Ribosomes associated with the RER produce proteins for export from the cell. The clusters or chains of free **polyribosomes** scattered in the cytoplasm form proteins for intracellular use.

Mitochondria

These ovoid or elongate organelles have an outer membrane enclosing an inner membrane that has numerous partition-like folds called cristae. The folds increase the surface area of the inner membrane to facilitate chemical reactions. Mitochondria can move about in the cytoplasm and can replicate themselves. They contain small amounts of RNA and DNA. They are sometimes called "powerhouses" of the cell since they are sites of ATP production by cellular respiration.

Lysosomes

These tiny oval sacs (label 7 in Figure 4.1) contain digestive enzymes and probably are formed by the Golgi apparatus. They release their enzymes into **vacuoles,** fluid-filled, membrane-enclosed sacs in the cytoplasm, to digest engulfed foreign particles or worn-out organelles. In fatally damaged cells, the lysosome membrane ruptures, and the released enzymes digest the entire cell.

Centrioles

Cells capable of cell division contain a pair of centrioles located close together and at right angles to each other. Together they constitute the **centrosome.** Each centriole is a short, rod-like bundle of microtubules that is involved in the formation of the spindle fibers during cell division.

Microvilli

These tiny cytoplasmic projections on the free surfaces of certain cells greatly increase the surface areas of the cells. Microvilli constitute the **brush border** on cells involved in the absorption of substances.

Cilia and Flagella

Both cilia and flagella are composed of microtubules derived from centrioles or their products. **Cilia** are numerous, short, hair-like projections on the surface of certain cells, which are used to move particles along the cell surface. They occur on cells lining the oviducts and respiratory passages. Spermatozoa are the only human cells with a **flagellum.** Each sperm has a single posteriorly projecting flagellum that propels it.

Assignment

1. Label Figure 4.1.
2. Locate the parts of a cell on the cell model.
3. Complete Sections A through C on the laboratory report.

Microscopic Study

The following microscopic observations of cells will aid your understanding of cellular structure. Complete the laboratory report as you make your observations.

The *Amoeba*

This common freshwater protozoan is large enough for you to observe the nucleus, cytoplasm, vacuoles and cytoplasmic granules. A sol-gel reversibility of the cytoplasm enables the flowing **amoeboid movement** and the capture of minute organisms by **phagocytosis** (engulfment of particles into vacuoles). The organisms are digested within the food vacuoles. In a similar way, your white blood cells can move among body tissues, and some of them engulf and digest bacteria as part of your defense against disease organisms.

Contractile (excretory) vacuoles maintain the intracellular water balance by collecting and pumping out excessive water that accumulates because water continuously diffuses into the *Amoeba*. As you observe an *Amoeba*, these vacuoles will appear as clear bubbles in the cytoplasm that disappear periodically as the water is forced out of the cell.

1. Place a drop from the bottom of the *Amoeba* culture on a clean slide and, before adding a cover glass, locate a specimen under low power. Use reduced light and observe the amoeboid movement.
2. Add a cover glass and study the cell under the high-dry objective. Locate the parts labeled in Figure 4.2.

Squamous Epithelial Cells

The epithelial cells lining the inside of your cheek are easily obtained for study. The cells that you will remove are dead and are continuously sloughed off normally.

1. Place a drop of methylene blue in the center of a clean slide.
2. *Gently* scrape the inside of your cheek with a clean toothpick, then swirl the end in the drop of methylene blue solution to dislodge the cells.

Figure 4.2 *Amoeba proteus.*

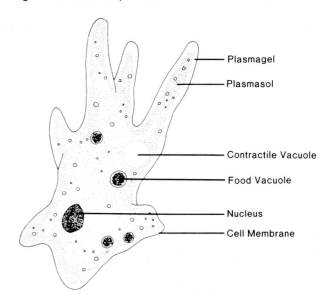

Plasmagel
Plasmasol
Contractile Vacuole
Food Vacuole
Nucleus
Cell Membrane

3. Add a cover glass and locate the cells with the low power objective.
4. Center the cells and rotate the high-dry objective into position for careful examination. Locate the nucleus and cytoplasm.

Human Sperm

Examine a prepared slide of human sperm set up under the oil immersion objective of a demonstration microscope. Note their small size. The head consists primarily of the sperm nucleus. Undulations of the posteriorly located flagellum enable forward movement by a sperm.

Exercise **5**

Mitotic Cell Division

Objectives

After completing this exercise, you should be able to:

1. List the stages of mitotic cell division and describe the characteristics of each stage.
2. Diagram and label cells in interphase and in stages of mitotic cell division.
3. Identify the mitotic stages when viewed microscopically.
4. Define all terms in bold print.

Materials

Prepared slides of whitefish mitosis

Mitotic cell division is characterized by the replication of **chromosomes** in the **parent cell** and their equal distribution to the new **daughter cells** that are formed. This orderly, controlled process produces millions of new cells in the body each day. These new cells enable either growth or replacement of damaged and worn out cells.

The life cycle of a cell typically consists of interphase and mitotic cell division. Cells that are not dividing are said to be in **interphase. Mitosis** is the orderly process that separates and equally distributes the replicated chromosomes to form the new nuclei of the daughter cells. This process assures that each new cell has the same genetic composition as the parent cell. The actual division of the parent cell is called **cytokinesis,** the division of the cytoplasm.

Cell division is normally a highly-controlled process. Some cells divide continuously, some occasionally, and some not at all. Uncontrolled cell division results in the formation of benign (nonspreading) or malignant (spreading) tumors.

Mitosis accounts for only 5–10% of the cell cycle and is a continuous process once it begins. For ease of understanding, mitosis is divided into four phases: prophase, metaphase, anaphase, and telophase.

Phases of the Cell Cycle

Refer to the photomicrographs and diagrams of mitotic division in *Ascaris* cells as you study the descriptions of the phases of the cell cycle. The cells of *Ascaris* possess only four chromosomes, which makes it easy to observe the individual chromosomes.

Interphase

Cells in interphase exhibit a distinct nucleus with an intact nuclear membrane. The chromosomes are extended and appear only as dark staining dots called **chromatin granules.** A **centrosome** containing a pair of **centrioles** is present near the nucleus. These cells are performing their normal metabolic functions. In cells that are destined to divide, replication of the centrioles and chromosomes occurs during interphase. Chromosome replication is a result of the replication of the **DNA** molecule forming the core of each chromosome.

Prophase

During **prophase,** the nuclear membrane disintegrates. The replicated chromosomes coil tightly and become visible as rod-shaped structures. Each replicated chromosome consists of two **chromatids** (sister chromatids) joined by the paired **centromeres.** The two pairs of centrioles move to opposite ends of the cell. By the end of prophase, the centrioles have formed **spindle fibers** (microtubules) that extend across the cell from one pole to the other. Each pole is formed by an **aster** that consists of a pair of centrioles and the astral rays that radiate from them. The chromosomes migrate toward the equator of the spindle.

Figure 5.1 Early stages of mitosis.

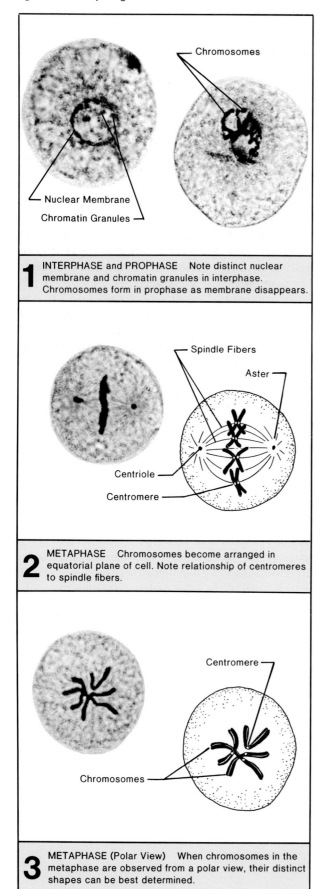

Chromosomes

Nuclear Membrane
Chromatin Granules

1 INTERPHASE and PROPHASE   Note distinct nuclear membrane and chromatin granules in interphase. Chromosomes form in prophase as membrane disappears.

Spindle Fibers
Aster
Centriole
Centromere

2 METAPHASE   Chromosomes become arranged in equatorial plane of cell. Note relationship of centromeres to spindle fibers.

Centromere
Chromosomes

3 METAPHASE (Polar View)   When chromosomes in the metaphase are observed from a polar view, their distinct shapes can be best determined.

Figure 5.2 Later stages of mitosis.

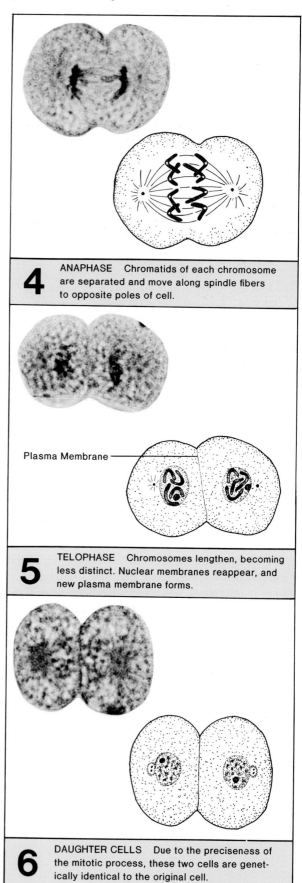

4 ANAPHASE   Chromatids of each chromosome are separated and move along spindle fibers to opposite poles of cell.

Plasma Membrane

5 TELOPHASE   Chromosomes lengthen, becoming less distinct. Nuclear membranes reappear, and new plasma membrane forms.

6 DAUGHTER CELLS   Due to the preciseness of the mitotic process, these two cells are genetically identical to the original cell.

Metaphase

During the brief **metaphase,** the replicated chromosomes line up on the equatorial plane of the spindle. Each centromere is attached to a spindle fiber that extends to a pole of the spindle.

Anaphase

In anaphase, the paired centromeres separate, and the sister chromatids migrate to opposite poles. They appear to be drawn to the poles by the spindle fibers. The separated sister chromatids are now called **daughter chromosomes.**

Telophase

During telophase, the daughter chromosomes extend and become less distinct. A new nuclear membrane forms around each set of chromosomes, forming the daughter nuclei. The spindle fibers gradually disappear. Cytokinesis begins with the formation of a **cleavage furrow** that constricts to divide the parent cell into two daughter cells.

Daughter Cells

These cells are again in the interphase stage of the cell cycle. The daughter cells have the identical DNA (genetic) composition as a result of chromosome replication, separation and equal distribution to their nuclei.

Assignment

Complete Sections A and B on the laboratory report.

Microscopic Study

In this section, you will examine mitotic cell division in whitefish cells. The stages of mitosis are readily distinguishable in whitefish cells, although they have more chromosomes than *Ascaris* cells. The slide that you will examine contains 3–6 sections sliced through a blastula, an early embryo. Each section contains many cells in various stages of cell division.

1. Use the low power objective to locate cells in interphase and the mitotic stages for study, and then switch to the high-dry objective for careful observation. Your study can be simplified by examining the stages as you locate them on the slide rather than searching for them as they occur in the mitotic sequence. You may need to examine all of the sections on your slide, or even another slide, in order to locate all of the mitotic stages.
2. Diagram and label whitefish cells in interphase and mitotic phases in Section C of the laboratory report.

Diffusion and Osmosis

Objectives

After completing this exercise you should be able to:

1. Explain the cause of Brownian movement, diffusion, and osmosis.
2. Describe the effect of molecular weight and temperature on the rate of diffusion.
3. Describe the relationship between osmotic equilibrium and the integrity of cell membranes.
4. Define all terms in bold print.

Materials

Compound microscopes
Hot plates
Beakers, 250 ml
Depression slides and cover glasses
Forceps
Mechanical pipetting devices
Petri dishes of agar-agar
Serological pipettes
Serological test tubes and test tube racks
Syringes and needles
Thermometers, Celsius
Thistle-tube osmometers
Fresh blood
Glucose solutions, 2.0%, 5.0%
India ink
Methylene blue crystals
NaCl solutions, 0.3%, 0.9% and 2.0%
Potassium permanganate crystals

Materials are constantly exchanged between body cells and the tissue fluid that bathes them. These substances, including organic nutrients, inorganic ions or water, pass into and out of cells through the plasma membrane. Some molecules, usually of small size, move passively through the membrane by diffusion along a concentration gradient. Others move either with or against a concentration gradient by active transport. In either case, the selective permeability of the plasma membrane plays an important role.

In this exercise you will investigate the passive transport of molecules.

Brownian Movement

Molecules in a liquid or gaseous state are in constant, random motion. Each molecule moves in a straight line until it bumps into another molecule, and then it bounces off in another straightline path. This molecular motion cannot be observed directly, but it may be observed indirectly by noting the random motion of microscopic particles suspended in water. The random bombardment of these particles by water molecules causes their random, vibratory movement that is called **Brownian movement** after Robert Brown, a Scottish botanist who first described it in 1827.

Diffusion

The net movement of the same kind of molecules from an area of higher concentration to an area of lower concentration is called **diffusion.** Molecules move away from the area of higher concentration because the area of lower concentration has fewer molecules to obstruct straightline movement of the molecules. For example, if a soluble substance such as a lump of sugar is placed in water, it will gradually diffuse throughout the liquid until it is equally distributed.

Diffusion occurs in liquids and gases, and is an important means of distributing materials within cells and of moving some materials into and out of cells. It is a passive process that results from the normal random motion of molecules and does not require an expenditure of energy by the cells. However, the rate of diffusion is affected by both temperature and molecular size.

Osmosis

Water is the most abundant substance in living cells. It is the **solvent** of living systems—the liquid medium in which the chemical reactions of life occur. Organic and inorganic materials constitute the **solutes.** Water molecules are small and move freely through cell membranes. Like other molecules, water molecules diffuse from an area of higher concentration to an area of lower concentration. The diffusion of water through a semipermeable or selectively permeable membrane is called **osmosis.**

The direction of a net movement of water across a membrane is dependent upon the concentration of water and thus, the concentration of solutes on each side of the membrane. If pure water is separated from a 10% sucrose solution by a semipermeable membrane, there will be a net movement of water into the sucrose solution. The force required to prevent this movement is called **osmotic pressure,** and its value is assigned to the sucrose solution. The greater the concentration of solute particles (molecules or ions), the greater is the osmotic pressure of the solution. Thus, the degree of ionization of solute molecules markedly affects the osmotic pressure of a solution.

If two solutions with the same osmotic pressure are separated by a semipermeable membrane, there will be no net movement of water from one solution into the other. Such solutions are said to be **isotonic,** and they are at **osmotic equilibrium.** The integrity of cell membranes is dependent upon osmotic equilibrium between the cells and the fluid that bathes them.

Now consider hypothetical sucrose solutions A and B that are separated by a semipermeable membrane. Solution A has a lower solute concentration, but a greater water concentration than B. Thus, A has a lower osmotic pressure than B, and the net movement of water will be from A to B. Solution A is said to be **hypotonic** to B, because it has a lower solute concentration and a lower osmotic pressure. Solution B is **hypertonic** to A, because it has a higher soluter concentration and a higher osmotic pressure. Whenever aqueous solutions with unequal osmotic pressures are separated by a semipermeable membrane, the net movement of water is always from the hypotonic solution into the hypertonic solution.

Solutions that are administered intravenously must be isotonic with blood cells to maintain the normal osmotic equilibrium of body fluids. Otherwise, osmosis will cause cells to swell or shrink, and they may be damaged.

Assignment

Complete Section A on the laboratory report.

Demonstrations and Experiments

The demonstrations and experiments in this section will aid your understanding of diffusion and osmosis. Record your observations and results on the laboratory report.

Brownian Movement

A hanging drop slide of diluted India ink has been set up under a demonstration microscope using the oil immersion objective. Observe the movement of the ink particles. What causes their movement?

Diffusion and Temperature

1. Fill a small beaker about two-thirds full with tap water.
2. Fill another beaker with the same amount of water that has been heated to 60° C.
3. Place the beakers on your table where they will not be disturbed and can remain motionless.
4. Drop a small crystal of potassium permanganate into each beaker without disturbing the water.
5. Observe the beakers at 5-minute intervals and record the time required for the molecules of potassium permanganate to diffuse throughout the water in each beaker.

Diffusion and Molecular Weight

1. Obtain a Petri dish containing about 12 ml of 1.5% agar-agar. The agar gel is 98.5% water, and water-soluble molecules readily diffuse through it.
2. Use forceps to place approximately equal-sized crystals of potassium permanganate and methylene blue on the agar about 5 cm apart as shown in Figure 6.1. Set the dish aside where it will not be disturbed.
3. After 1 hour, measure and record the diameter (mm) of the colored area around each crystal to determine any difference in the rate of diffusion. The molecular weight of potassium permanganate is 158; the molecular weight of methylene blue is 320.

Osmosis

1. Examine the two thistle-tube osmometers, like the one shown in Figure 6.2, that were set up at the start of the laboratory session. Each osmometer was

Figure 6.1 Comparing rates of diffusion.

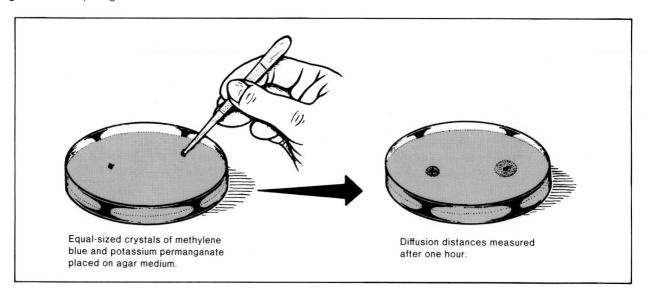

Equal-sized crystals of methylene blue and potassium permanganate placed on agar medium.

Diffusion distances measured after one hour.

Figure 6.2 Osmosis setup.

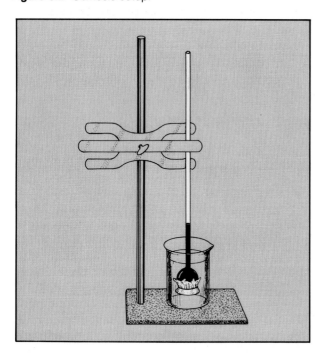

Figure 6.3 Routine for tube preparations.

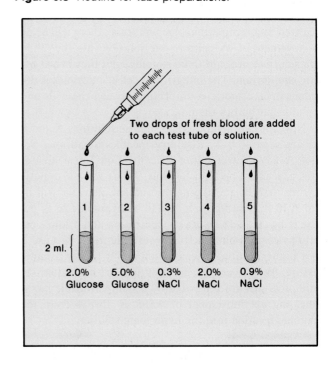

Two drops of fresh blood are added to each test tube of solution.

2 ml.

1	2	3	4	5
2.0% Glucose	5.0% Glucose	0.3% NaCl	2.0% NaCl	0.9% NaCl

prepared with a different concentration of sucrose in the thistle-tube and immersed in a beaker of distilled water. A semipermeable cellulose membrane separates the sucrose solution from the distilled water.

2. Measure and record the change in the height of the water-sucrose column in osmometers A and B during a 30-minute interval.

Osmosis and Cell Membrane Integrity

In this experiment, you will determine the effect of the osmotic pressure of four solutions on red blood cells. Figure 6.3 shows the protocol for the experiment. Recall that osmotic equilibrium between cells and the fluid bathing them is required for the normal functioning of cells. Figure 6.4 shows the effect of hypertonic, isotonic, and hypotonic solutions on red blood cells. Hypertonic solutions cause water

Figure 6.4 Effects of solutions on red blood cells.

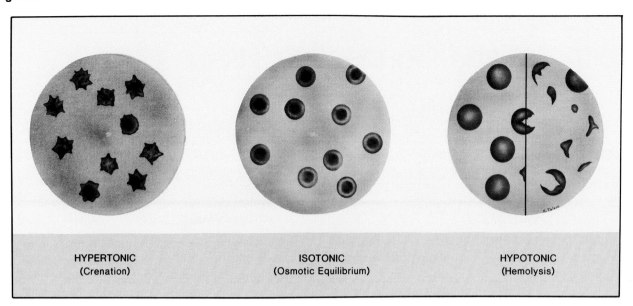

| HYPERTONIC (Crenation) | ISOTONIC (Osmotic Equilibrium) | HYPOTONIC (Hemolysis) |

to leave the cells, and they shrivel (crenation). Isotonic solutions cause no net movement of water. Hypotonic solutions cause water to enter the cells, and they ultimately burst (lyse), due to the excessive internal pressure that is generated. The rupture of red blood cells is called **hemolysis.**

1. Label 5 clean serological test tubes 1 to 5 and place them in a test tube rack.
2. Use a 5 ml pipette to place 2 ml of solution in each tube:
   Tube 1—2.0% glucose
   Tube 2—5.0% glucose
   Tube 3—0.3% NaCl
   Tube 4—0.9% NaCl
   Tube 5—2.0% NaCl
   Use the mechanical pipetting devise as shown in Figure 6.5 to dispense the solutions. Rinse the pipette with distilled water between delivery of each solution.
3. Use a syringe to add 2 drops of blood to each tube. Shake the tubes from side to side to mix thoroughly and return them to the test tube rack.
4. After 5 minutes, examine the tubes by holding them up to the light to determine the relative transparency of the three solutions. Interpret the clarity of the solutions as follows:
   Hypotonic solution: transparent due to hemolysis.
   Isotonic solution: cloudy due to the intact red blood cells.
   Hypertonic solution: clarity intermediate between transparent and cloudy due to the crenation of the cells.

Figure 6.5 Mechanical pipetting device works best when controlled with thumb.

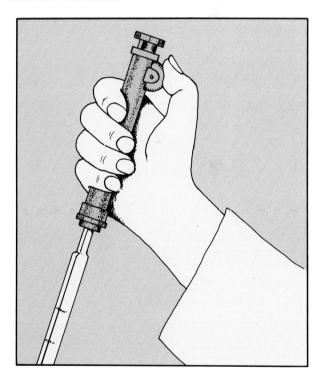

5. Examine the demonstration microscope set-ups utilizing hanging drop slides and oil immersion objectives, which show the effect of these solutions on the shape of red blood cells.
6. Complete the laboratory report.

Epithelial and Connective Tissues

Objectives

After completing this exercise, you should be able to:

1. List the types of epithelial and connective tissues, and for each describe its structure, functions and locations.
2. Recognize each type of epithelial and connective tissue when viewed with a microscope.
3. Define all terms in bold print.

Materials

Prepared slides of epithelial tissues.
   columnar, simple ciliated
   columnar, simple nonciliated
   columnar, stratified
   columnar, pseudostratified ciliated
   columnar, pseudostratified nonciliated
   cuboidal
   squamous, simple
   squamous, stratified
   transitional
Prepared slides of connective tissues.
   areolar tissue
   adipose tissue
   bone, compact, 1.s., x.s.
   dense fibrous tissue
   elastic cartilage
   fibrocartilage
   hyaline cartilage
   reticular tissue

Although body cells share many common structures and functions, they may differ considerably in size, shape, and structure in accordance with their specialized functions. Cells of a similar type usually occur in groups. An aggregation of cells that are similar in structure and function is a **tissue.** The scientific study of tissues is **histology.**

There are four basic types of tissues in the body: epithelial, connective, muscle, and nerve. In this exercise, you will study only epithelial and connective tissues. Muscle and nerve tissues will be studied in Exercise 18.

Epithelial Tissues

Epithelial tissues cover surfaces of the body and internal organs, line cavities, and form glands. They may be protective, absorptive, secretory, or excretory in function. They are characterized by: (1) closely packed cells without intercellular substances or blood vessels, and (2) the presence of a noncellular **basilar lamina** (basement membrane) that attaches the tissue to underlying connective tissue.

Epithelial tissues are categorized on the basis of: (1) single or multiple cell layers, (2) cell shape, and (3) presence or absence of cilia. The three basic categories are simple, stratified, and pseudostratified. Figure 7.1 depicts the morphological classification of epithelial tissues.

Figure 7.1 A morphologic classification of epithelial types.

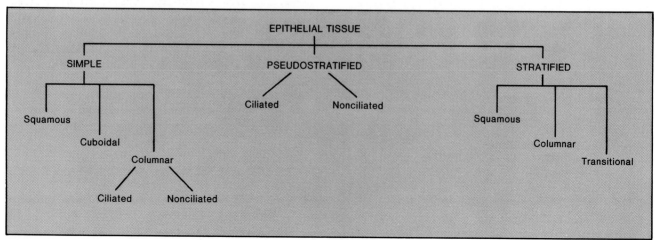

Figure 7.2 Simple epithelia.

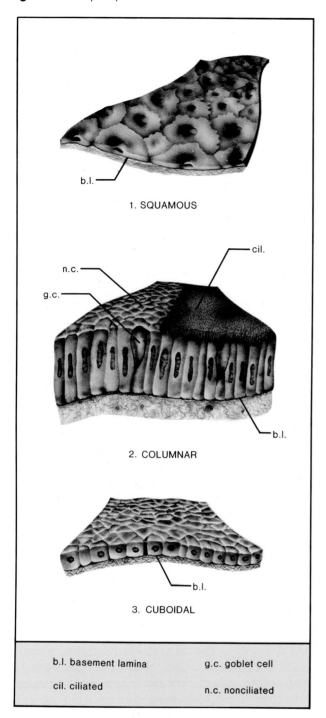

1. SQUAMOUS

2. COLUMNAR

3. CUBOIDAL

b.l. basement lamina   g.c. goblet cell

cil. ciliated   n.c. nonciliated

Simple Epithelia

Epithelia in this category are composed of a single layer of cells. Figure 7.2 illustrates the basic types.

Simple Squamous   These cells are thin and flat, and appear as irregular polygons in a surface view. The peritoneum, pleurae, and inner linings of arteries and veins, capillaries and kidney tubules are formed of simple squamous epithelium. This epithelium is involved in filtering and exchange processes.

Simple Columnar   The cells of this tissue are elongate and appear polygonal on their free surfaces. Scattered among the columnar cells are special **goblet cells** that secrete **mucus,** which flows over the surfaces of adjacent cells and protects the tissue.

Ciliated columnar epithelium lines the small bronchi of the lungs and oviducts. **Nonciliated (plain) columnar epithelium** lines the digestive tract from stomach to anus. These intestinal cells have **microvilli** on their free surface, which greatly increase the absorptive surface area of the cells.

Cuboidal   The cells of this tissue are block-like in cross section and hexagonal in a surface view. Secretion and absorption are its chief functions. Cuboidal epithelium occurs in kidney tubules, and in both endocrine and exocrine glands. For example, it is found in the thyroid gland, pancreas, ovaries, sweat glands, and salivary glands.

Stratified Epithelia

Three types of stratified epithelia are shown in Figure 7.3. Note that the cells are arranged in layers. The tissues are classified according to the shape of the surface cells.

Stratified Squamous   The superficial cells are distinctly squamous in shape, but the deepest cells are cuboidal or columnar. Cells change in shape as they migrate to the surface of the tissue. Protection is the chief function of this tissue found in regions where abrasions or friction occur, such as the outer layer of the skin, oral cavity, vagina, cornea of the eye, and esophagus.

Stratified Columnar   The columnar cells at the surface of this tissue are formed from the deeper rounded cells as they migrate upward. This tissue occurs in only a few areas of the body, such as the conjunctiva, parts of the pharynx and epiglottis, and the anal canal. Protection and secretion are its primary functions.

Figure 7.3 Stratified and pseudostratified epithelia.

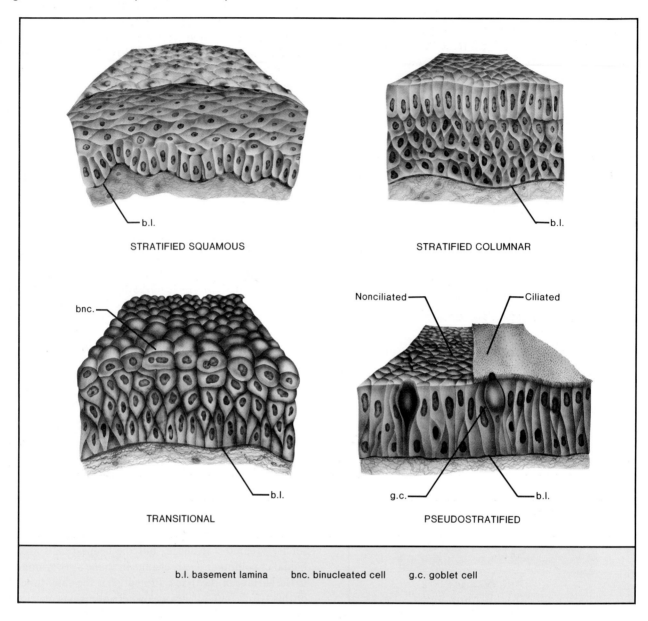

STRATIFIED SQUAMOUS

STRATIFIED COLUMNAR

TRANSITIONAL

PSEUDOSTRATIFIED

b.l. basement lamina   bnc. binucleated cell   g.c. goblet cell

Transitional This tissue occurs in organs of the body where stretching of the tissue may occur, such as the urinary bladder and ureters. Elasticity is a key characteristic. The deepest cells are columnar and loosely held together, while the surface cells may be dome-shaped and cuboidal or squamous, depending upon the degree of stretching.

Pseudostratified Epithelia

These tissues consist of a single layer of columnar cells but appear to have more than one layer. This illusion occurs because: (1) each cell is in contact with the basal lamina, but all cells do not reach the free surface of the tissue; and (2) the nuclei occur at different levels in the cells. Ciliated

forms of this tissue line the nasal cavity, trachea, primary bronchi, and Eustachian tubes. Nonciliated types are found in parts of the male urethra and in large ducts of the parotid salivary gland. Mucus-secreting goblet cells occur in both ciliated and nonciliated types.

About 90% of all cancers are **carcinomas,** malignant tumors originating from epithelial cells. They usually occur in epithelium exposed to the external environment: skin, respiratory passages, and digestive tract. **Sarcomas,** cancers of connective tissues, are especially hazardous since they often metastasize.

Connective Tissues

These tissues provide support and attachment for various organs and fill spaces in the body. They are characterized by an abundance of nonliving intercellular substance, the **matrix,** and few cells. Connective tissue may be classified in several ways. The system used here is based on the nature of the matrix. See Figure 7.4.

Connective Tissue Proper

These tissues are pliable and have a matrix that is semi-fluid, fibrous, or both. Three types of protein fibers produced by **fibroblasts** may be present. The strong, nonelastic white fibers are composed of collagen, which gives them great tensile strength. The elasticity of **yellow (elastic) fibers** is due to the presence of a protein called elastin. These fibers are often branched and not as strong as white fibers. The thin, branching **reticular fibers** form a network of very fine collagenous threads.

Areolar (Loose) Connective Tissue   This tissue, which is widespread in the body, has all three types of fibers scattered throughout the loosely organized matrix. Fibroblasts are the most common cells, although mast cells and macrophages are present. See Figure 7.5. Areolar tissue provides a flexible framework within and between organs,

Figure 7.4 Types of connective tissue.

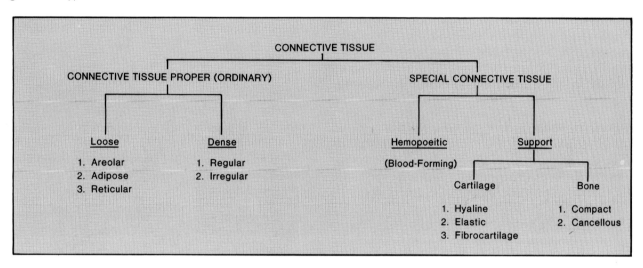

Figure 7.5 Areolar and adipose tissues.

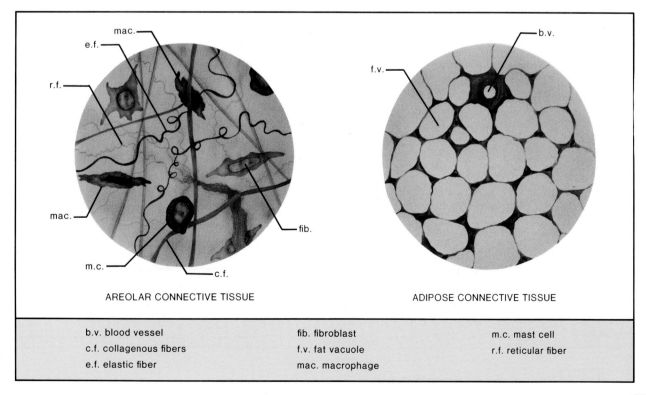

contains blood vessels that nourish surrounding tissues, and is a site of immune reactions. It occurs beneath epithelial tissues, around and within muscles and nerves, in serous membranes, and composes the superficial fascia.

Adipose Tissue   Small groups of fat or adipose cells are scattered throughout the body. Large accumulations of fat cells form adipose tissue that serves as a protective cushion and insulation for the body. The vacuoles of fat cells are filled with lipids and usually compress the cytoplasm and nucleus to the edge of the cell.

Reticular Tissue   This tissue is characterized by a network of fine reticular fibers. It provides a supporting framework for many vascular organs such as the liver, hemopoietic tissue, and lymph nodes.

Dense (Fibrous) Connective Tissue   This tissue differs from loose connective tissue in having a predominance of fibers in the matrix and relatively few cells, mostly fibroblasts. There are two subcategories of dense connective tissue based on the arrangement of the fibers. See Figure 7.6.

Dense regular connective tissue consists of closely-packed bundles of white fibers that are roughly parallel to each other and a few elastic fibers. The few cells present are nearly all fibroblasts. There are relatively few blood vessels present. This tissue is noted for its tensile strength.

It composes tendons that attach muscles to bones, and ligaments that join bones together, and forms the deep fascia.

Scar tissue forms from the accumulation of collagen produced by fibroblasts. Excessive collagen produces a **keloid scar** that is raised above the tissue surface. Scar tissue formed at sites of inflammation or after surgery may result in **adhesions,** the abnormal fusion of tissues, that may have to be surgically removed.

Dense irregular connective tissue consists of bundles of white fibers and a few elastic fibers that are not parallel but interwoven in a random fashion. This tissue forms much of the dermis of the skin, and the sheaths around nerves and tendons.

Special Connective Tissue

The three main types of special connective tissues are hemopoietic, cartilage, and bone. See Figure 7.4. This section will deal with cartilage and bone. The formed elements of the blood will be covered in Exercise 25.

Cartilage and bone possess: (1) a matrix that is solid and strong, (2) cells located in **lacunae,** tiny cavities in the matrix, and (3) an external membrane capable of producing new tissue.

Figure 7.6 Reticular and dense connective tissue.

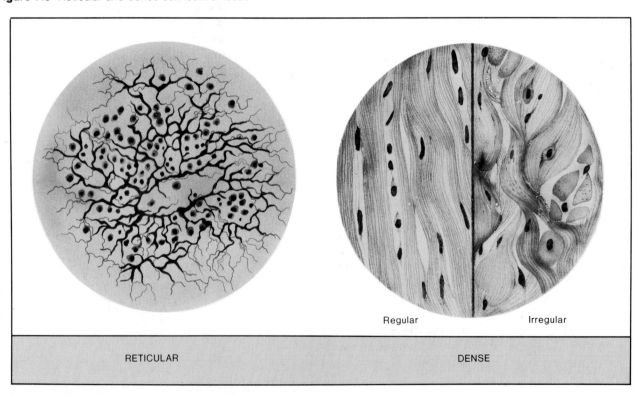

Regular   Irregular

RETICULAR   DENSE

Cartilage

All three types of cartilage are surrounded by a fibrous membrane, the **perichondrium,** and have cartilage cells, **chondrocytes,** in the lacunae of the matrix. Unlike other connective tissue, cartilage lacks blood vessels, so its cells derive nourishment by diffusion from capillaries in the perichondrium. See Figure 7.7.

Hyaline Cartilage   This is the most common type of cartilage. It has a white, glassy appearance due to the nature of the matrix. Hyaline cartilage composes much of the fetal skeleton prior to ossification and covers the ends of long bones in adults. It also forms the cartilaginous portion of the nose and the cartilaginous supports of the respiratory passages.

Elastic Cartilage   The presence of yellow elastic fibers makes this cartilage more flexible and elastic than hyaline cartilage. It is located in such places as the external ear and the epiglottis.

Fibrocartilage   This very tough cartilage contains bundles of collagenous fibers that separate short rows of chondrocytes. It forms the intervertebral discs and parts of the knee joint and pelvic girdle where it serves as a cushioning shock absorber.

Bone

Bone, the most rigid connective tissue, is hard, strong, and lightweight. It protects vital organs, provides an internal support for the body, and serves as a site for the attachment of muscles. It contains **red marrow** that forms blood cells.

In embryonic and fetal development, bones are preformed in either hyaline cartilage or fibrous connective tissue. These tissues provide a framework for the subsequent deposition of calcium salts by cells called **osteoblasts,** which form the bone matrix. Like other connective tissues, bone is continually modified and reconstructed.

There are two basic types of bone tissue as shown in Figure 7.8. Solid, dense **compact bone** (label 2) forms the outer portion of the bone, and porous **spongy** or **cancellous bone** (label 1) forms the inner portion of the bone.

Compact Bone   The structural unit of compact bone is the **Haversian system** or **osteon,** label 12 in Figure 7.8. The **Haversian canal** occupies the center of each osteon and contains one or more capillaries. Bone matrix is formed around each Haversian canal in concentric rings called **lamellae.** The lamellae are secreted by **osteoblasts** that have numerous extensions radiating from the cell. Once osteoblasts are trapped in lacunae (label 10), by their own depositions, they are called **osteocytes.** The radiating microcanals between lacunae are the **canaliculi** (label 9). They contain the cellular extensions of the osteocytes and serve as passageways for materials moving between the Haversian canals and the lacunae.

Figure 7.7 Types of cartilage.

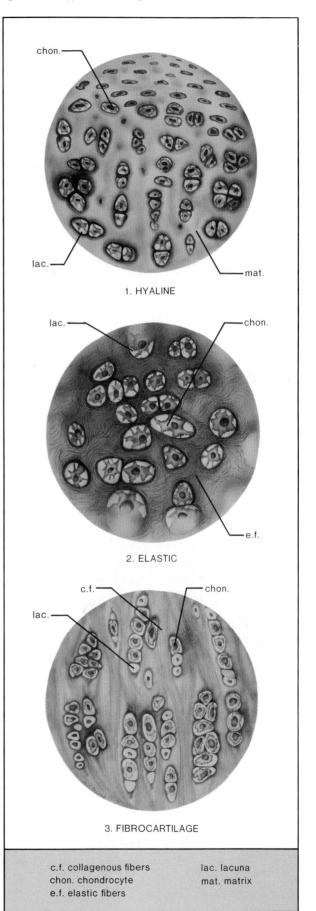

1. HYALINE

2. ELASTIC

3. FIBROCARTILAGE

c.f. collagenous fibers	lac. lacuna
chon. chondrocyte	mat. matrix
e.f. elastic fibers	

Figure 7.8 Bone tissue.

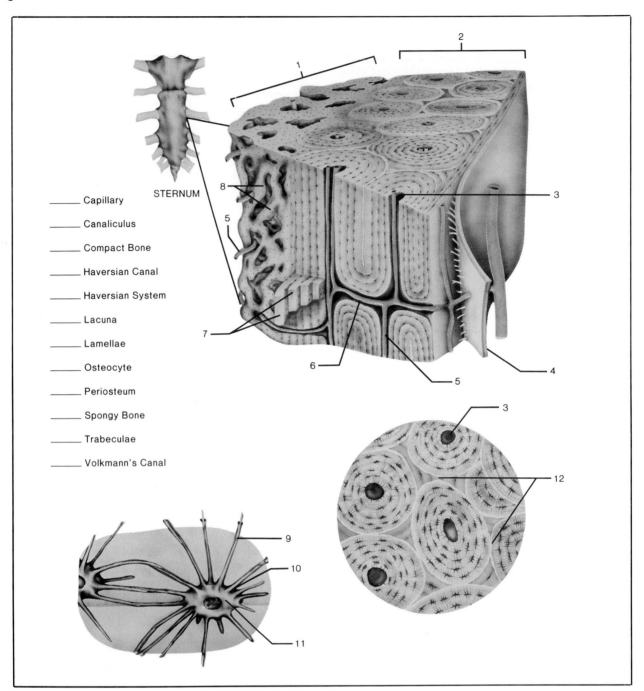

_____ Capillary

_____ Canaliculus

_____ Compact Bone

_____ Haversian Canal

_____ Haversian System

_____ Lacuna

_____ Lamellae

_____ Osteocyte

_____ Periosteum

_____ Spongy Bone

_____ Trabeculae

_____ Volkmann's Canal

STERNUM

Adjacent osteons are fused together at their outer lamellae and by bone deposits between them to yield a solid, hard matrix of calcium salts. Haversian canals of adjacent osteons are connected by **Volkmann's canals** that penetrate the osteons perpendicularly or obliquely to the osteon axis. These canals contain capillaries derived from the periosteum to nourish the osteocytes.

The outer surface of compact bone is covered by the **periosteum.** It consists of: (1) an outer, fibrous membrane, and (2) an inner layer that is a source of new bone-forming cells and capillaries. The periosteum is firmly attached to the compact bone by fibers that penetrate the outer lamellae.

Spongy Bone   This tissue is interior to and continuous with compact bone. There is no distinct line of demarcation between the two tissues. The distinctive feature is the presence of numerous branching bony plates called **trabeculae** with interconnecting spaces between them. Osteons and Haversian canals are absent. Instead, osteocytes are scattered through the trabeculae and are nourished by diffusion from blood vessels that meander through the spaces. The spaces are filled with red marrow and serve to decrease the weight of the bone.

Assignment

1. Label Figure 7.8.
2. Complete Sections A, B, and C on the laboratory report.

Microscopic Study

Perform a careful microscopic study of epithelial and connective tissues using the prepared slides provided. Consult Figures HA–1 through HA–9 in the Histology Atlas that starts on page 191. Complete the laboratory report.

The Integument

Objectives

After completing this exercise, you should be able to:

1. Describe the structure and function of the integument.
2. Identify the components of the skin when viewed microscopically.
3. Define all terms in bold print.

Materials

Prepared slides of human skin showing hair follicles, glands and receptors

The **integument** or **skin** consists of epithelial and connective tissues with associated nerve and muscle tissues. The functions of the skin include: (1) protection of underlying tissues from dehydration, microorganisms, and ultraviolet radiation; (2) excretion; (3) maintenance of body temperature; and (4) detection of certain stimuli. The skin consists of an outer, multilayered **epidermis** and an underlying **dermis.**

The Epidermis

The epidermis, label 8 in Figure 8.1, consists of stratified squamous epithelium. Prepared slides of thick skin reveal four distinct layers: an outer **stratum corneum,** a thin translucent **stratum lucidum,** a darkly stained **stratum granulosum,** and an inner **stratum malpighi.** The malpighian layer is formed of two sublayers: a multilayered **stratum spinosum** and an innermost **stratum germinativum** or **basale,** a single layer of columnar cells. Cells of all four layers originate from cells produced by the stratum germinativum. The new cells change in distinct ways and finally die as they are pushed to the surface.

The brown-black skin pigment **melanin** is produced by **melanocytes** in the stratum germinativum and is incorporated into the cells of this layer and the cells formed by this layer. Inherited differences in skin color are due to the amount of melanin produced rather than the number of melanocytes. Melanin protects underlying tissues from harmful ultraviolet radiation. In light-skinned persons, exposure to ultraviolet radiation increases the production of melanin, resulting in a "tan."

The primary cause of skin cancer is overexposure to ultraviolet light. Usually, the cancer involves non-pigmented cells in the stratum germinativum and is curable. **Melanoma,** a cancer of the melanocytes, is an especially lethal skin cancer because it rapidly metastasizes into other tissues. And, its incidence of occurrence is rapidly increasing.

The granules in cells of the stratum granulosum are precursors of **keratin,** a water-repellent protein. As the cells are pushed toward the surface, they become the keratinized, flattened, dead cells of the stratum corneum. Surface cells of this outer layer are constantly sloughed off and replaced by underlying cells. The thin stratum lucidum is apparent only in very thick skin.

The Dermis

Often called the true skin, the dermis varies in thickness from less than 1 mm to over 6 mm. It is composed of fibrous connective tissue; its numerous capillaries provide nourishment not only for itself but also for the epidermis. The dermis consists of two layers: an outer papillary layer and an inner reticular layer.

The **papillary layer** is in contact with the epidermis and has numerous projections or **papillae** that extend into the epidermis. These papillae are most abundant and pronounced in skin of the finger tips, palms of the hands, and soles of the feet where they produce epidermal ridges that improve the frictional characteristics of these areas. The pattern of these epidermal ridges is unique in each person.

The **reticular layer** contains more collagenous fibers than the papillary layer and provides much of the strength of the skin.

Figure 8.1 Skin structure.

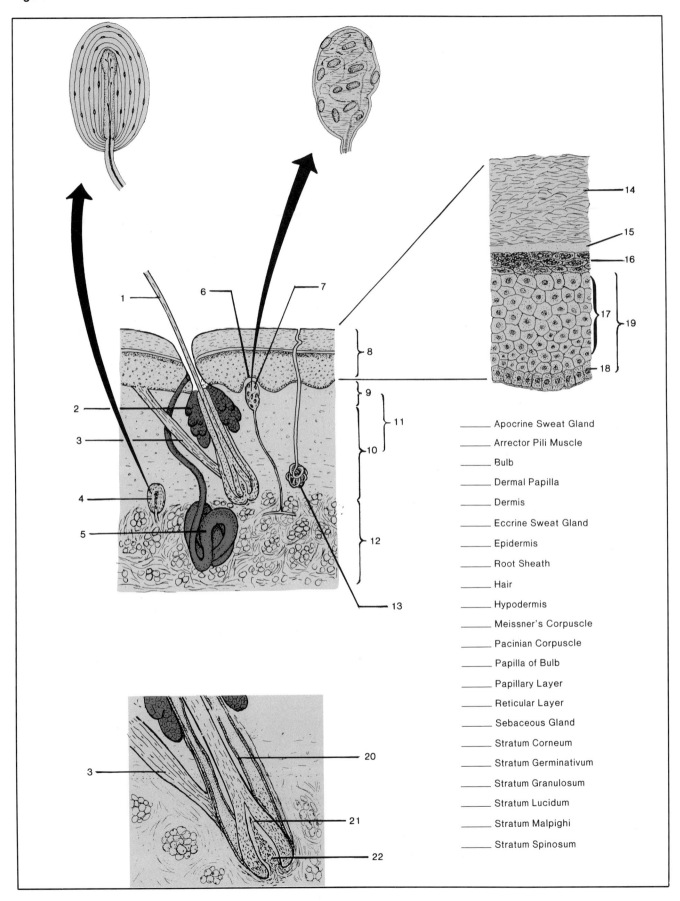

_____ Apocrine Sweat Gland

_____ Arrector Pili Muscle

_____ Bulb

_____ Dermal Papilla

_____ Dermis

_____ Eccrine Sweat Gland

_____ Epidermis

_____ Root Sheath

_____ Hair

_____ Hypodermis

_____ Meissner's Corpuscle

_____ Pacinian Corpuscle

_____ Papilla of Bulb

_____ Papillary Layer

_____ Reticular Layer

_____ Sebaceous Gland

_____ Stratum Corneum

_____ Stratum Germinativum

_____ Stratum Granulosum

_____ Stratum Lucidum

_____ Stratum Malpighi

_____ Stratum Spinosum

Subcutaneous Tissue

The subcutaneous tissue or **hypodermis** is also called the **superficial fascia.** It consists of loose connective tissue, adipose tissue, nerves, and blood vessels. It attaches the skin to the underlying tissues. The adipose tissue provides insulation for underlying tissues as well as storing energy reserves.

Hair

A hair consists of keratinized epidermal cells that are compactly cemented together. The **root** of a hair is surrounded by a **root sheath** of epidermal cells that extend into the dermis to form the **hair follicle.** The **shaft** of a hair extends above the epidermis. A hair is formed by division of the stratum germinativum cells that compose the **bulb** at the base of the follicle. The blood vessels of the dermal papilla, that extends into the base of the bulb, provide nourishment for hair growth.

An **arrector pili muscle,** formed of smooth muscle fibers, extends diagonally from the hair follicle upward to the epidermis. Contraction of this muscle raises the hair into a more upright position and produces "goose bumps."

Glands

Two kinds of glands are present in the skin: sebaceous glands and sweat glands.

Sebaceous Glands   These glands are located within the epithelial tissue of the hair follicle. They secrete an oily secretion called **sebum** into the follicle, which serves to keep the hair soft, pliable, and waterproof.

Excessive production of sebum tends to produce **blackheads,** whose color results from melanin and oils. Certain bacteria deriving nutrients from the sebum may produce inflammation, resulting in **pimples** and **boils.**

Sweat Glands   There are two types of sweat glands: eccrine and apocrine. The smaller **eccrine sweat glands** carry their secretions directly to the surface of the skin.

These are simple tubular glands that have a coiled basilar portion located in the dermis. They are widely distributed over the body and are abundant on the head, neck, back, palms of the hands, and soles of the feet. They are absent on the lips, glans penis, and clitoris. Eccrine glands located in the axillae and palms may be stimulated by psychic factors such as fear and stress, but most eccrine glands are activated primarily by thermal stimuli. The watery perspiration secreted by these glands serves to remove waste materials from the blood and cools the body surface by evaporation.

The larger **apocrine sweat glands** empty their secretions into a hair follicle. The coiled basilar portion is located in the hypodermis. These glands are essentially scent glands producing a white or yellow odiferous secretion that contributes to body odors. They are most abundant in the axillae, scrotum of the male, and perigenital region of the female. Apocrine secretions are stimulated more by psychic factors than thermal factors.

Receptors

The skin plays an important role in sensory perception because it contains numerous sensory receptors for touch, pressure, heat, cold, and pain stimuli. Touch and pressure receptors are shown in Figure 8.1. The touch receptors, **Meissner's corpuscles,** are located in dermal papillae just under the epidermis. Pressure receptors, **Pacinian corpuscles,** are spherical or ovoid laminated structures that lie deep in the reticular layer of the dermis.

Assignment

1. Label Figure 8.1.
2. Complete Sections A and B on the laboratory report.

Microscopic Study

1. Examine prepared slides of human skin and identity as many structures as you can. Compare your observations with Figure 8.1 and also with Figure HA-10 in the Histology Atlas.
2. Complete the laboratory report.

The Skeletal Plan

Objectives

After completing this exercise, you should be able to:

1. Identify the parts of a longitudinally-sectioned long bone.
2. Identify the major types of bone fractures.
3. Identify the major bones of the body and their relationships.
4. Define all terms in bold print.

Materials

Fresh beef bones sawed longitudinally
Human femur sawed longitudinally
Articulated human skeleton
Colored pencils

In this exercise, you will study the structure of a typical long bone, certain types of bone fractures, and the skeleton as a whole.

Long Bone Structure

Figure 9.1 depicts a femur that has been sectioned to reveal its internal structure. The bone consists of an elongated shaft, the **diaphysis,** and expanded terminal portions, the **epiphyses.** During the growing years, a plate of hyaline cartilage, the **epiphyseal disk,** is present at the junction of the epiphyses and the diaphysis. As new cartilage is formed on the epiphyseal side of the disk, cartilage is replaced by bone on the diaphyseal side. Linear growth ceases when the cartilage ceases to reproduce and is completely replaced by bone. The lines of fusion that form between the epiphyses and the diaphysis are called **epiphyseal lines.**

The interior of the epiphyses and ends of the diaphysis consist of **cancellous bone. Compact bone** forms the rigid tube of the diaphysis and is reduced in thickness where it covers the underlying cancellous bone.

The hollow chamber in the diaphysis is the **medullary cavity.** It is lined with a thin epithelial membrane, the **endosteum,** that extends into the spaces in cancellous bone and into the Haversian canals. The medullary cavity and the adjoining spaces in cancellous bone are filled with fatty **yellow marrow. Red marrow,** which forms blood cells, occurs in the epiphyses of the femur and humerus, but the epiphyses of other long bones contain yellow marrow. Most red marrow is found in cancellous bone of the ribs, sternum, and vertebrae.

Except for the articular surfaces, the bone is covered with the **periosteum,** a tough, vascular, fibrous membrane that tightly adheres to the bone surface. The articular surfaces are covered with a smooth **articular cartilage** (hyaline type) that reduces friction and protects the ends of the bone.

Assignment

1. Label Figure 9.1.
2. In Figure 9.1, color the articular cartilages and the epiphyseal disks blue, and the yellow marrow yellow to distinguish these parts.
3. Examine the cut surface of a beef bone. Locate the parts shown in Figure 9.1. Feel the surface of the articular cartilages. Note the texture of the yellow marrow. Verify that the periosteum is continuous with tendons and ligaments.
4. Examine the split human femur. Locate the epiphyseal lines and note the distribution of compact and cancellous bone.

Bone Fractures

Figure 9.2 shows some common types of fractures. If a broken bone pierces the skin or mucous membrane, it is a **compound fracture;** if not, it is a **simple fracture. Complete fractures** are those in which the bone is broken completely

Figure 9.1 Long bone structure.

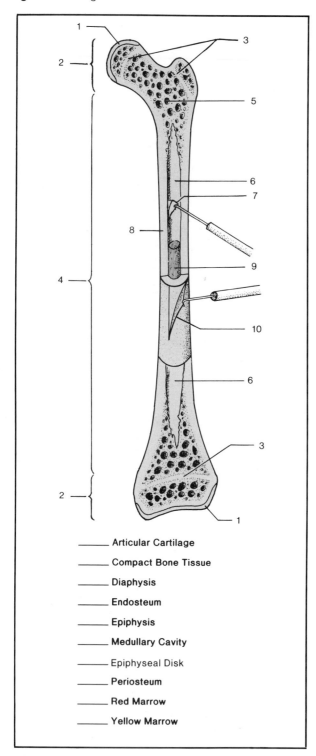

_____ Articular Cartilage

_____ Compact Bone Tissue

_____ Diaphysis

_____ Endosteum

_____ Epiphysis

_____ Medullary Cavity

_____ Epiphyseal Disk

_____ Periosteum

_____ Red Marrow

_____ Yellow Marrow

Figure 9.2 Types of bone fractures.

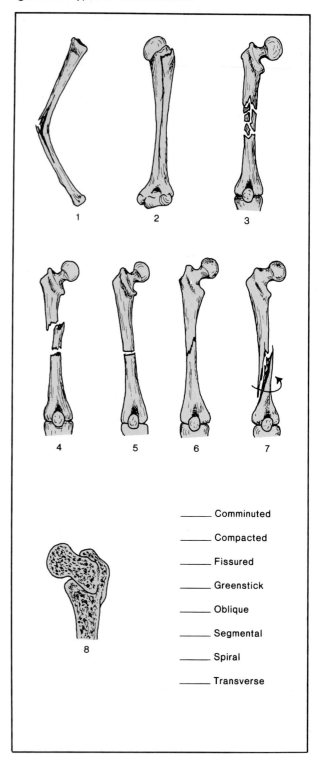

_____ Comminuted

_____ Compacted

_____ Fissured

_____ Greenstick

_____ Oblique

_____ Segmental

_____ Spiral

_____ Transverse

through. In **incomplete fractures,** the bone is only partially broken or splintered.

There are several types of complete fractures. If the break is at right angles to the long axis of the bone, it is a **transverse fracture.** Fractures at some other angle are **oblique fractures.** A **spiral fracture** results from excessive twisting of the bone. Several pieces are broken out of the bone in a **comminuted fracture.** If only one piece is broken out of the bone, it is a **segmental fracture.**

Two types of incomplete fractures are: **greenstick fractures,** where the break is on one side only due to bending of the bone, and **fissured fractures,** where a linear split occurs.

When one piece of the bone is forced into another part of the same bone, a **compacted fracture** (illustration 8) results. **Compression fractures** (not shown) of the vertebrae (crushed vertebrae) sometimes occur due to excessive vertical forces.

Assignment

Label Figure 9.2.

Parts of the Skeleton

The adult skeleton consists of 206 named bones and several small unnamed ones. Bones vary considerably in size and shape, and are categorized into two main groups: (1) bones of the **axial skeleton,** and (2) bones of the **appendicular skeleton.** Only the major bones of the skeleton will be considered in this section. Refer to Figure 9.3.

The Axial Skeleton

This part of the skeleton consists of the **skull, hyoid bone, vertebral column,** and **rib cage.** The hyoid bone is a horseshoe-shaped bone located under the lower jaw. The rib cage is formed by 12 pairs of **ribs** and the **sternum,** or breastbone.

The Appendicular Skeleton

This portion of the skeleton is formed by the bones of the shoulder and pelvic girdles, and the upper and lower limbs.

The **shoulder girdle** consists of the **clavicles** (collar bones) and **scapulae** (shoulder blades), which attach the upper limbs to the trunk. An upper limb consists of the **humerus** (upper arm bone), **radius,** and **ulna** (lower arm bone), and the bones of the hand. The radius is lateral to the ulna.

The pelvic girdle is formed by two **os coxae** bones that unite with the sacrum at the base of the vertebral column and with each other anteriorly at the **symphysis pubis** joint. A lower limb consists of the **femur** (thighbone), **tibia** (shinbone), and **fibula** (smaller bone of the lower leg), **patella** (kneecap), and the bones of the foot.

Assignment

1. Label Figure 9.3.
2. In Figure 9.3, color the appendicular skeleton red.
3. Complete the laboratory report.
4. Identify the major bones on an articulated skeleton.

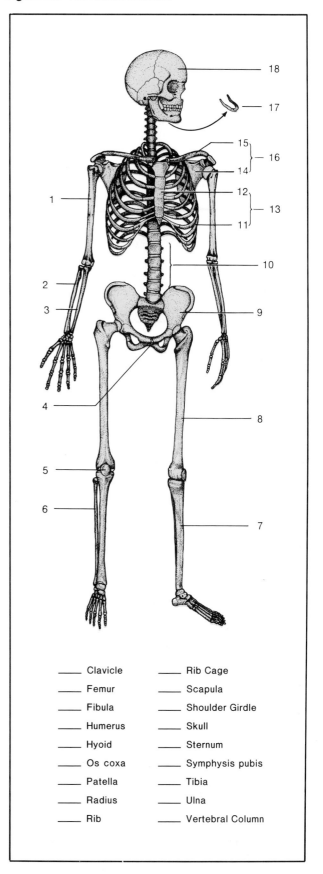

Figure 9.3 The human skeleton.

_____ Clavicle	_____ Rib Cage
_____ Femur	_____ Scapula
_____ Fibula	_____ Shoulder Girdle
_____ Humerus	_____ Skull
_____ Hyoid	_____ Sternum
_____ Os coxa	_____ Symphysis pubis
_____ Patella	_____ Tibia
_____ Radius	_____ Ulna
_____ Rib	_____ Vertebral Column

The Skull

Objectives

After completing this exercise, you should be able to:

1. Identify the skull bones and their distinctive features, sutures, paranasal sinuses, and major foramina.
2. Describe the relationships of the bones forming the skull.
3. Identify the fontanels and unossified sutures of the fetal skull.
4. Define all terms in bold print.

Materials

Human skulls with removable top of cranium
Fetal Skulls
Pipe cleaners

The adult skull consists of twenty-one bones, which are firmly joined together by immovable joints called **sutures,** plus the movable lower jaw—twenty-two bones in all. Eight of the interlocked bones form the **cranium** encasing the brain, and thirteen compose the face.

While studying the laboratory skulls, be very careful not to damage them. Some parts are delicate and easily broken. *Never use a pencil or pen as a pointer or probe.* Use a pipe cleaner instead.

The Cranium

The cranium is formed of the following bones: one **frontal,** two **parietals,** one **sphenoid,** two **temporals,** one **occipital,** and one **ethmoid.** Locate the bones and their distinctive features in Figures 10.1 through 10.4 as you study this section.

Figure 10.1 Lateral view of skull.

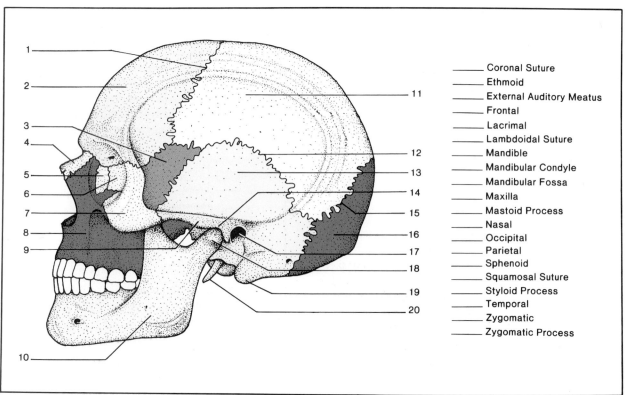

_____	Coronal Suture
_____	Ethmoid
_____	External Auditory Meatus
_____	Frontal
_____	Lacrimal
_____	Lambdoidal Suture
_____	Mandible
_____	Mandibular Condyle
_____	Mandibular Fossa
_____	Maxilla
_____	Mastoid Process
_____	Nasal
_____	Occipital
_____	Parietal
_____	Sphenoid
_____	Squamosal Suture
_____	Styloid Process
_____	Temporal
_____	Zygomatic
_____	Zygomatic Process

Figure 10.2 Bottom view of skull.

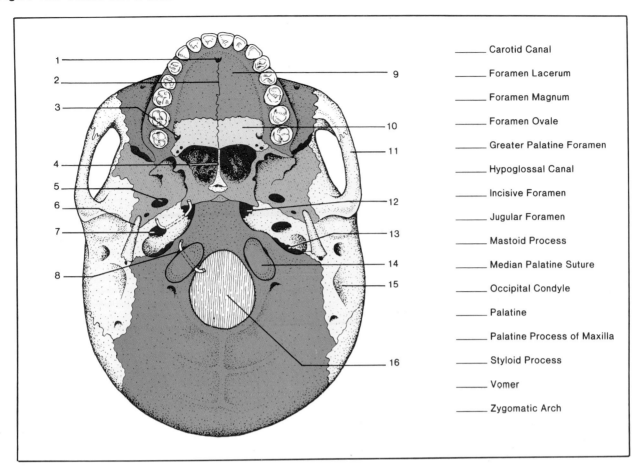

1	
2	
3	
4	
5	
6	
7	
8	
9	
10	
11	
12	
13	
14	
15	
16	

_____ Carotid Canal

_____ Foramen Lacerum

_____ Foramen Magnum

_____ Foramen Ovale

_____ Greater Palatine Foramen

_____ Hypoglossal Canal

_____ Incisive Foramen

_____ Jugular Foramen

_____ Mastoid Process

_____ Median Palatine Suture

_____ Occipital Condyle

_____ Palatine

_____ Palatine Process of Maxilla

_____ Styloid Process

_____ Vomer

_____ Zygomatic Arch

Frontal Bone

The frontal bone forms the anterior superior portion of the cranium including the forehead, upper parts of the eye orbits, and the roof of the nasal cavity. Above each eye orbit is a **supraorbital foramen** that is reduced in some skulls to a **supraorbital notch.** Foramina are passageways in bones for blood vessels and nerves.

Parietal Bones

The parietal bones form the roof and sides of the cranium posterior to the frontal bone to which they are joined by **coronal sutures.** The parietals are joined at the midline by the **sagittal suture.**

Occipital Bone

The posterior inferior portion of the skull consists of the occipital bone. It joins to the parietals by the **lambdoidal suture.** The floor of the occipital contains the large **foramen magnum,** which surrounds the brain stem. On each side of this opening are the **occipital condyles** that articulate with the first vertebra of the vertebral column. Condyles are raised knucklelike processes that fit into fossa (depressions) of adjacent bones. The **hypoglossal canals**

(label 8 in Figure 10.2 and label 15 in Figure 10.3) occur on each side of the foramen magnum and pass through the base of the occipital condyles. They are passageways for the 11th and 12th cranial nerves.

Temporal Bones

The temporal bones are located just below the parietal bones on each side of the skull. Each joins to the parietal bone above it by a **squamosal suture** and articulates posteriorly with the occipital bone by the lambdoidal suture. The ear components are located within the temporal bones. The **external auditory meatus,** which leads inward toward the middle ear, is located near the inferior margin. Just anterior to the auditory meatus is the **mandibular fossa** (depression). The mandibular condyle fits into this fossa to form the **temporomandibular joint.**

There are three major processes on each temporal bone. The **zygomatic process** is an anterior extension that articulates with the cheekbone (zygomatic). The **styloid process** is a slender spinelike process that extends downward below the auditory meatus. The **mastoid process** is a rounded eminence inferior and posterior to the auditory canal.

The Skull   **49**

Figure 10.3 Floor of cranium.

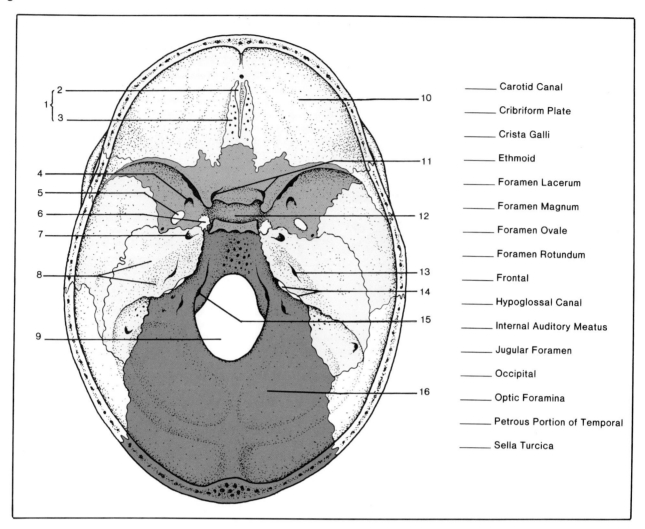

_____	Carotid Canal
_____	Cribriform Plate
_____	Crista Galli
_____	Ethmoid
_____	Foramen Lacerum
_____	Foramen Magnum
_____	Foramen Ovale
_____	Foramen Rotundum
_____	Frontal
_____	Hypoglossal Canal
_____	Internal Auditory Meatus
_____	Jugular Foramen
_____	Occipital
_____	Optic Foramina
_____	Petrous Portion of Temporal
_____	Sella Turcica

The mastoid process contains small spaces lined with mucous membranes, which communicate with the middle ear. The spaces are separated from the cranial cavity by a thin partition of bone. If a middle ear infection spreads to the mastoid air cells, a serious condition develops because of the danger of the infection spreading to the membranes covering the brain.

Figure 10.2 depicts three important paired foramina of the temporal bones. The **jugular foramina** are located just medial to the styloid processes at the junction of the temporal and occipital bones. They allow passage of the jugular veins from the brain to the neck. Just anterior to the jugular foramina are the **carotid canals** through which the carotid arteries pass to the brain. The **foramina lacerum** are medial and anterior to the carotid canals at the juncture of the temporal, occipital, and sphenoid bones.

In Figure 10.3, the **petrous portion** (label 8) of the temporal bone contains the inner ear and is probably the hardest portion of the skull. The **internal auditory meatus** is the opening on the medial sloping surface of the petrous portion. It contains facial and acoustic nerves. A jugular foramen is just posterior to each internal auditory meatus at the junction of the temporal and occipital bones; a carotid canal (label 7) appears anterior to each internal auditory meatus near the junction of the temporal and sphenoid bones. Insert a pipe cleaner or straightened paper clip through these foramina to clarify their external and internal openings.

Sphenoid Bone

The sphenoid is colored light pink in the figures. It forms part of the floor of the cranium and extends laterally to each side of the cranium, where it forms part of the lateral cranial wall and the posterior wall of the orbits. A ventral view of the skull (Figure 10.2) shows an oval opening, the

Figure 10.4 Anterior aspect of skull.

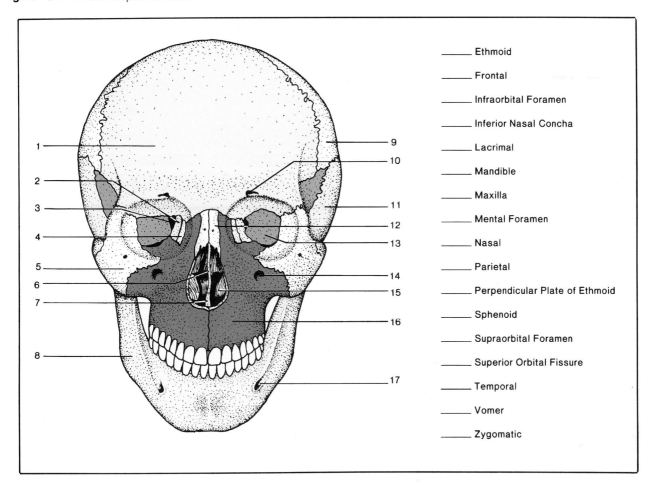

_____ Ethmoid

_____ Frontal

_____ Infraorbital Foramen

_____ Inferior Nasal Concha

_____ Lacrimal

_____ Mandible

_____ Maxilla

_____ Mental Foramen

_____ Nasal

_____ Parietal

_____ Perpendicular Plate of Ethmoid

_____ Sphenoid

_____ Supraorbital Foramen

_____ Superior Orbital Fissure

_____ Temporal

_____ Vomer

_____ Zygomatic

foramen ovale, on each side of the sphenoid bone lateral to the foramen lacerum. The mandibular nerve passes through this opening.

In the cranial floor (Figure 10.3), the medial portion of the sphenoid forms a saddle-like structure, the **sella turcica** (label 12). The hypophysis (pituitary gland) hangs downward from the brain and occupies the depression in the sella turcica. There are four major paired foramina near the sella turcica. Closest to the posterior margin of the sella turcica are the jagged internal openings of the foramina lacerum. The foramina ovale are more lateral. Anterior to the foramina ovale are the **foramina rotundum** (label 4), which are passageways for the maxillary nerves. The **optic foramina** (label 11), passageways for the optic nerves, are located on each side in the anterior portion of the sella turcica.

Figure 10.4 shows a part of the sphenoid forming the posterior portion of the eye orbits. The **superior orbital fissure** is the opening medial to the sphenoid. The **inferior orbital fissure** is below the sphenoid.

Ethmoid Bone

The ethmoid forms part of the roof of the nasal cavity and part of the medial surface of the orbit (label 6 in Figure 10.1), and closes the anterior portion of the cranium. The **perpendicular plate** of the ethmoid (label 6 in Figure 10.4 and label 3 in Figure 10.7) extends downward to form most of the nasal septum. It articulates anteriorly with the nasal and frontal bones, and posteriorly with the sphenoid and vomer. Delicate scroll-shaped processes, the **superior and middle nasal conchae** (not shown), project from the lateral portions of the ethmoid toward the perpendicular plate.

The superior portion of the ethmoid forms the roof of the nasal cavity and a small part of the anterior floor of the cranium articulating with the frontal and sphenoid bones. This portion consists of (1) two **cribriform plates** that contain numerous **olfactory foramina** for the passage of olfactory nerves from the brain to the nasal cavity, and (2) the **crista galli** (label 2 in Figure 10.3 and label 4 in Figure 10.7) that projects upward into the cranial cavity between the cribriform plates. Membranes enclosing the brain are attached to the crista galli.

Figure 10.5 The mandible.

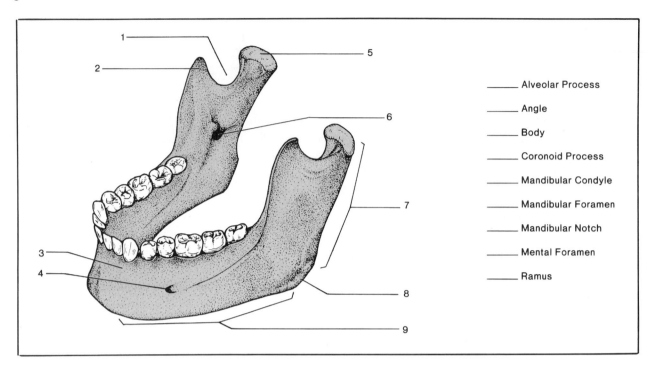

Alveolar Process

Angle

Body

Coronoid Process

Mandibular Condyle

Mandibular Foramen

Mandibular Notch

Mental Foramen

Ramus

Assignment

Label the cranial bones and their pertinent parts in Figures 10.1 through 10.4.

The Face

The paired bones of the face are: **maxillae, palatines, zygomatics, lacrimals, nasals,** and **inferior nasal conchae.** The **vomer** and **mandible** are single bones. Refer to Figures 10.1, 10.2, 10.4, and 10.5 as you study this section.

Maxillae

The two maxillary bones are joined by a suture at the midline to form the upper jaw. The anterior portion of the roof of the mouth (hard palate) is formed by the **palatine processes** of the maxillae that are joined by the **median palatine suture.** The portion of each maxillary bone containing teeth is the **alveolar process.** Together these processes form the **alveolar** or **dental arch.** Each tooth occupies an **alveolus** (socket) in the dental arch.

Three foramina are shown in the maxillae. The **incisive foramen** is located anteriorly on the midline of the hard palate, and a large **infraorbital foramen** is present below each eye orbit.

Palatine Bones

The posterior portion of the hard palate is formed by the palatine bones that are joined at the midline. A large **greater palatine foramen** is located near the lateral margin of each bone. The superior extension (not shown) of each palatine bone helps to form the lateral walls of the nasal cavity.

Zygomatic Bones

These cheekbones form the prominences of the cheeks and the lateral walls of the orbits. A posterior extension, the **temporal process,** unites with the zygomatic process of the temporal bone to form the **zygomatic arch.**

Lacrimal Bones

A small lacrimal bone is located between the ethmoid and maxilla on the medial wall of each eye orbit. Each lacrimal bone has a small groove for a tear duct that carries tears from the eye into the nasal cavity.

Nasal Bones

These thin bones are fused at the midline to form the bridge of the nose.

Vomer

This thin, flat bone (label 4 in Figure 10.2 and label 7 in Figure 10.4) is located on the midline of the nasal cavity. Its anterior superior margin articulates with the perpendicular plate of the ethmoid to form the nasal septum. Its posterior superior margin joins to the sphenoid, and its inferior margin articulates with the maxillae and palatines.

Figure 10.6 The paranasal sinuses.

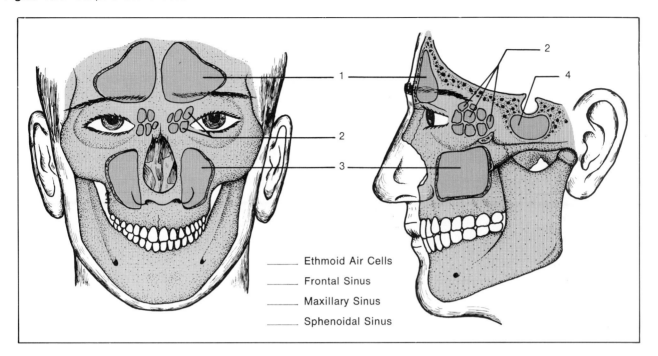

_____ Ethmoid Air Cells

_____ Frontal Sinus

_____ Maxillary Sinus

_____ Sphenoidal Sinus

Inferior Nasal Conchae

These scroll-like bones (label 15 in Figure 10.4) extend from the lateral walls of the nasal cavity inferior to the middle nasal conchae of the ethmoid bone. Like all nasal conchae, they support mucous membranes in the nasal cavity.

Assignment

Complete the labeling of Figures 10.1, 10.2, and 10.4.

Mandible

This lower jawbone consists of a horseshoe-shaped, horizontal **body** with an upward-projecting **ramus** at each end. The posterior inferior junction of each ramus with the body is called the **angle** (label 8 in Figure 10.5) of the mandible. Each ramus ends in an anterior **coronoid process** and a posterior **mandibular condyle** that are separated by the **mandibular notch.** The **alveolar process** contains the teeth.

The prominent **mandibular foramina** are located on the medial surfaces of the rami. Blood vessels and nerves supplying the teeth enter these openings and later emerge through the smaller **mental foramina,** which are located on the anterolateral external surfaces of the mandibular body, to supply the lips and chin.

The Paranasal Sinuses

Some of the skull bones contain cavities, the **paranasal sinuses,** that reduce the weight of the skull without appreciably weakening it. The sinuses are lined with a mucous membrane and have passageways leading into the nasal cavity. The sinuses are named after the bones in which they are located and are shown in Figure 10.6. The **frontal sinuses** occur in the forehead above the eyes. The large **maxillary sinuses** are below the eyes. The **sphenoid sinus** is centrally located under the sella turcica (Figure 10.7). The **ethmoid sinuses** consist of a number of small, air-filled spaces.

Inflammation of the mucous membranes lining the paranasal sinuses may block the drainage into the nasal cavity. If this occurs, the pressure increase often results in a sinus headache.

Assignment

1. Label Figures 10.5, 10.6, and 10.7.
2. Complete Sections B and C on the laboratory report.
3. Examine a human skull and locate the parts illustrated and described above. _Remember to use only pipe cleaners as pointers._

The Fetal Skull

The incompletely ossified fetal skull is shown in Figure 10.8. The unossified sutures and **fontanels,** membranous areas at the junction of several cranial bones, allow the compression of the skull during childbirth and growth of the brain after birth.

Figure 10.7 Sagittal section of skull.

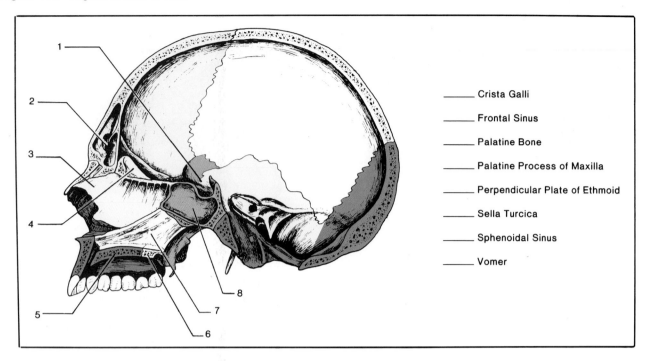

_____ Crista Galli

_____ Frontal Sinus

_____ Palatine Bone

_____ Palatine Process of Maxilla

_____ Perpendicular Plate of Ethmoid

_____ Sella Turcica

_____ Sphenoidal Sinus

_____ Vomer

Figure 10.8 The fetal skull.

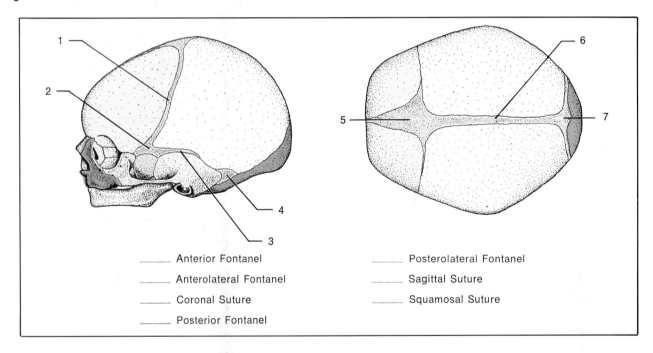

_____ Anterior Fontanel

_____ Anterolateral Fontanel

_____ Coronal Suture

_____ Posterior Fontanel

_____ Posterolateral Fontanel

_____ Sagittal Suture

_____ Squamosal Suture

There are six fontanels. The large **anterior fontanel** lies on the midline at the junction of the frontal and parietal bones. The smaller **posterior fontanel** is on the midline at the junction of the parietal and occipital bones. An **anterolateral fontanel** occurs on each side of the skull at the junction of the frontal, parietal, zygomatic, sphenoid, and temporal bones. A **posterolateral fontanel** is on each side of the skull at the junction of the temporal, parietal, and occipital bones.

Assignment

1. Label Figure 10.8.
2. Complete Section A on the laboratory report.
3. Examine a fetal skull and locate the parts shown in Figure 10.8.
4. Complete the laboratory report.

The Vertebral Column and Thorax

Objectives

After completing this exercise, you should be able to:

1. Identify the bones and their significant parts that form the vertebral column and thorax.
2. Define all terms in bold print.

Materials

Skeletons, articulated and disarticulated
Vertebral column, mounted
Pipe cleaners

The Vertebral Column

The vertebral column forms a flexible but sturdy vertical axis extending from the skull to the pelvis. It is composed of twenty-four movable vertebrae, the sacrum, and the coccyx. They are all bound together by ligaments and muscles to form a unified structure. The vertebral column exhibits four defined **spinal curvatures** that are shown in Figure 11.1. From top to bottom, they are: the **cervical, thoracic, lumbar,** and **pelvic curvatures.**

Vertebrae share many common features, although they vary in size and shape in different regions of the vertebral column. Each has an anterior structural mass, the **body,** that is the major load-bearing contact between adjacent vertebrae. Between the bodies of adjacent vertebrae are fibrocartilagenous **intervertebral disks** that serve as protective cushions.

As a person ages, the intervertebral disks become less compressible and resilient. Excessive vertical force on the vertebral column may result in a herniated (slipped) disk, usually between lumbar vertebrae 4 and 5, or between lumbar 5 and the sacrum. If pressure on the spinal nerves results, it causes considerable pain and also may cause numbness or paralysis in the area innervated.

Projecting posteriorly from the vertebral body are two processes, the **pedicles,** (label 9) that form the anterolateral portion of the **neural arch.** The posterolateral portion is formed by the **laminae** (label 7). A **transverse process** projects laterally from each side of the neural arch, and a **spinous process** projects posteriorly. Note that each vertebra has two **inferior articulating facets** (label 6) that articulate with the **superior articulating facets** (label 4) of the adjacent inferior vertebra.

The **vertebral foramen** is in the center of the neural arch. The spinal cord descends from the brain through the vertebral foramina of the vertebrae and is protected by the surrounding neural arches. The small openings between the pedicles of adjacent vertebrae, the **intervertebral foramina,** contain spinal nerves emerging from the spinal cord.

Cervical Vertebrae

The first seven vertebrae are the **cervical vertebrae** of the neck. The first vertebra is the **atlas.** Illustration A shows its superior surface. Its vertebral foramen is larger than other vertebrae to accommodate the brain stem. The superior articular surfaces articulate with the occipital condyles of the skull. A small **transverse foramen** occurs in each transverse process. These foramina are found only in cervical vertebrae and serve as passageways for the vertebral arteries and veins.

The second cervical vertebra, the **axis,** is shown in Illustration B. It is unique in having an upward projecting **odontoid process** arising from the body, which serves as a pivot for the rotation of the atlas. When the head is turned, the atlas rotates on the axis.

Thoracic Vertebrae

Below the cervical vertebrae are the twelve **thoracic vertebrae** (numbers 8–19). A superior view of one is shown in Illustration C. These vertebrae are distinguished in having ribs attached to them. They have facets (label 12) on their transverse processes and on the sides of their bodies that articulate with the ribs. Also, their spinous processes project downward.

Figure 11.1 The vertebral column.

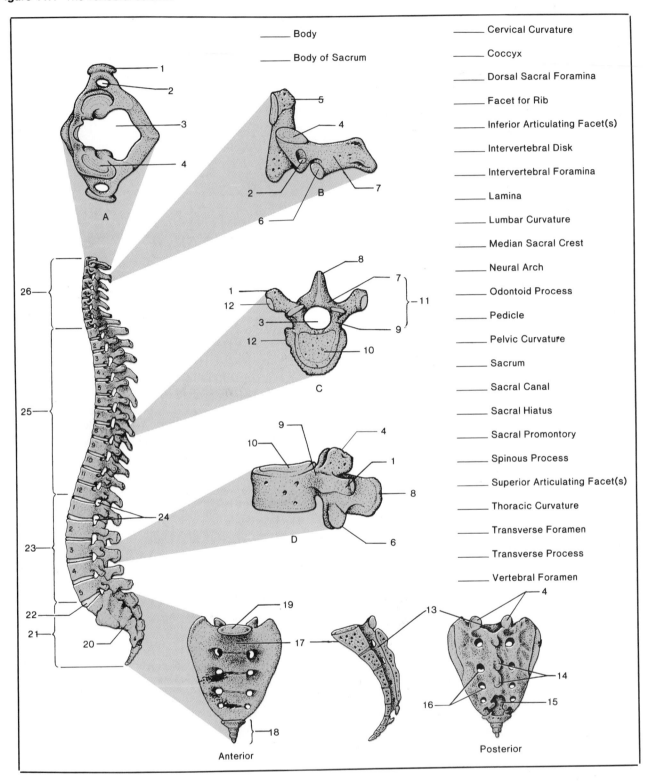

_____ Body

_____ Body of Sacrum

_____ Cervical Curvature

_____ Coccyx

_____ Dorsal Sacral Foramina

_____ Facet for Rib

_____ Inferior Articulating Facet(s)

_____ Intervertebral Disk

_____ Intervertebral Foramina

_____ Lamina

_____ Lumbar Curvature

_____ Median Sacral Crest

_____ Neural Arch

_____ Odontoid Process

_____ Pedicle

_____ Pelvic Curvature

_____ Sacrum

_____ Sacral Canal

_____ Sacral Hiatus

_____ Sacral Promontory

_____ Spinous Process

_____ Superior Articulating Facet(s)

_____ Thoracic Curvature

_____ Transverse Foramen

_____ Transverse Process

_____ Vertebral Foramen

Anterior

Posterior

Figure 11.2 The thorax.

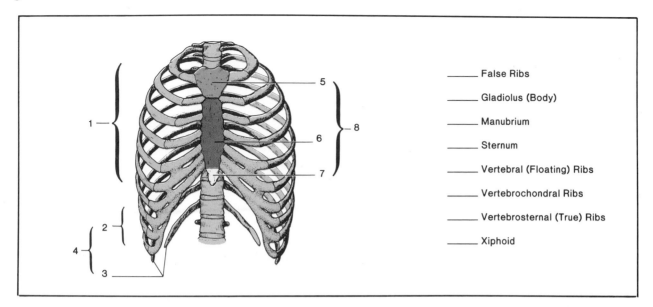

_____ False Ribs

_____ Gladiolus (Body)

_____ Manubrium

_____ Sternum

_____ Vertebral (Floating) Ribs

_____ Vertebrochondral Ribs

_____ Vertebrosternal (True) Ribs

_____ Xiphoid

Lumbar Vertebrae

The next five vertebrae (numbers 20–24) are the **lumbar vertebrae** (Illustration D). They have larger bodies than other vertebrae due to the greater stress that occurs in this region of the vertebral column.

The Sacrum

Inferior to the fifth lumbar vertebra is the **sacrum,** which composes the posterior wall of the pelvic cavity. It consists of five fused vertebrae. The transverse ridges where the vertebrae have fused are visible. The sacrum curves posteriorly causing an anterior protrusion of the body of the first sacral vertebra. This protrusion is the **sacral promontory.**

On the posterior surface, the spinous processes have fused to form the tubercles of the **median sacral crest.** On each side are rows of **dorsal sacral foramina** through which nerves and blood vessels pass. The neural arches of the fused vertebrae form the **sacral canal,** which continues to an inferior opening, the **sacral hiatus.**

The Coccyx

The last four or five vertebrae fuse together to form the **coccyx** or tailbone.

The Thoracic Cage

The thoracic vertebrae, ribs, costal cartilages, and sternum form the skeleton of the thorax. It protects the heart and lungs, and provides a support for the shoulder girdles. See Figure 11.2.

Ribs

There are twelve pairs of **ribs** attached to thoracic vertebrae. Each rib attaches to the transverse process and body of its own vertebra, and also to the body of the vertebra just above it.

The first seven pairs of ribs attach directly to the sternum by **costal cartilages.** They are the **vertebrosternal** or **true ribs.** The remaining five pairs are **false ribs.** The cartilages of the first three pairs of false ribs are fused to the costal cartilages of ribs above them. These are the **vertebrochondral ribs.** The last two pairs of false ribs are not attached anteriorly and are called **vertebral** or **floating ribs.**

Sternum

The **sternum,** or breastbone, is notched along the sides where the costal cartilages attach. It consists of three parts: an upper **manubrium,** a middle **body** or **gladiolus,** and a lower **xiphoid process.**

A red marrow biopsy is usually obtained from the sternum because of its accessibility. A **sternal puncture** is made by inserting a large-bore needle into the sternum under local anesthetic, and aspirating a sample of red marrow.

Assignment

1. Label Figures 11.1 and 11.2.
2. Complete the laboratory report.
3. Examine the skeletal material available in the laboratory and locate the parts discussed above. _Use only pipe cleaners as pointers._

The Appendicular Skeleton

Objectives

Upon completion of this exercise, you should be able to:

1. Identify the bones, including their significant parts, that compose the appendicular skeleton.
2. Describe the relationships of the bones involved.
3. Define all terms in bold print.

Materials

Skeletons, articulated and disarticulated
Pelvic girdles, male and female
Pipe cleaners

The appendicular skeleton consists of the shoulder and pelvic girdles, and the extremities. Refer to the figures to locate the bones and their parts as you study this exercise.

The Shoulder Girdle

The **shoulder** or **pectoral girdle** is formed by the collar bones and shoulder blades. The **clavicle,** or collar bone, is a slender, S-shaped bone that articulates with the sternum medially and with the acromion process of the scapula laterally.

The **scapula,** or shoulder blade, is a flat, triangular-shaped bone that does not articulate directly with the axial skeleton. Instead, it is held in place by muscles, thus giving greater mobility to the shoulder. On its **axillary margin** is the shallow **glenoid cavity** that articulates with the head of the humerus. Above the glenoid cavity are two large processes. The **coracoid process** projects anteriorly under the clavicle. The **acromion process** projects posteriorly and articulates with the clavicle. On its posterior surface, the scapular **spine** extends from the medial **vertebral margin** to the acromion process. A **scapular notch** occurs on the **superior margin** at the base of the coracoid.

Figure 12.1 The scapula.

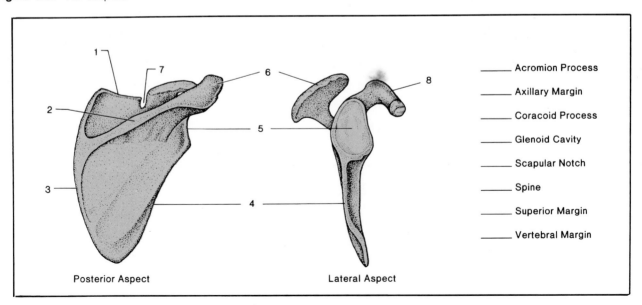

Posterior Aspect

Lateral Aspect

_____ Acromion Process

_____ Axillary Margin

_____ Coracoid Process

_____ Glenoid Cavity

_____ Scapular Notch

_____ Spine

_____ Superior Margin

_____ Vertebral Margin

Figure 12.2 The upper extremity and pectoral girdle.

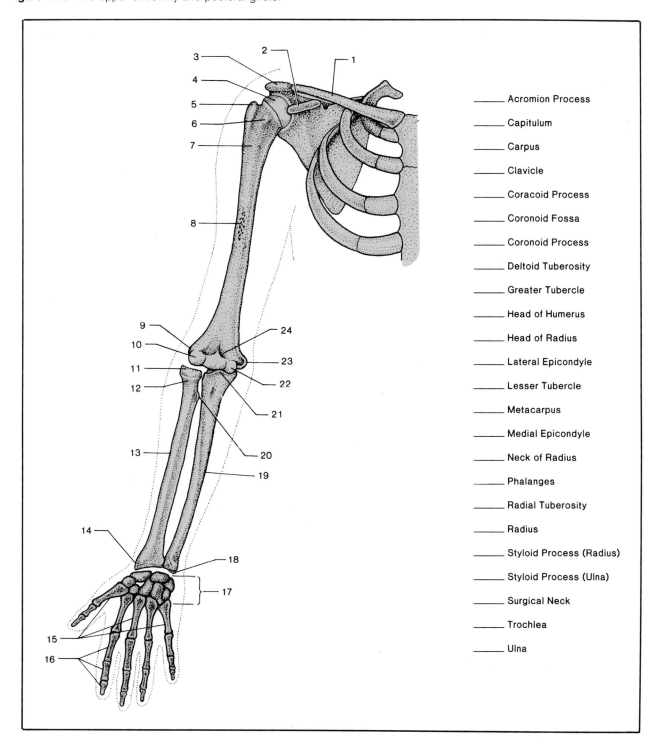

_____ Acromion Process

_____ Capitulum

_____ Carpus

_____ Clavicle

_____ Coracoid Process

_____ Coronoid Fossa

_____ Coronoid Process

_____ Deltoid Tuberosity

_____ Greater Tubercle

_____ Head of Humerus

_____ Head of Radius

_____ Lateral Epicondyle

_____ Lesser Tubercle

_____ Metacarpus

_____ Medial Epicondyle

_____ Neck of Radius

_____ Phalanges

_____ Radial Tuberosity

_____ Radius

_____ Styloid Process (Radius)

_____ Styloid Process (Ulna)

_____ Surgical Neck

_____ Trochlea

_____ Ulna

The Appendicular Skeleton   **59**

Falling on an outstretched arm often provides sufficient force to fracture the clavicle. It is the most frequently broken bone in the body because it transmits forces from the upper extremity to the trunk.

The Upper Extremity

The rounded **head** of the **humerus** fits into the glenoid cavity of the scapula. Just inferior to the head are two processes: a **greater tubercle** on the lateral surface and a **lesser tubercle** on the anterior surface. The **surgical neck** is inferior to the tubercles and is so named due to the frequency of fractures in this area. The **deltoid tuberosity** is a rough, raised area near the midpoint of the lateral surface of the shaft.

The distal end of the humerus has two condyles. The **capitulum** is the lateral condyle articulating with the radius. The **trochlea** is the medial condyle that articulates with the ulna. Superior to these condyles are the **lateral** and **medial epicondyles.** The depression on the anterior surface just superior to the trochlea is the **coronoid fossa.** The **olecranon fossa** (not shown in Figure 12.2) is in a similar location on the posterior surface of the humerus.

Two bones occur in the forearm: a lateral **radius** and a medial **ulna.** The disk-like **head** of the radius articulates with the capitulum of the humerus and enables the head to rotate with the supination and pronation of the hand (see Illustrations H and I, Figure 14.2). A short distance below the head is the **radial tuberosity.** The **neck** is located between the head and the tuberosity. At the distal end of the radius, a lateral **styloid process** is present at the articulation with the wrist.

The proximal posterior prominence of the ulna, the **olecranon process** (not shown), forms the point of the elbow and fits into the olecranon fossa when the arm is extended. It has a depression, the **trochlear** or **semilunar notch,** that articulates with the trochlea of the humerus. A small eminence at the anterior margin of the trochlear notch is the **coronoid process** (label 21). At the distal end of the ulna, the knob-like head articulates with the radius and a fibrocartilagenous disk that separates it from the wrist. The **styloid process** is the distal medial prominence.

The skeleton of each hand consists of carpus, metacarpus, and phalanges. Eight small bones form the **carpus** or wrist. The **metacarpus** or palm consists of five metacarpal bones numbered 1–5 starting with the thumb. The **phalanges** are the bones of the fingers; two in the thumb and three in each finger.

The Pelvic Girdle

The **pelvic girdle** or **pelvis** is formed by two **os coxae** (hipbones), the sacrum, and the coccyx. The coxal bones unite with the sacrum posteriorly at the **sacroiliac joints** and with each other anteriorly at the **symphysis pubis.** See Figure 12.4.

Each coxal bone is formed of three fused bones as shown in Figure 12.3. The **ilium** is the broad upper bone whose superior margin forms the **iliac crest,** the prominence of the hip. The **ischium** is the lower posterior portion. The **pubis** is the lower anterior part. A large lateral fossa, the **acetabulum,** is the socket that receives the head of the femur.

Figure 12.3 The os coxa.

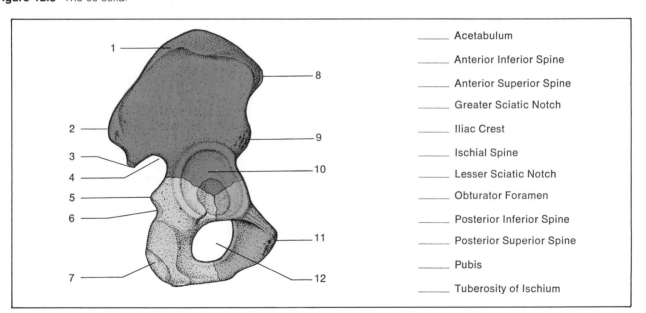

_____ Acetabulum

_____ Anterior Inferior Spine

_____ Anterior Superior Spine

_____ Greater Sciatic Notch

_____ Iliac Crest

_____ Ischial Spine

_____ Lesser Sciatic Notch

_____ Obturator Foramen

_____ Posterior Inferior Spine

_____ Posterior Superior Spine

_____ Pubis

_____ Tuberosity of Ischium

Figure 12.4 The lower extremity.

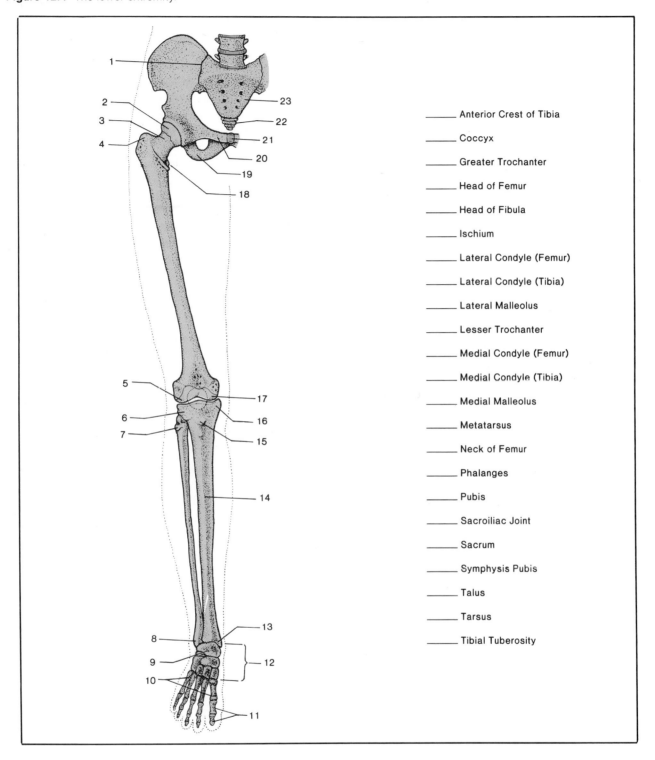

_____ Anterior Crest of Tibia

_____ Coccyx

_____ Greater Trochanter

_____ Head of Femur

_____ Head of Fibula

_____ Ischium

_____ Lateral Condyle (Femur)

_____ Lateral Condyle (Tibia)

_____ Lateral Malleolus

_____ Lesser Trochanter

_____ Medial Condyle (Femur)

_____ Medial Condyle (Tibia)

_____ Medial Malleolus

_____ Metatarsus

_____ Neck of Femur

_____ Phalanges

_____ Pubis

_____ Sacroiliac Joint

_____ Sacrum

_____ Symphysis Pubis

_____ Talus

_____ Tarsus

_____ Tibial Tuberosity

The iliac crest extends between the **anterior superior spine** (label 8) and the **posterior superior spine** (label 2). Just inferior to each of these prominences are smaller iliac spines: the **anterior inferior spine** and the **posterior inferior spine.** Between the posterior inferior spine and the **ischial spine** is the **greater sciatic notch.** The lesser **sciatic notch** is just inferior to the ischial spine. The **tuberosity of the ischium** is located at the posterior inferior angle of the ischium. The large opening surrounded by the ischium and pubis is the **obturator foramen.**

The female pelvis is modified for childbearing. The size of the pelvic outlet is largely determined by the curvature of the coccyx and size of the ischial spines. Physicians carefully measure the pelvic outlet to be sure that it is ample for the passage of the fetus. If it is not, a Caesarean delivery is indicated.

The Lower Extremity

The rounded **head** of the **femur** fits into the acetabulum of an os coxa. Two large processes occur at the base of the **neck** (label 3 in Figure 12.4): the lateral **greater trochanter** and the medial **lesser trochanter.** The enlarged lower end of the femur terminates with the **lateral** and **medial condyles,** which have **lateral** and **medial epicondyles** just above them.

The bones of the leg are the **tibia** (shinbone) and the smaller **fibula.** The upper end of the tibia consists of **lateral** and **medial condyles** that articulate with the corresponding condyles of the femur. The **tibial tuberosity** is located on the anterior surface just inferior to the condyles. A sharp **anterior crest** is evident on the shaft. At the distal end a process called the **medial malleolus** forms the medial prominence of the ankle.

The superior end or **head** of the fibula articulates with the tibia but does not form part of the knee joint. The **neck** lies just below the head and an **anterior crest** is present. At the inferior end, the **lateral malleolus** forms the lateral prominence of the ankle.

The skeleton of each foot consists of tarsus, metatarsus, and phalanges. There are seven tarsal bones forming the **tarsus** or ankle. The most prominent tarsal bones are the **talus,** which articulates with the tibia and fibula, and the **calcaneous** or heelbone. The instep or **metatarsus** is formed by five metatarsal bones numbered 1–5 starting on the medial side. The bones of the toes are the **phalanges;** two in the great toe and three in each of the other toes.

Assignment

1. Label Figures 12.1 through 12.4.
2. Complete Sections A and B on laboratory report.
3. Examine the skeletal material and locate the parts described above.
4. Compare the differences in male and female pelves in accordance with Section C on the laboratory report.

Articulations

Objectives

After completing this exercise, you should be able to:

1. State the basic functional types of joints and describe their characteristics.
2. Classify the joints of the body as to functional type.
3. Identify the components of diarthrotic joints.
4. Define all terms in bold print.

Materials

Skeleton, articulated
Fresh knee joint of cow or lamb, sectioned
  longitudinally

The Basic Functional Types

All joints of the body are classified according to three functional categories as shown in Figure 13.1.

Immovable Joints

The lack of movement of these **synarthrotic joints** is due to the bonding of the bones with fibrous connective tissue or cartilage. There are two basic types of immovable joints.

Sutures are irregular joints between bones of the cranium that are bonded together by fibrous connective tissue. This tissue is continuous with the **periosteum** (label 2) on the outside and the **dura mater** on the inside.

Synchondroses have cartilage as the bonding tissue. An example is the bonding of the epiphyses to the diaphysis by the epiphyseal cartilages in children.

Figure 13.1 Types of articulations.

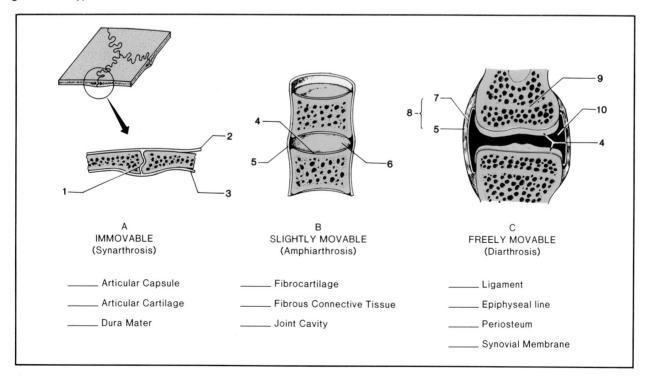

A	B	C
IMMOVABLE	SLIGHTLY MOVABLE	FREELY MOVABLE
(Synarthrosis)	(Amphiarthrosis)	(Diarthrosis)

_____ Articular Capsule   _____ Fibrocartilage   _____ Ligament

_____ Articular Cartilage   _____ Fibrous Connective Tissue   _____ Epiphyseal line

_____ Dura Mater   _____ Joint Cavity   _____ Periosteum

_____ Synovial Membrane

Slightly Movable Joints

The bones of these **amphiarthrotic joints** are also bound by fibrous connective tissue or cartilage but not as tightly as in immovable joints.

Symphyses have a pad of fibrocartilage between the bones. The articulations of the bodies of vertebrae and the symphysis pubis are examples. Note in Figure 13.1 that the bone surfaces adjacent to the fibrocartilage are covered with a hyaline type **articular cartilage** and that the joint is enveloped by an **articular capsule.**

Syndesmoses are held together by fibrous connective tissue, forming interosseous ligaments. The connections between the distal ends of the ulna and radius are of this type.

Freely Movable Joints

Bones forming these **diarthrotic** or **synovial joints** are bound together by a fibrous **articular capsule** formed of ligaments. A **synovial membrane** lines the inside of the capsule and secretes **synovial fluid** that lubricates the joint. The articular surfaces of the bones are covered by protective, friction-reducing **articular cartilages** of the hyaline type. These joints may also contain **bursae,** sacs of synovial fluid, that reduce friction. There are six kinds of synovial joints.

Gliding joints occur between small bones with flat or slightly convex surfaces such as the carpal and tarsal bones.

Hinge joints allow movement in only one plane such as in the elbow and knee.

Condyloid joints allow movement in two planes. They are formed by a rounded condyle articulating with an elliptical depression such as between the carpus and the radius.

Saddle joints occur where the ends of both bones are saddle-shaped, convex in one direction and concave in the other. An example is the joint between the trapezium (a carpal bone) and the metacarpal bone of the thumb, which permits a variety of movements.

Pivot joints allow rotational movement around a pivot point such as the atlas around the odontoid process of the axis.

Ball–and–socket joints allow angular movement in all directions such as in the shoulder and hip joints.

Joints are subject to a variety of disorders. **Bursitis** is the inflammation of a bursa or synovial membrane due to trauma, stress, or infection. The joint is painful and often swollen. **Dislocations** involve the displacement of the articulating bones, which produces damage to ligaments and tendons. Pain, swelling, deformity, and immobility of the joint are characteristic. **Sprains** result from excessive stretching or tearing of ligaments and tendons without dislocating the articulating bones. Pain, swelling, and restricted movement are characteristic.

The Shoulder Joint

The shoulder joint is the most freely movable joint in the body. Figure 13.2 shows its structure. Locate the **articular cartilages** that cover the head of the humerus and line the glenoid cavity. The cut ends of the acromion process and clavicle are seen above the joint. Note the **subdeltoid bursa** located between the **deltoid muscle** (label 1) and the insertion of the **supraspinatus muscle** (label 5). The **subacromial bursa** is seen below the epiphysis of the humerus.

Assignment

Label Figures 13.1 and 13.2.

The Knee Joint

The structure of this complex hinge joint is shown in Figure 13.3. The knee is probably the most highly stressed joint in the body. Some of this stress is absorbed by two **semilunar cartilages** or **menisci.** These lateral and medial fibrocartilagenous pads are thicker at the periphery and thinner at the center of the joint providing a recess for the condyles. The menisci are wrapped together peripherally by a **transverse ligament** (label 20).

Figure 13.2 The shoulder joint.

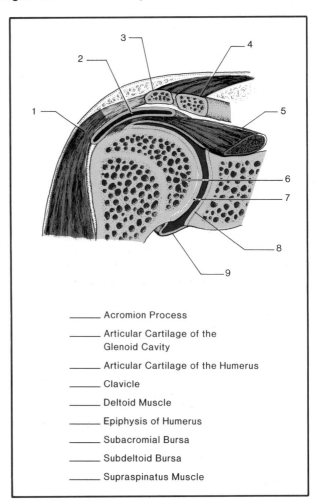

_____ Acromion Process

_____ Articular Cartilage of the Glenoid Cavity

_____ Articular Cartilage of the Humerus

_____ Clavicle

_____ Deltoid Muscle

_____ Epiphysis of Humerus

_____ Subacromial Bursa

_____ Subdeltoid Bursa

_____ Supraspinatus Muscle

Figure 13.3 The knee joint.

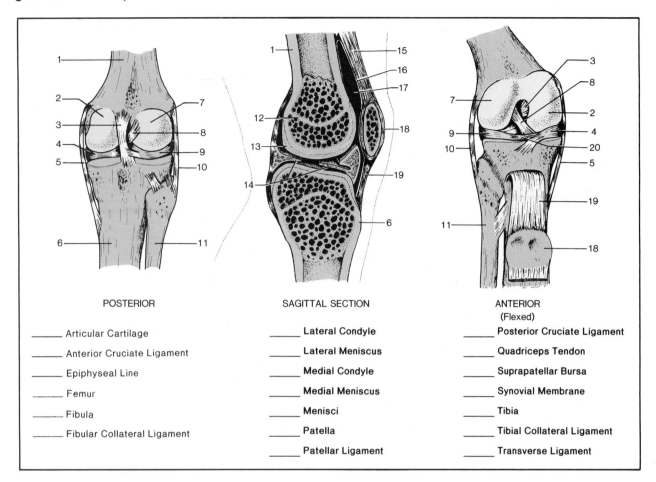

POSTERIOR

_____ Articular Cartilage

_____ Anterior Cruciate Ligament

_____ Epiphyseal Line

_____ Femur

_____ Fibula

_____ Fibular Collateral Ligament

SAGITTAL SECTION

_____ Lateral Condyle

_____ Lateral Meniscus

_____ Medial Condyle

_____ Medial Meniscus

_____ Menisci

_____ Patella

_____ Patellar Ligament

ANTERIOR
(Flexed)

_____ Posterior Cruciate Ligament

_____ Quadriceps Tendon

_____ Suprapatellar Bursa

_____ Synovial Membrane

_____ Tibia

_____ Tibial Collateral Ligament

_____ Transverse Ligament

The joint is enveloped by a fibrous capsule composed of a complex of several layers of ligaments and muscle tendons. Innermost are the **posterior** and **anterior cruciate ligaments** that form an X on the midline of the posterior surface. The anterior cruciate ligament is the outermost one (label 8) in the anterior view. The **fibular collateral ligament** is the lateral ligament extending from the fibula to the lateral epicondyle of the femur. The **tibial collateral ligament** extends from the tibia to the medial epicondyle of the femur. The **oblique** and **arcuate ligaments** are not shown. Note that there is no bony reinforcement to prevent dislocation, only the tough articular capsule. Thus, the knee joint is vulnerable to excessive lateral and rotational forces.

The sagittal section shows the kneecap held in place by the **quadriceps tendon** (label 15) and the **patellar ligament** (label 19). A **suprapatellar bursa** is posterior to the quadriceps tendon, and an **infrapatellar bursa** lies posterior to the patellar ligament. The **synovial membranes** of these bursae are evident.

A common football injury results from a lateral blow to knee. This usually results in tearing the tibial collateral ligament. It may also involve damage to the anterior cruciate ligament and the medial meniscus (torn cartilage).

Assignment

1. Label Figure 13.3.
2. Complete Sections A through C on the laboratory report.
3. Examine an articulated skeleton and classify the joints in accordance with Section D on the laboratory report.
4. Examine a fresh cow or sheep hinge joint that has been sectioned longitudinally. Compare it with Figure 13.3 and identify as many structures as possible.

Muscle Organization and Body Movements

Objectives

1. Identify and describe the structure of a skeletal muscle.
2. Describe the arrangement of muscles in the body.
3. Identify the various types of body movements in diarthrotic joints.
4. Define all terms in bold print.

Movement of limbs and other structures by skeletal muscles results from the contraction of the muscle fibers. The type of movement is determined by the sites of muscle attachment and the type of articulation involved. Typically, a muscle is attached to the skeleton at each end and spans across a joint where the movement occurs. In this exercise, you will study the structure of skeletal muscles and types of body movements at freely movable joints.

Figure 14.1 shows the orientation of two muscles of the upper arm as examples. The **triceps brachii** (label 7) extends the forearm, and the **brachialis** flexes the forearm.

Muscle Attachments

Most muscles attach to bones, but a few attach to other muscles or to soft tissues like the skin. Muscles may be attached to a bone in three ways: (1) directly to the periosteum, (2) by a tendon, or (3) by an aponeurosis. A **tendon** is a band or cord of white, fibrous connective tissue, which is an extension of the connective tissue within the body of the muscle. It intertwines with the periosteum at its terminus. An **aponeurosis** is a broad sheet of white fibrous connective tissue that attaches a muscle to a bone or to another muscle (not shown in Figure 14.1).

The immovable end of a muscle is called the **origin,** and the movable end—the end where action occurs—is called the **insertion.** Figure 14.1 shows that the triceps has three origins or **heads,** two on the humerus and one on the scapula, and one insertion on the olecranon of the ulna. Contraction of the muscle causes it to shorten and exert a forceful pull on the olecranon, which extends the forearm. The brachialis has a single origin on the shaft of the humerus and a single insertion on the coronoid process of the ulna. Contraction shortens the muscle causing a pull on the ulna, which results in flexion of the forearm.

The two heads of the triceps that are attached to the humerus and the origin of the brachialis are attached directly to the periosteum by the connective tissue of the muscle; no tendon is involved. In contrast, the scapular head of the triceps and the insertions of both muscles consist of tendons.

Microscopic Structure

Illustrations A, B and C in Figure 14.1 show the microscopic structure of a skeletal muscle. A muscle is composed of a number of **fasciculi** (label 11), like the single fasciculus shown in Illustration C. Each fasciculus consists of several muscle fibers separated by a thin layer of connective tissue, the **endomysium.** A fibrous connective tissue sheath, the **perimysium,** surrounds each fasciculus. Groups of fasciculi are enveloped by a layer of coarser connective tissue, the **epimysium.** External to the epimysium is the **deep fascia,** which covers the entire muscle and forms the outer layer of a tendon or aponeurosis. Note in Illustration A how the connective tissue of the muscle body extends to form the tendon that unites with the periosteum.

Assignment

Label Figure 14.1.

Figure 14.1 Muscle anatomy and attachment.

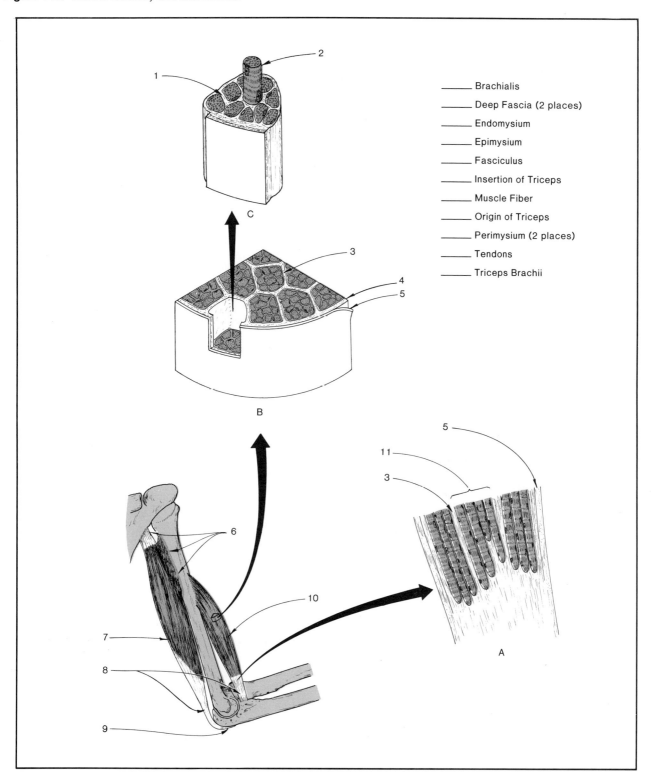

_____ Brachialis
_____ Deep Fascia (2 places)
_____ Endomysium
_____ Epimysium
_____ Fasciculus
_____ Insertion of Triceps
_____ Muscle Fiber
_____ Origin of Triceps
_____ Perimysium (2 places)
_____ Tendons
_____ Triceps Brachii

Body Movements

The arrangement of muscles in the body allows them to produce specific body movements upon contraction. The common types of movements are noted below.

Flexion and Extension

Flexion is the decrease of the angle formed by the bones at a joint. **Extension** is the increase of this angle. Thus, flexion occurs when the movement of the forearm brings the hand closer to the shoulder of the same arm. The opposite movement is extension.

The flexing of the foot upwards is called **dorsiflexion.** Extension of the foot is called **plantar flexion.** When extension is excessive, such as when leaning backwards, it is called **hyperextension.**

Abduction and Adduction

The movement of a limb away from the midline of the body is **abduction.** Movement toward the midline is **adduction.** These terms may be applied to parts of a limb, such as fingers and toes, by using the axis of the limb as the point of reference.

Pronation and Supination

The palm of the hand faces anteriorly in the anatomical position. Rotating the hand so the palm faces posteriorly is **pronation. Supination** rotates the palm to again face anteriorly. These terms also apply when the arm is not in the anatomical position. Turning the palm downward is pronation, and turning it upward is supination.

Inversion and Eversion

These terms apply to foot movements. **Inversion** is when the sole of the foot is turned inward toward the midline of the body. **Eversion** is turning the sole outward away from the midline.

Rotation and Circumduction

The movement of a body part around its longitudinal axis is **rotation.** If the body part can be rotated so that it describes a circle, the movement is called **circumduction.** For example, the head, arm, and leg can be rotated, but only the arm and leg can be circumducted.

Constriction and Dilation

Sphincter muscles are circular muscles that surround openings of the body. Contraction causes **constriction** of the opening; relaxation causes **dilation** of the opening.

Muscle Arrangements

Muscles usually function in groups to bring about the body movements. The **prime mover** is the primary muscle in the group that produces the desired action. The assisting muscles, the **synergists,** impart steadiness and prevent unnecessary action. Other muscles that hold structures in position for action are called **fixators.**

Muscles are usually arranged in **antagonistic groups** so that the contraction of one muscle, the **agonist,** causes a specific action and the contraction of the other muscle, the **antagonist,** causes the opposite action. Smooth movement requires that each muscle relaxes while the opposing muscle contracts.

Assignment

1. Identify the types of movements in Figure 14.2.
2. Complete the laboratory report.

Figure 14.2 Body movements.

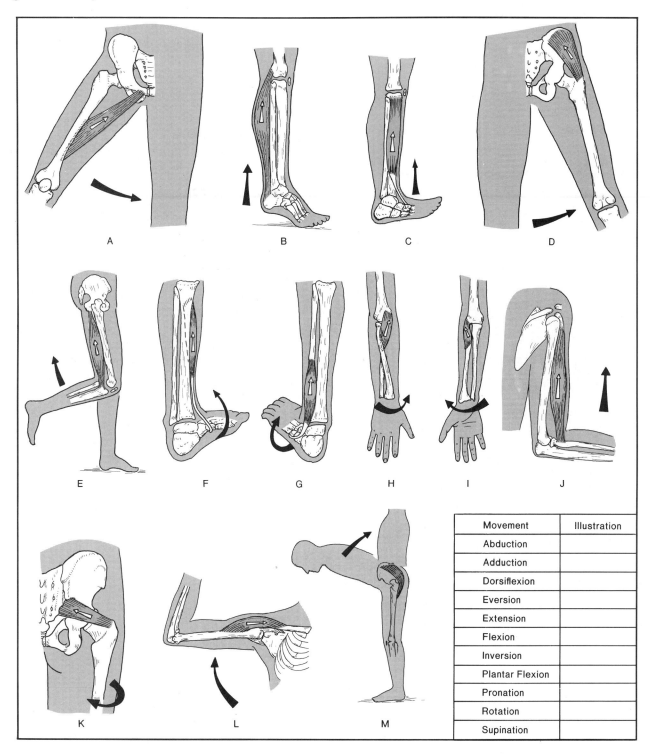

Movement	Illustration
Abduction	
Adduction	
Dorsiflexion	
Eversion	
Extension	
Flexion	
Inversion	
Plantar Flexion	
Pronation	
Rotation	
Supination	

Head and Trunk Muscles

Objectives

Upon completion of this exercise, you should be able to:

1. Identify the major superficial muscles of the head and trunk on charts or manikin.
2. Describe the action of the muscles studied.
3. Define all terms in bold print.

Materials

Skeleton, articulated
String
Colored pencils

Head Muscles

Refer to Figure 15.1 as you study this section.

Orbicularis Oris   This sphincter muscle encircles the mouth. Its *origin* is on several facial muscles, maxilla, mandible, and nasal septum. The *insertion* is on the lips.
   Action: Closes and puckers the lips.

Orbicularis Oculi   This is the sphincter muscle encircling the eye. Its *origin* is on the frontal and maxilla bones. Its *insertion* is on the eyelids.
   Action: Closes the eyelids.

Epicranius   This muscle is composed of two muscular parts that lie over the frontal and occipital bones. They are joined by the **epicranial aponeurosis,** which covers the top of the skull. The **frontalis** has its *origin* on the aponeurosis and its *insertion* on the soft tissue under the eyebrows. The **occipitalis** has its *origin* on the mastoid process and occipital bone. Its *insertion* is on the aponeurosis.
   Action: The frontalis elevates the eyebrows and wrinkles the forehead. The occipitalis pulls the scalp backward.

Zygomaticus   This muscle has its *origin* on the zygomatic arch and extends diagonally to its *insertion* on the orbicularis oris at the corner of the mouth.
   Action: Draws the corner of the mouth upward and backward as in smiling and laughing.

Triangularis   This muscle has its *origin* on the mandible and its *insertion* on the orbicularis oris at the corner of the mouth. It is an antagonist of the zygomaticus.
   Action: Draws the corner of the mouth downward.

Platysma   This is a broad sheet-like muscle that covers the front and side of the neck. Its *origin* is on the fascia of the upper chest and shoulder. The *insertion* is on the mandible and the muscles around the mouth. It is involved in many facial expressions.
   Action: Pulls the angle of the mouth downward and backward, and assists in opening the mouth.

Buccinator   This horizontal muscle is located in the walls of the cheeks. Its *origin* is on the mandible and maxilla, and its *insertion* is on the orbicularis oris at the corner of the mouth.
   Action: Compresses the cheek, holds food between teeth during chewing, and interacts with other muscles in facial expressions.

Masseter   This is the primary chewing muscle. Its *origin* is on the zygomatic arch, and its *insertion* is on the lateral surface of the ramus and angle of the mandible.
   Action: Raises the mandible.

Temporalis   This is a large, fan-shaped muscle on the side of the head. Its *origin* is on the temporal, parietal, and frontal bones. The *insertion* is on the coronoid process of the mandible.
   Action: Acts synergistically with the masseter to raise the mandible.

Figure 15.1 Head muscles.

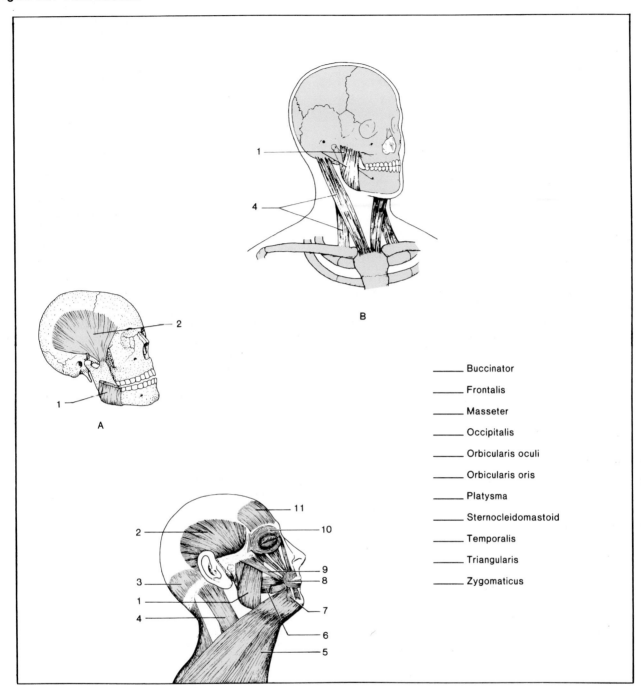

_____ Buccinator

_____ Frontalis

_____ Masseter

_____ Occipitalis

_____ Orbicularis oculi

_____ Orbicularis oris

_____ Platysma

_____ Sternocleidomastoid

_____ Temporalis

_____ Triangularis

_____ Zygomaticus

Sternocleidomastoid   This muscle is located on the side of the neck and is partially covered by the platysma. Its _origin_ is on the sternum and medial end of the clavicle, and its _insertion_ is on the mastoid process.

Action: Contraction of both muscles flexes the head down toward the chest. Contraction of one muscle only turns the face away from the side of the contracting muscle.

Assignment

Label Figure 15.1.

Anterior Upper Trunk Muscles

Refer to Figure 15.2 as you study this section.

Pectoralis Major   This fan-shaped muscle occupies the upper quadrant of the chest. Its _origin_ is on the clavicle, sternum, costal cartilages, and aponeurosis of the external oblique. Its _insertion_ is in a groove between the greater and lesser tubercles of the humerus.

Action: Adduction and medial rotation of the upper arm.

Figure 15.2 Shoulder and trunk muscles, anterior.

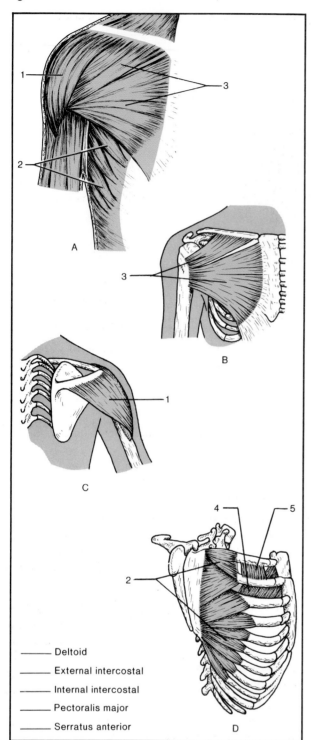

Figure 15.2 Shoulder and trunk muscles, anterior.

——— Deltoid

——— External intercostal

——— Internal intercostal

——— Pectoralis major

——— Serratus anterior

Deltoid   This primary muscle of the shoulder *originates* on the lateral third of the clavicle, and the acromion and spine of the scapula. It *inserts* on the deltoid tuberosity of the humerus.

Action: Abduction and extension of the upper arm.

Serratus Anterior   This muscle covers the upper lateral surface of the ribs. Its *origin* is on the upper eight or nine ribs, and its *insertion* is on the anterior surface of the scapula.

Action: Pulls scapula downward and forward toward the chest.

External Intercostals   Intercostal muscles are found between the ribs. The external intercostals (label 5) *originate* on the lower border of the upper rib and *insert* on the upper edge of the lower rib. The fibers are directed obliquely forward.

Action: Elevates rib cage and increases volume of the thorax, resulting in inspiration of air in breathing.

Internal Intercostals   These muscles *originate* on the upper border of the lower rib and *insert* on the lower edge of the upper rib. The muscle fibers are arranged in a direction opposite to the external intercostals.

Action: Lowers rib cage and draws ribs closer together. This decreases volume of the thorax, causing expiration of air in breathing.

Assignment

Label Figure 15.2.

Posterior Upper Trunk Muscles

Refer to Figure 15.3 as you study this section.

Trapezius   This large triangular muscle of the upper back has its *origin* on the occipital bone and the spinous processes of the cervical and thoracic vertebrae. It *inserts* on the spine and acromion of the scapula, and the lateral third of the clavicle.

Action: Adducts and elevates scapula; pulls head backward (hyperextension) if scapulae are fixed.

Latissimus Dorsi   This large muscle covers the lower back. Its *origin* is a large aponeurosis attached to thoracic and lumbar vertebrae, spine of the sacrum, iliac crest, and lower ribs. Its *insertion* is on the intertubercular groove of the humerus.

Action: Extends, adducts, and rotates the arm medially; pulls shoulder downward and backward.

Figure 15.3 Shoulder and back muscles.

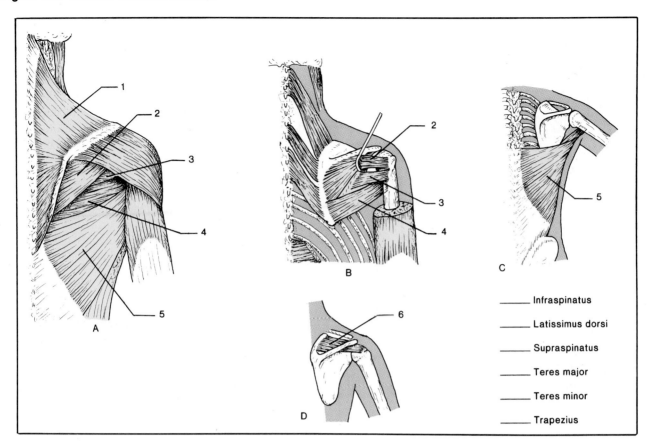

_____ Infraspinatus

_____ Latissimus dorsi

_____ Supraspinatus

_____ Teres major

_____ Teres minor

_____ Trapezius

Infraspinatus This muscle (label 2) *originates* on the inferior margin of the spine and posterior surface of the scapula. Its *insertion* is on the greater tubercle of the humerus. It is partially covered by the trapezius and deltoid in Illustration A. The small muscle just below it, the **teres minor,** synergistically assists the infraspinatus.

Action: Rotates upper arm laterally.

Supraspinatus This muscle (label 6) is covered by the trapezius and deltoid. It *originates* from the fossa above the scapular spine and *inserts* on the greater tubercle of the humerus.

Action: Assists deltoid in abducting upper arm.

Teres Major This muscle is the most inferior of the three muscles that *originate* on the posterior surface of the scapula. Its *insertion* is on the lesser tubercle of the humerus.

Action: Rotates upper arm medially.

Assignment

Label Figure 15.3.

Abdominal Muscles

The abdominal wall consists of four pairs of thin muscles that have a common collective action. The right side of the abdominal wall has been removed in Figure 15.4 to show the three layers composed of the first three muscles described below, starting with the outermost muscle.

External Oblique The *origin* of this superficial muscle is on the lower eight ribs. Its *insertion* is on the iliac crest and the **linea alba** (label 6), the white line at the midline of the abdomen where the aponeuroses (label 5) of the right and left external obliques meet. The lower margin of each aponeurosis forms the **inguinal ligament** that extends from the anterior superior spine of the ilium to the pubic tubercle. The muscle fibers run diagonally downward from the ribs toward the linea alba.

Internal Oblique This muscle lies just under the external oblique. Its *origin* is on the iliac crest and inguinal ligament. Its *insertion* is on the costal cartilages of the lower three ribs, the linea alba, and the pubic crest. The muscle fibers also run diagonally.

Figure 15.4 Abdominal muscles.

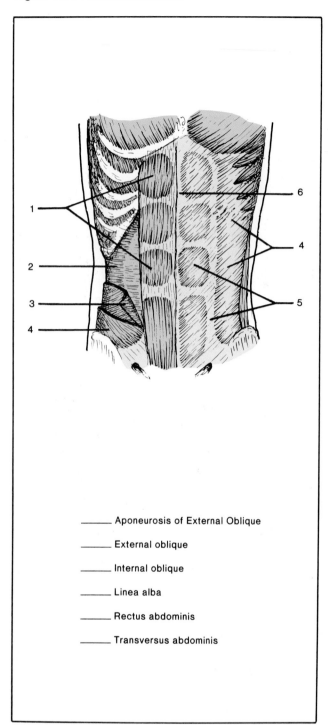

_____ Aponeurosis of External Oblique

_____ External oblique

_____ Internal oblique

_____ Linea alba

_____ Rectus abdominis

_____ Transversus abdominis

Transversus Abdominis This innermost muscle *originates* from the inguinal ligament, iliac crest, and the costal cartilages of the lower six ribs. It *inserts* on the linea alba and the pubic crest. The muscle fibers run horizontally across the abdomen.

Rectus Abdominis The right rectus abdominis is the narrow segmented muscle extending from the rib cage to the pubic bone in Figure 15.4. It is embedded within the aponeurosis formed by the above three muscles. Its *origin* is on the pubic bone, and its *insertion* is on the cartilages of the fifth, sixth, and seventh ribs. The rectus abdominis aids in flexion of the trunk in the lumbar region.

Collective Action: These four muscles compress the abdominal organs and maintain or increase intra-abdominal pressure aiding in urination, defecation, and childbirth. They are antagonists to the diaphragm and aid in forcing air from the lungs.

Assignment

1. Label Figure 15.4.
2. On the figures, color the origin blue and the insertion red for each muscle wherever they are exposed.
3. Complete the laboratory report.
4. Locate these muscles on a wall chart or manikin.
5. Study the action of these muscles by locating their origins and insertions on an articulated skeleton. Place one end of a piece of string on the origin and the other end on the insertion to help you visualize the effect of muscle contraction.

Muscles of the Upper Extremity

Objectives

After completing this exercise, you should be able to:

1. Identify the muscles studied on a chart or manikin.
2. Describe the action of the muscles studied.
3. Define all terms in bold print.

Materials

Skeleton, articulated
String
Colored pencils

The major muscles moving the upper arm were studied in Exercise 15 except for the coracobrachialis. It is included here since it is primarily located on the upper arm instead of on the trunk. See Illustration D, Figure 16.1.

   Coracobrachialis   This muscle covers the upper medial surface of the humerus. Its *origin* is on the coracoid process of the scapula, and its *insertion* is on the middle medial surface of the humerus.
   Action: Flexes and adducts upper arm.

Forearm Movements

The first four muscles described occur on the upper arm and are shown in Figure 16.1. The last two are forearm muscles and are depicted in Illustrations A and B in Figure 16.2.

   Biceps Brachii   This large muscle on the anterior surface of the upper arm bulges when the forearm is flexed. It *originates* at two sites on the scapula: the coracoid process and the tubercle above the glenoid cavity. Its *insertion* is on the radial tuberosity.
   Action: Flexes forearm; rotates forearm laterally (supination).

   Brachialis   This muscle lies just under the biceps. Its *origin* is on the lower half of the humerus. Its *insertion* is on the coronoid process of the ulna.
   Action: Flexes the forearm.

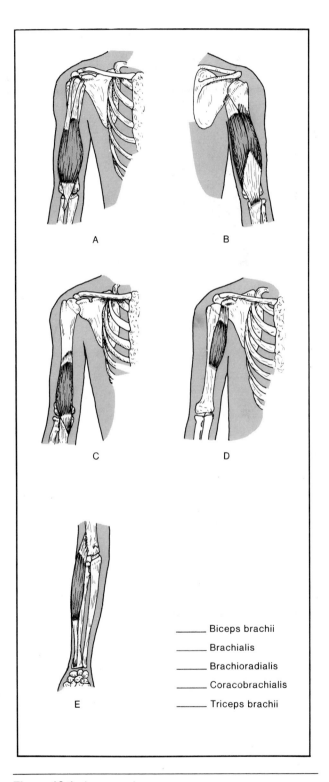

_____ Biceps brachii

_____ Brachialis

_____ Brachioradialis

_____ Coracobrachialis

_____ Triceps brachii

Figure 16.1 Arm muscles.

Brachioradialis This is the most superficial muscle on the lateral side of the forearm. Its *origin* is on the lateral epicondyle of the humerus. Its *insertion* is on the lateral surface of the radius just above the styloid process.
Action: Flexes the forearm.

Triceps Brachii This muscle covers the posterior surface of the upper arm and is the antagonist of the brachialis. Its *origin* occurs at three sites: the tubercle below the glenoid cavity, and the posterior and medial surfaces of the humerus. It *inserts* on the olecranon process of the ulna.
Action: Extends the forearm.

Supinator This small muscle (label 1, Figure 16.2) near the elbow originates from the lateral epicondyle of the humerus and proximal crest of the ulna. It curves around the radius and *inserts* on the lateral margin of the radial tuberosity.
Action: Rotates forearm laterally (supination).

Pronator Teres This muscle originates from the medial epicondyle of the humerus and *inserts* on the proximal lateral surface of the radius.
Action: Rotates forearm medially (pronation).

Pronator Quadratus This small muscle located near the wrist is a synergist of the pronator teres. Its *origin* is on the distal part of the ulna, and its *insertion* is on the distal lateral part of the radius.
Action: Pronates forearm.

Hand Movements

Muscles moving the hand are shown in Figure 16.2. All illustrations are of the right arm. Anterior views are shown in Illustrations C, D, and E; posterior views are in Illustrations F, G, and H.

Flexor Carpi Radialis This muscle shown in Illustration C *originates* from the medial epicondyle of the humerus and extends diagonally across the arm where its tendon divides to *insert* on the proximal ends of the second and third metacarpals.
Action: Flexes and abducts the hand.

Flexor Carpi Ulnaris This muscle also *originates* on the medial epicondyle of the humerus, but it *inserts* on the base of the fifth metacarpal.
Action: Flexes and adducts the hand.

Flexor Digitorum Superficialis This muscle (label 6) *originates* on the humerus, ulna, and radius. Its tendon divides to *insert* on the middle phalanges of fingers two through five. It is partially covered by the flexor carpi radialis and flexor carpi ulnaris.
Action: Flexes middle phalanges of fingers two through five.

Flexor Digitorum Profundus This muscle (label 8) *originates* on the anterior surface of the ulna. Its tendon divides to *insert* on the distal phalanges of fingers two through five.
Action: Flexes the distal phalanges of fingers two through five.

Flexor Pollicis Longus This smaller muscle in Illustration E *originates* on the radius and ulna, and *inserts* on the distal phalanx of the thumb.
Action: Flexes the thumb.

Extensor Carpi Radialis Longus and Brevis These two muscles are shown in Illustration F. Both *originate* on the lateral epicondyle of the humerus; the longus is attached in the more proximal position. The longus *inserts* on the base of the second metacarpal, while the brevis *inserts* on the base of the third metacarpal.
Action: Both muscles extend and abduct the hand.

Extensor Carpi Ulnaris This muscle lies on the medial side of the arm in Illustration H. It *originates* on the lateral epicondyle of the humerus, crosses over, and *inserts* on the base of the fifth metacarpal.
Action: Extends and adducts the hand.

Extensor Digitorum Communis This muscle (label 14) also *originates* from the lateral epicondyle of the humerus. Its tendon divides to *insert* on the posterior surfaces of the distal phalanges of fingers two through five.
Action: Extends fingers two through five.

Extensor Pollicis Longus This muscle (label 11) *originates* on the radius and ulna, and *inserts* on the distal phalanx of the thumb.
Action: Extends the thumb.

Abductor Pollicis This is the superior muscle in Illustration G. It *originates* on the interosseus ligament between the radius and ulna, and *inserts* on the lateral surface of the first metacarpal.
Action: Abducts the thumb.

Assignment

1. Label Figures 16.1 and 16.2.
2. For each muscle in the figures, color the origin blue and the insertion red.
3. Complete the laboratory report.
4. Locate the muscles on a wall chart or manikin.
5. Study the action of the muscles by locating their origins and insertions on an articulated skeleton. Place one end of a piece of string on the origin and the other end on the insertion to help you visualize the effect of muscle contraction.

Figure 16.2 Forearm muscles.

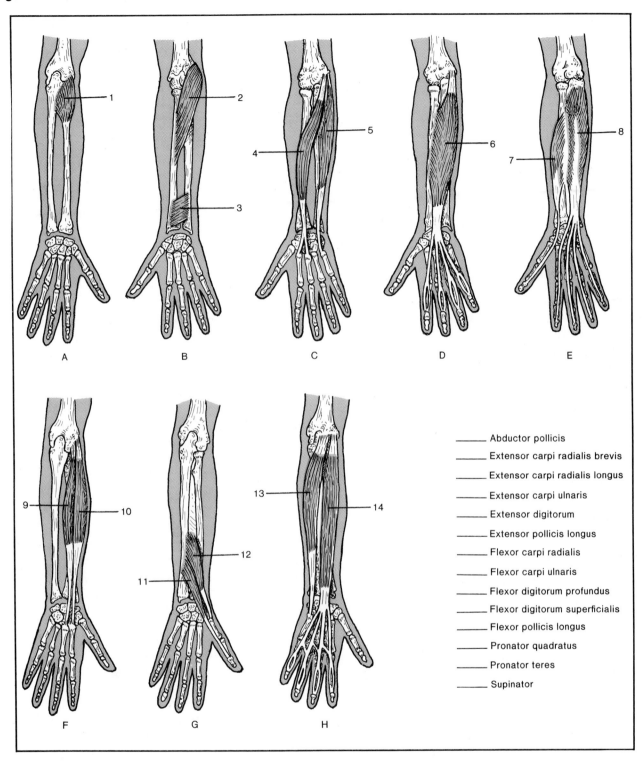

A   B   C   D   E

F   G   H

_____ Abductor pollicis
_____ Extensor carpi radialis brevis
_____ Extensor carpi radialis longus
_____ Extensor carpi ulnaris
_____ Extensor digitorum
_____ Extensor pollicis longus
_____ Flexor carpi radialis
_____ Flexor carpi ulnaris
_____ Flexor digitorum profundus
_____ Flexor digitorum superficialis
_____ Flexor pollicis longus
_____ Pronator quadratus
_____ Pronator teres
_____ Supinator

Muscles of the Upper Extremity   **77**

Muscles of the Lower Extremity

Objectives

After completing this exercise, you should be able to:

1. Identify the major leg muscles on a chart or manikin.
2. Describe the action of the major leg muscles.
3. Define all terms in bold print.

Materials

Skeleton, articulated
String
Colored pencils

The major muscles of the leg are described in this exercise according to functional groups.

Thigh Movements

These muscles originate on the pelvis and insert on the femur. See Figure 17.1.

Gluteus Maximus   This superficial buttocks muscle is covered by the deep fascia that invests the thigh muscles. A broad tendon, the **iliotibial tract** (label 2), extends downward from the deep fascia and attaches to the tibia. The gluteus maximus *originates* from the ilium, sacrum and coccyx and *inserts* on the posterior surface of the femur and the iliotibial tract.

Action: Extends and rotates the thigh laterally.

Figure 17.1   Muscles that move the thigh.

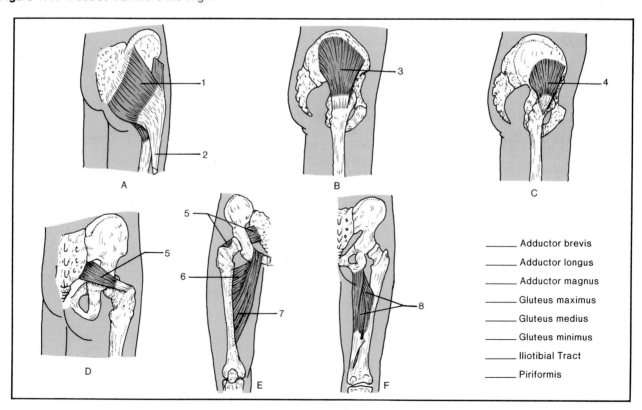

_____ Adductor brevis

_____ Adductor longus

_____ Adductor magnus

_____ Gluteus maximus

_____ Gluteus medius

_____ Gluteus minimus

_____ Iliotibial Tract

_____ Piriformis

Figure 17.2 Muscles that move the leg.

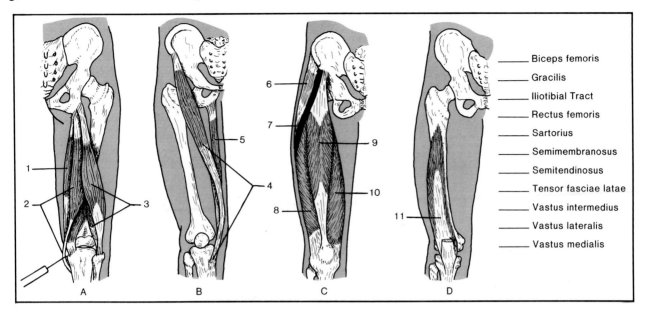

The figure shows four views (A, B, C, D) of leg muscles with the following labels:

- _____ Biceps femoris
- _____ Gracilis
- _____ Iliotibial Tract
- _____ Rectus femoris
- _____ Sartorius
- _____ Semimembranosus
- _____ Semitendinosus
- _____ Tensor fasciae latae
- _____ Vastus intermedius
- _____ Vastus lateralis
- _____ Vastus medialis

Gluteus Medius and Gluteus Minimus These muscles lie under the gluteus maximus with the minimus being the innermost. They _originate_ from the ilium and _insert_ on the greater trochanter of the femur.
Action: Abduct and rotate the thigh medially.

Piriformis This small muscle (label 5) _originates_ from the anterior surface of the sacrum and _inserts_ on the greater trochanter of the femur.
Action: Abducts and rotates the thigh laterally.

Adductor Longus and Adductor Brevis Both muscles _originate_ from the pubis and _insert_ on the posterior surface of the femur. The adductor brevis (label 6) is shorter and inserts superior to the adductor longus.
Action: Adduct, flex and rotate the thigh laterally.

Adductor Magnus This muscle, the strongest of the adductors, _originates_ from the ischium and pubis and _inserts_ on the posterior surface and the medial epicondyle of the femur.
Action: Adducts, flexes and rotates the thigh laterally.

Assignment
Label Figure 17.1.

Thigh and Leg Movements

These muscles, shown in Figure 17.2, are primarily involved in the flexion and extension of the leg.

Hamstrings These three muscles, shown in Illustration A, flex the leg.

Biceps Femoris This is the most lateral of the three hamstring muscles. It _originates_ from the ischial tuberosity and the posterior surface of the femur and _inserts_ on the head of the fibula and lateral epicondyle of the tibia.

Semimembranosus This is the most medial muscle of the three hamstrings. Its _origin_ is on the ischial tuberosity.
Insertion is on the posterior surface of the medial epicondyle of the tibia.

Semitendinosus Located between the other two hamstrings, this muscle _originates_ on the ischial tuberosity and _inserts_ on the upper medial surface of the tibia.
Collective Action: Flex lower leg; biceps rotates thigh laterally; the other two rotate the thigh medially.

Quadriceps Femoris This large muscle forms the anterior part of the thigh. It is composed of four muscles that unite to form a common tendon. The tendon attaches to the patella and continues as the patellar ligament to _insert_ on the tibial tuberosity.

Figure 17.3 Leg muscles.

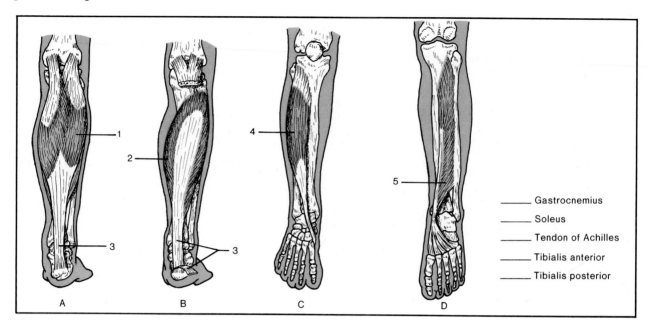

_____ Gastrocnemius
_____ Soleus
_____ Tendon of Achilles
_____ Tibialis anterior
_____ Tibialis posterior

Rectus Femoris   This is the anterior central portion of the quadriceps. It _originates_ from the inferior ischial spine and a site just above the acetabulum.

Vastus Lateralis   This lateral part of the quadriceps _originates_ from the greater trochanter and posterior surface of the femur.

Vastus Medialis   This media portion _originates_ from the posterior surface of the femur.

Vastus Intermedius   This portion (label 11) is obscured by the other portions of the qaudriceps. It _originates_ from the anterior and lateral surfaces of the femur.
Collective Action: Extends the lower leg; the rectus femoris also flexes the thigh.

Sartorius   This long muscle _originates_ from the anterior superior spine of the ilium and _inserts_ on the proximal medial surface of the tibia.
Action: Flexes the lower leg and the thigh; rotates leg medially.

Tensor Fasciae Latae   This muscle _originates_ from the anterior iliac crest and _inserts_ in the iliotibial tract (label 7).
Action: Flexes and abducts thigh.

Gracilis   This muscle _originates_ from the lower edge of the pubis and _inserts_ on the proximal medial surface of the tibia.
Action: Adducts the thigh and flexes the leg.

Assignment
Label Figure 17.2.

Leg and Foot Movements

Muscles that move the leg and foot are shown in Figures 17.3 and 17.4.

Gastrocnemius   This large superficial muscle covers the calf of the leg. It _originates_ from the lateral and medial condyles of the femur and _inserts_ on the posterior surface of the calcaneous.
Action: Flexes leg; extends the foot (plantar flexion).

Soleus   This muscle lies under the gastrocnemius and _originates_ from the head and posterior surface of the fibula and shaft of the tibia. Its tendon unites with the tendon of the gastrocnemius to form the **tendon of Achilles** that _inserts_ on the calcaneous.
Action: Extends the foot (plantar flexion).

Tibialis Anterior   This muscle _originates_ from the anterior surface of the lateral condyle and upper two-thirds of the tibia. It crosses over at the ankle to _insert_ on the inferior surface of a tarsal bone and the first metatarsus.
Action: Produces dorsiflexion and inversion of the foot.

Tibialis Posterior   This deep muscle _originates_ from the posterior surfaces of the tibia and fibula and _inserts_ on the inferior surfaces of several tarsal bones and metatarsal bones two through four.
Action: Extends (plantar flexion) and inverts the foot.

Figure 17.4 Leg muscles.

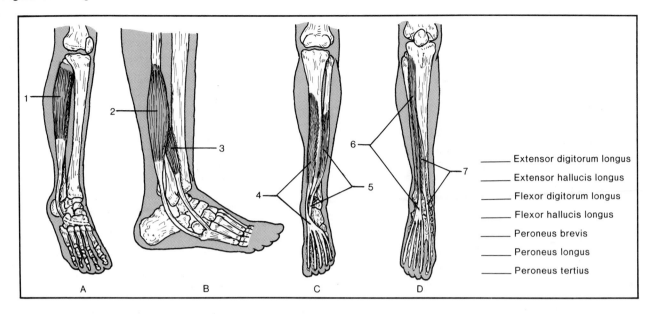

Extensor digitorum longus

Extensor hallucis longus

Flexor digitorum longus

Flexor hallucis longus

Peroneus brevis

Peroneus longus

Peroneus tertius

Peroneus Longus   This muscle (label 1, Figure 17.4) *originates* from the head and proximal portion of the fibula. Its tendon passes under the arch of the foot to *insert* on the first metatarsal and a tarsal bone.

Action: Extends and everts the foot; supports transverse arch.

Peroneus Brevis   This muscle (label 2) lies under the peroneus longus and *originates* on the middle portion of the fibula and *inserts* on the lateral proximal margin of the fifth metatarsal.

Action: Extends and everts the foot.

Peroneus Tertius   This is the smallest of the three peroneus muscles. It *originates* from the lower portion of the fibula and *inserts* on the medial proximal margin of the fifth metatarsal.

Action: Dorsiflexes and everts the foot.

Flexor Hallucis Longus   This muscle *originates* from the lower two-thirds of the fibula. Its tendon runs diagonally under the foot to *insert* on the distal phalanx of the great toe.

Action: Flexes the great toe.

Flexor Digitorum Longus   This muscle *originates* on the lower two-thirds of the media posterior surface of the tibia. Its tendon passes along the bottom of the foot and divides to *insert* on the distal phalanges of toes two through five.

Action: Flexes toes two through five; aids in extension of the foot.

Extensor Digitorum Longus   This muscle (label 6) *originates* on the lateral condyle of the tibia and the anterior surface of the fibula. Its tendon divides to *insert* on the upper surfaces of the second and third phalanges of toes two through five. It lies lateral and posterior to the tibialis anterior.

Action: Extends toes two through five; aids in dorsiflexion of the foot.

Extensor Hallucis Longus   This muscle (label 7) *originates* on the anterior surface of the fibula and *inserts* on the upper surface of the distal phalanx of the great toe.

Action: Extends the great toe; aids in dorsiflexion of the foot.

Assignment

1. Label Figures 17.3 and 17.4.
2. For each muscle in the figures, color the origin blue and the insertion red.
3. Complete Sections A through D on the laboratory report.
4. Locate the muscles on a wall chart or manikin.
5. Study the actions of the muscles by locating their origins and insertions on an articulated skeleton. Place one end of a piece of string on the origin and the other end on the insertion to help you visualize the effect of muscle contraction.
6. Review your understanding of superficial muscles by labeling Figure 17.5.
7. Complete the laboratory report.

Figure 17.5 Major surface muscles of the body.

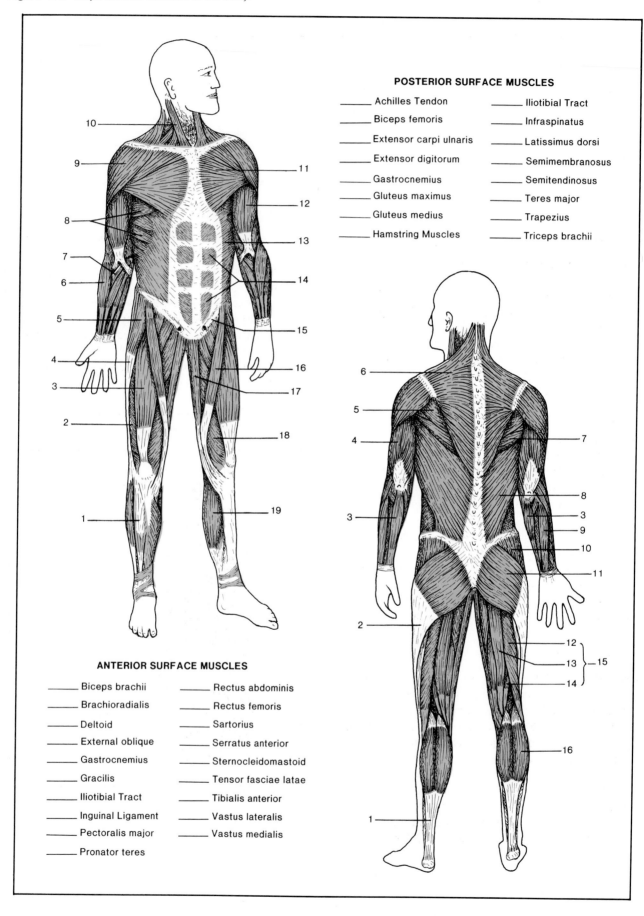

POSTERIOR SURFACE MUSCLES

_____ Achilles Tendon
_____ Biceps femoris
_____ Extensor carpi ulnaris
_____ Extensor digitorum
_____ Gastrocnemius
_____ Gluteus maximus
_____ Gluteus medius
_____ Hamstring Muscles

_____ Iliotibial Tract
_____ Infraspinatus
_____ Latissimus dorsi
_____ Semimembranosus
_____ Semitendinosus
_____ Teres major
_____ Trapezius
_____ Triceps brachii

ANTERIOR SURFACE MUSCLES

_____ Biceps brachii
_____ Brachioradialis
_____ Deltoid
_____ External oblique
_____ Gastrocnemius
_____ Gracilis
_____ Iliotibial Tract
_____ Inguinal Ligament
_____ Pectoralis major
_____ Pronator teres

_____ Rectus abdominis
_____ Rectus femoris
_____ Sartorius
_____ Serratus anterior
_____ Sternocleidomastoid
_____ Tensor fasciae latae
_____ Tibialis anterior
_____ Vastus lateralis
_____ Vastus medialis

Muscle and Nerve Tissue

Objectives

After completing this exercise, you should be able to:

1. Describe the characteristics of smooth, skeletal and cardiac muscle tissue and identify each type when viewed microscopically.
2. Describe the characteristics of motor, sensory and internuncial neurons and identify neurons when viewed microscopically.
3. Define all terms in bold print.

Materials

Prepared slides of:
cardiac muscle tissue
skeletal muscle tissue, teased and unteased
muscle tissue, teased and unteased
giant multipolar neurons

Muscle and nerve cells exhibit **excitability,** the ability to transmit electrochemical impulses along their membranes. The adaptive use of excitability results in (1) the contraction of muscle cells to enable movement, and (2) the transmission of impulses by nerve cells to coordinate body functions.

Muscle Tissues

Contractility, the ability to shorten or contract, is a major characteristic of muscle tissue. During contraction the elongate **muscle cells** or **fibers,** shorten and exert a pull at their attached ends that causes movement of body parts. Muscle tissues may be classified according to location, structure and function.

Skeletal Muscle

Skeletal muscle tissue occurs in muscles attached to bones or other muscles. The actions of skeletal muscles are consciously controlled. Thus, this tissue is classified functionally as **voluntary.**

Skeletal muscle fibers are elongate and cylindrical. They are grouped together in bundles called **fasciculi.** Each fasciculus is surrounded by fibrous connective tissue, the **perimysium,** and the individual fibers are separated by a thin layer of connective tissue, the **endomysium.** Each fiber is distinguished by numerous peripherally located nuclei and alternating dark and light bands called **striae,** or **striations,** in the **sarcoplasm** (cytoplasm). The entire fiber is covered by a very thin membrane, the **sarcolemma.** See Figures 14.1 and 18.1.

Figure 18.1 A fascicle of skeletal muscle fibers.

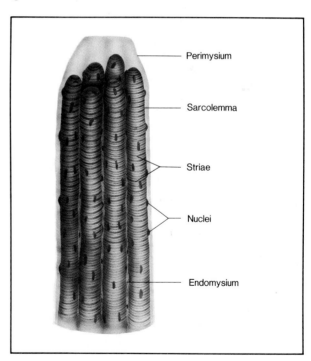

- Perimysium
- Sarcolemma
- Striae
- Nuclei
- Endomysium

Figure 18.2 Cardiac muscle fibers.

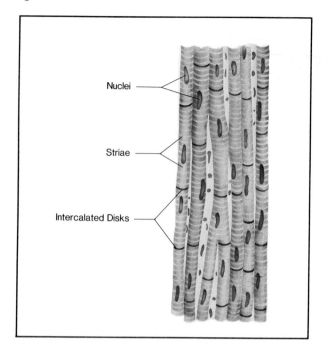

Figure 18.3 Smooth muscle fibers, teased.

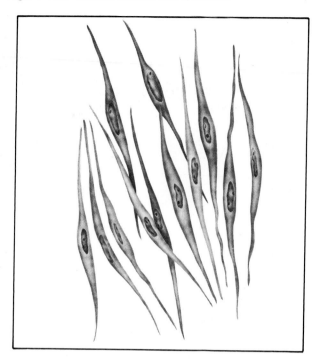

Cardiac Muscle

The muscular walls of the heart contain **cardiac muscle tissue** that contracts rhythmically without conscious control or neural stimulation. Thus, it is classified functionally as **involuntary.**

The fibers of cardiac muscle are composed of cells joined end-to-end and separated from each other by **intercalated disks.** The interconnected, branching fibers form a complex network. Each cell has a single nucleus, and striae are present in the sarcoplasm. See Figure 18.2.

Smooth Muscle

Smooth muscle tissue occurs in the walls of hollow internal organs and blood vessels. Individual cells are spindle-shaped, have a single centrally located nucleus and lack striae in the sarcoplasm. Since this tissue is not under conscious control, it is **involuntary** in function. See Figure 18.3.

Nerve Tissue

Nerve tissue occurs in the central nervous system (brain and spinal cord) and in peripheral nerves. The basic cells of this tissue are nerve cells or **neurons.** The associated **neuroglial cells** in the central nervous system provide structural support and prevent interneuron contact except at particular sites. The structure of neurons varies in accordance with their particular roles, but all neurons possess common features.

Figure 18.4 illustrates the structure of a **sensory neuron** and a **motor neuron.** It also shows the relationships between the three fundamental types of neurons: sensory, motor and internuncial neurons. These three types of neurons plus a sensory receptor and an effector, compose a **reflex arc** as shown in the cross section of the spinal cord.

Neuron Structure

The **cell body** (perikaryon) of a neuron is the portion containing the nucleus. The nucleus is quite large and has a conspicuous nucleolus. Numerous **Nissl bodies** are visible in the cytoplasm of the cell body. They are aggregations of rough endoplasmic reticulum. **Neurofibrils,** slender protein filaments that extend into the cell processes, are also visible.

There are two types of **neuron processes** or **fibers:** dendrites and axons. A **dentrite** or a **dendritic zone** receives impulses from receptors or from axon tips of other neurons. An **axon** is a single elongated process of the cytoplasm that carries the impulses from the dendrites to another neuron or an effector. Axons may have branches called **collaterals.** Neurons may have several dendrites, but there is only one axon. The cell body may be located anywhere along the conduction pathway, depending on the morphological type.

The processes of peripheral (i.e. outside the central nervous system) neurons are enclosed by a covering of **Schwann cells.** In larger peripheral fibers as shown in Figure 18.4, the inner multiple wrappings of Schwann cell

Figure 18.4 Sensory and motor neurons.

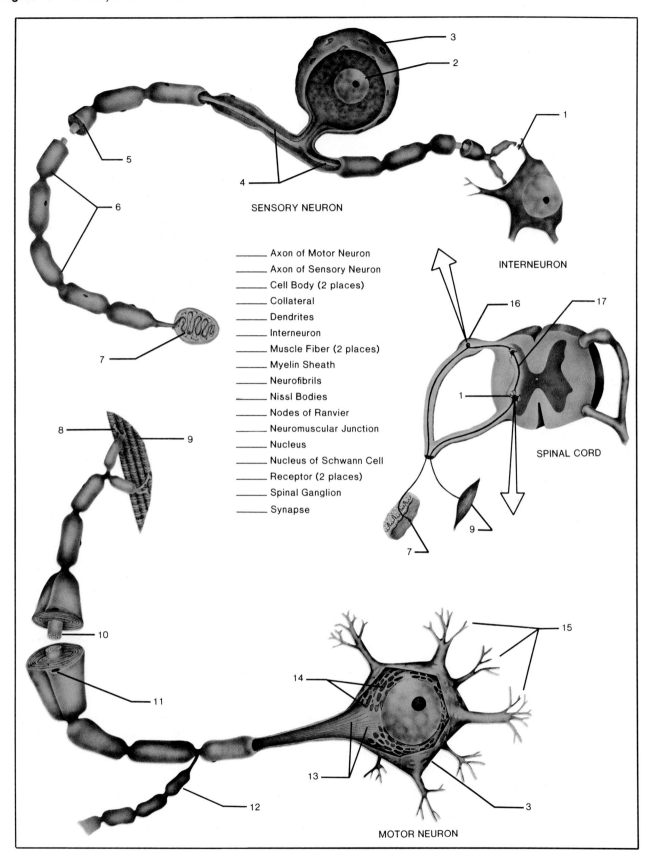

SENSORY NEURON

INTERNEURON

_____ Axon of Motor Neuron
_____ Axon of Sensory Neuron
_____ Cell Body (2 places)
_____ Collateral
_____ Dendrites
_____ Interneuron
_____ Muscle Fiber (2 places)
_____ Myelin Sheath
_____ Neurofibrils
_____ Nissl Bodies
_____ Nodes of Ranvier
_____ Neuromuscular Junction
_____ Nucleus
_____ Nucleus of Schwann Cell
_____ Receptor (2 places)
_____ Spinal Ganglion
_____ Synapse

SPINAL CORD

MOTOR NEURON

membranes form the **myelin sheath.** The outer layer of the Schwann cell, the part containing the nucleus and most of the cytoplasm, constitutes the **neurilemma.** Neuron fibers with a myelin sheath are said to be **myelinated;** those enclosed in Schwann cells but lacking a myelin sheath are **unmyelinated.** Schwann cells are essential for the regeneration of damaged neuron fibers. They provide a pathway for the regrowth of the neuron processes.

The myelin sheath of neurons in the central nervous system is formed by a type of neuroglial cell. Since Schwann cells are absent, damaged neuron processes in the brain and spinal cord do not regenerate, which results in permanent impairment of function.

The myelin sheath and neurilemma are interrupted at regular intervals along the neuron fiber by the **nodes of Ranvier.** These nodes are tiny spaces between the Schwann cells where the neuron fiber is exposed. They play an important role in increasing the speed of impulse transmission.

Types of Neurons

Neurons vary considerably in size, shape and function. There are three functional types of neurons: sensory, internuncial and motor neurons. Refer to Figure 18.4 as you study this section.

Sensory Neurons These neurons carry impulses from **receptors** (label 7) toward the central nervous system (brain and spinal cord). Thus, sensory neurons are also called **afferent neurons.** The cell bodies of sensory neurons in spinal nerves are found within **spinal ganglia** (label 16) that are located outside the spinal cord. The entire myelinated fiber (label 4) is classified as an axon, although

part of it carries impulses toward the cell body. A **synapse** (label 1) is shown where the tip of an axon contacts a dendritic zone of an adjacent neuron. Note that only the tip of the axon lies within the spinal cord.

Internuncial Neurons Large numbers of these unmyelinated, multipolar **interneurons** occur in the gray matter of the central nervous system. They serve as links between other neurons and transmit impulses from one part of the central nervous system to another. In Figure 18.4, the interneuron is the connecting link between the sensory and motor neurons. A synapse is indentified between the axon tip of the interneuron and a dendrite of the motor neuron.

Motor Neurons These neurons carry impulses away from the central nervous system and terminate at an **effector,** a muscle or gland. Thus, motor neurons are also called **efferent neurons.** In Figure 18.4, the effector is a muscle, and the union of the axon tip and the muscle constitutes the **neuromuscular junction.** Motor neurons are multipolar with only the myelinated axon located outside the central nervous system.

Assignment

1. Label Figure 18.4.
2. Complete Sections A, C, D, and E on the laboratory report.
3. Examine slides of each type of muscle tissue using Figures 18.1, 18.2, 18.3, and Figures HA-11 and HA-12 in the histology atlas as references.
4. Examine a slide of giant multipolar neurons using Figure HA-13 as a reference.
5. Complete the laboratory report.

The Nature of Muscle Contraction

Objectives

After completing this exercise, you should be able to:

1. Describe the basic structure and function of the neuromuscular junction.
2. Describe the orientation of myosin and actin filaments in resting and contracted muscle fibers.
3. Describe the basic mechanism of muscle contraction.
4. Define all terms in bold print.

Materials

Test tube of glycerinated rabbit psoas muscle
Dropping bottles of:
glycerol
    0.25% ATP in triple distilled water
    0.05 M KCl
    0.001 M MgCl₂
Microscope slides and cover glasses, 3 each
Dissecting instruments
Plastic ruler, clear and flat
Microscopes, compound and dissecting

Contraction of a muscle fiber is a complex physico-chemical process that involves both microscopic structural changes and chemical changes within the muscle fiber. Furthermore, contraction is dependent upon impulses from a motor neuron to initiate the process. In this exercise you will study the interaction of a motor neuron and a muscle fiber in the process of contraction, as well as the mechanism of contraction.

Skeletal muscle fibers are innervated by motor neurons that emerge from the spinal cord through the anterior roots of spinal nerves. A large motor neuron may branch many times to innervate from 3 to 2,000 skeletal muscle fibers.

Contraction of a skeletal muscle fiber involves the interaction of (1) a motor neuron, (2) the neuromuscular junction, and (3) a muscle fiber. As an impulse (action potential) passes along an axon to the neuromuscular junction, it causes chemical changes at the neuromuscular junction that form an action potential in the muscle fiber. This action potential, in turn, initiates physicochemical changes in the muscle fiber that culminate in contraction.

The Neuromuscular Junction

The union of the tip of a motor neuron axon with the sarcolemma of a muscle fiber forms the **neuromuscular junction.** See Figure 19.1.

The **axon terminal branches** (label 6) fit into invaginations of the **sarcolemma** (label 5). The minute space between the terminal branches and the sarcolemma is the **synaptic cleft.** Numerous folds of the sarcolemma adjacent to the synaptic cleft greatly increase its surface area.

When a neural impulse or action potential (depolarization of the neurilemma) reaches the axon terminal branches, **synaptic vesicles** (label 3) in the axon tip release the neurotransmitter **acetylcholine** into the synaptic cleft. Acetylcholine quickly reacts with its specific receptors on the sarcolemma, initiating an action potential (depolarization of the sarcolemma) in the muscle fiber. As the action potential penetrates the **transverse tubule system** (label 1) and then the **sarcoplasmic reticulum** (label 2), calcium ions are released, which enable contraction of the myofibrils.

Contraction is powered by ATP supplied by numerous **mitochondria** (label 8). Within 1/500 of a second, acetylcholine is deactivated and the sarcolemma is repolarized. The muscle fiber is now ready for another neural impulse to initiate another contraction.

Figure 19.1 The neuromuscular junction.

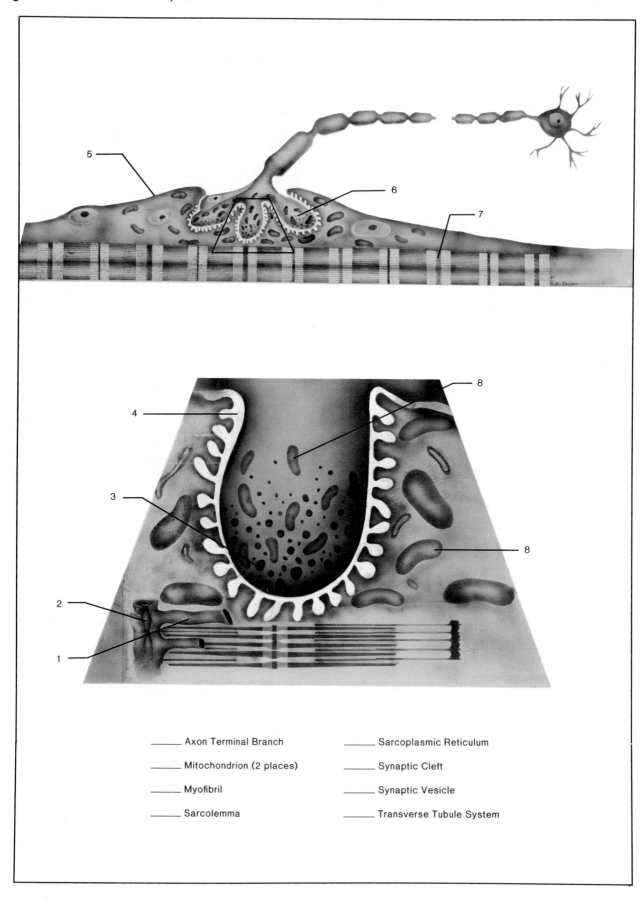

_____ Axon Terminal Branch		_____ Sarcoplasmic Reticulum	
_____ Mitochondrion (2 places)		_____ Synaptic Cleft	
_____ Myofibril		_____ Synaptic Vesicle	
_____ Sarcolemma		_____ Transverse Tubule System	

Myofibril Ultrastructure

The upper illustration in Figure 19.2 depicts the structure of a muscle fiber. Note the enveloping sarcolemma, the sarcoplasmic reticulum, and the transverse tubule system encircling the numerous **myofibrils.**

The striations of a muscle fiber are due to the arrangement of two kinds of **myofilaments** within each myofibril. See the middle illustration of Figure 19.2. **Myosin** myofilaments are the thicker of the two and are the main components of the **A Band.** The thinner **actin** myofilaments exend into the A Band from the **Z Lines.** The light-colored **H Zone** lies in the center of the A Band where only myosin is present. The light-colored **I Band** lies between the A Band and a Z Line. The Z lines form the boundaries of a **sarcomere,** the contractile unit of a myofibril.

Compare the relationships of actin and myosin myofilaments during the relaxed state (middle illustration) and contracted state (lower illustration) in Figure 19.2. Note that the actin filaments and Z lines are drawn toward the center of the A band in the contracted state.

Mechanics of Contraction

The shortening of a muscle fiber by contraction results from the interaction of actin and myosin myofilaments in the presence of calcium and magnesium ions. ATP provides the necessary energy. The most accepted explanation for myofibril contraction is known as the ratchet theory. The sequence of events is described below.

1. As the action potential moves along the sarcolemma, transverse tubule system, and sacroplasmic reticulum, large quantities of calcium ions are released.
2. Calcium ions cause changes that expose the chemically active sites on the actin myofilaments.
3. The cross-bridges of the myosin myofilaments attach to the active sites on the actin molecule. The cross-bridges bend, exerting a power stroke that pulls the actin filaments toward the center of the A band.
4. After the first power stroke, the cross-bridges separate from the first actin active sites, bond with next active sites, and produce another power stroke.
5. This ratchet-like process repeats itself until maximum contraction is attained.

Assignment

1. Label Figure 19.1.
2. Complete Sections A through D on the laboratory report.

Figure 19.2 Skeletal muscle fiber ultrastructure.

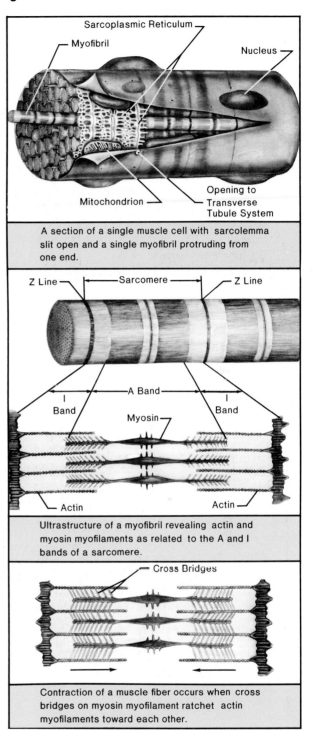

A section of a single muscle cell with sarcolemma slit open and a single myofibril protruding from one end.

Ultrastructure of a myofibril revealing actin and myosin myofilaments as related to the A and I bands of a sarcomere.

Contraction of a muscle fiber occurs when cross bridges on myosin myofilament ratchet actin myofilaments toward each other.

Experimental Muscle Contraction

Perform the following experiment in which a glycerinated rabbit psoas muscle will be induced to contract using ATP, and magnesium and potassium ions.

Your instructor has removed the muscle from its test tube and cut it into 2 cm segments that have been placed in a Petri dish of glycerol. These muscle segments have been teased apart to obtain thin muscle strands composed of very few fibers. The strands should be no more than 0.2 mm thick for use in the experiment.

1. Place a muscle strand in a small drop of glycerol on a clean microscope slide and add a cover glass. Examine it microscopically with the high dry or oil immersion objective. Draw the pattern of striations of the relaxed muscle fibers in Section E of the laboratory report.
2. Place 3–4 of the thinnest strands on another slide in a minimal amount of glycerol. Arrange the strands straight, parallel, and close to each other.
3. Using a dissecting microscope, measure the length of the strands by placing a clear plastic ruler under the slide. Record the lengths on the laboratory report. Estimate the width of the strands.
4. While observing through the dissecting microscope, add one drop from each of the solutions: ATP, magnesium chloride, and potassium chloride. Observe the contraction.
5. After 30 seconds, remeasure and record the length of the strands. Calculate and record the percentage of contraction. Has the width of the strands changed?
6. Remove one of the contracted strands to another slide, add a cover glass, and observe the pattern of the striations using a compound microscope. Diagram the pattern of striations on the laboratory report.
7. Clean your work station.
8. Complete the laboratory report.

The Physiology of Muscle Contraction

Objectives

After completing this exercise, you should be able to:

1. Explain the basis of a simple twitch and summation as studied in the frog gastrocnemius.
2. Describe the parts of a myogram.
3. Define all terms in bold print.

Materials

Grass frog, double pithed
Dissecting instruments and pins
Dissecting pan, wax bottom, or frog board
Ringer's solution in a squeeze bottle
Cotton thread
Physiograph IIIS with transducer coupler
SM-1 electronic stimulator
Myograph
Myograph tension adjuster
Stimulus switch
Transducer stand
Event marker cable
Sleeve electrode and pin electrode

The physiology of muscle contraction will be studied using a gastrocnemius muscle of a frog. The results will be similar to those obtained when using mammalian muscle tissue.

The experiments require careful preparation of the frog muscle, and correct set-up and operation of a considerable amount of equipment. Because of the complexity of the experiment, it will be performed by groups of 4–6 students who will perform the roles as outlined in Table 20.1.

Table 20.1. Team Member Responsibilities.

Equipment Engineer
Hooks up the various components (recorder, stimulator, transducer, etc.). Balances the transducer and calibrates recorder. Operates the stimulator. Responsible for correct use of equipment to prevent damage. Dismantles equipment and returns all components to proper places of storage at end of period.

Engineer's Assistant
Works with Engineer in setting up equipment to ensure procedures are correct. Makes necessary tension adustments before and during experiment. Applies electrode to nerve and muscle for stimulation.

Recorder
Labels events produced on tracings. Keeps a log of the sequence of events as experiment proceeds. Sees that each member of team is provided with chart records for attachment to Laboratory Report.

Surgeon
Removes skin on leg and exposes sciatic nerve. Responsible for maintaining viability of tissue. Attaches thread to nerve and muscle end.

Head Nurse
Assists surgeon as needed during dissection. Keeps exposed tissues moist with Ringer's solution. Helps to keep work area clean.

Coordinator
Oversees entire operation. Communicates among team members. Reports to instructor any problems that seem to be developing that seem insurmountable. Double checks the set-up and clean-up procedures. Keeps things moving.

Figure 20.1 Equipment setup with Narco Bio-Systems equipment.

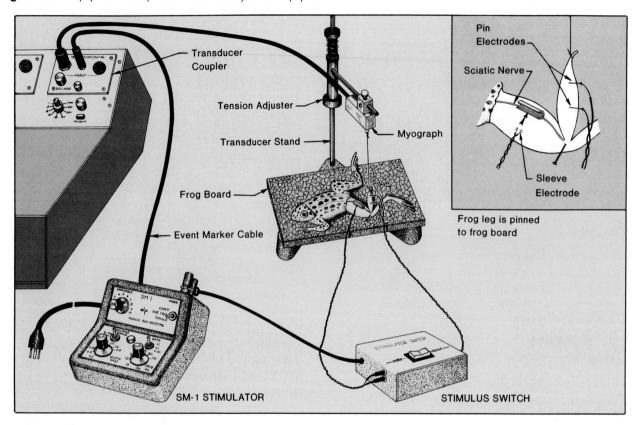

The exercise describes procedures for use of Narco Bio-Systems equipment. The set-up is shown in Figure 20.1. However, your equipment may be different. Your instructor will indicate the type and arrangement of the equipment available for your use.

Preparations

While two team members prepare the frog gastrocnemius muscle for study, the other members are to set up the equipment.

Frog Preparation

Obtain a double-pithed or decapitated and pithed frog. Such a frog has had its brain destroyed and is "brain dead," but its peripheral nerves, muscles, and other tissues are still living. Remove the skin from the left hind leg as shown in photos 1–3 in Figure 20.2.

Refer to photos 4–6 of this figure for the next steps. Use a dissecting needle to tear the fascia and separate the thigh muscles to expose the **sciatic nerve.** Be careful not to break the blood vessel near the nerve or to traumatize the nerve. The less it is touched with metal, the better. Pull a 10 in. length of thread under the nerve with forceps. The thread will be used to manipulate the nerve later; *don't tie it to the nerve. Keep the exposed tissues moist with Ringer's solution at all times.*

Equipment Preparation

Set up the equipment as shown in Figure 20.1. A **stimulus switch** is used between the **stimulator** and the frog. This switch makes it possible to switch from direct muscle stimulation to nerve stimulation without disturbing the setup. Note that an **event marker cable** connects the stimulator with the **Physiograph.** The receptacle for this cable is on the back of the stimulator and labeled MARKER OUTPUT. The other end of the cable is inserted into the middle receptacle of the **transducer coupler.**

After affixing the **myograph tension adjuster** to the transducer stand, attach the **myograph** to the support rod, and tighten the thumb screw to hold it in place. Insert the free end of the **myograph cable** into the left receptacle of the transducer coupler.

Plug the power cords for the Physiograph and SM-1 stimulator into grounded 110V outlets. Insert the jacks for the **pin electrodes** into the *right side* of the stimulus switch and the jacks for the **sleeve electrodes** into the *left side* of the stimulus switch. Now, attach the cable from the stimulus switch to the output posts of the SM-1 stimulator.

With the completion of all hookups, balance the channel and calibrate the myograph according to the instructions in Appendix A.

Figure 20.2 Preparation of the frog.

1. The first incision through the skin is made at the base of the thigh.

2. After the skin is cut around the leg, it is pulled away from the muscles.

3. While the body is held firmly with the left hand, the skin is stripped off the entire leg.

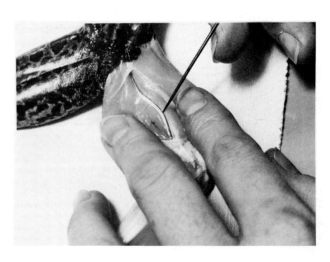

4. The muscles of the thigh are separated with a dissecting needle to expose the sciatic nerve.

5. Forceps are inserted under the nerve to grasp the end of the string on the other side.

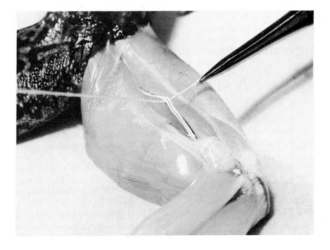

6. The thread used for manipulating the nerve is pulled through.

The Threshold Stimulus

Before severing the Achilles tendon and attaching the muscle to the myograph with thread as shown in Figure 20.1, you will determine the threshold stimulus. A **threshold stimulus** is the minimal voltage that elicits a muscle twitch. A stimulus that does not yield a muscle twitch is a **subliminal stimulus.** Proceed as follows to determine the threshold stimulus for both muscle and nerve stimulation. *Keep the tissues moist with Ringer's solution.*

1. Set the **Width** control at 2 msec and the **Voltage** at 0.1 volt (range switch at 0–10, variable control at lowest position).
2. Insert the **pin electrodes** into the gastrocnemius muscle about 1 cm apart.
3. Press the MUSCLE side of the button on the stimulus switch to assure that the muscle will receive stimuli.
4. Deliver a single stimulus to the muscle by depressing the **Mode** switch on the stimulator to SINGLE. If the muscle twitches, reduce the **Width** control to the lowest value that produces a twitch. Record 0.1 volt as the threshold stimulus for muscle stimulation.
   If the muscle does not twitch at 0.1 volts and 2 msec, increase the voltage in increments of 0.1 volt until a twitch is seen. Record this voltage as the threshold stimulus.
5. Carefully lift the nerve with the thread and position the rubber tubing sleeve around the nerve; insert the sleeve electrode within the tubing next to the nerve. After the sleeve electrode is in place, cut the sciatic nerve at the thigh-body junction to eliminate any extraneous impulses.
6. Press the NERVE side of the button on the stimulator switch to enable nerve stimulation.
7. Return the **Voltage** to 0.1 volt and stimulate the nerve by pressing the **Mode** switch to SINGLE. If no twitch occurs, increase the voltage in increments of 0.1 volt until a twitch is seen. Record this voltage as the threshold stimulus for nerve stimulation.
8. Place the foot in a flexed position and determine the minimum voltage that causes contraction to extend the foot. Record this voltage.

Myogram of a Muscle Twitch

Now, you will produce a **myogram** similar to the one shown in Figure 20.3 for the human soleus muscle. Note that there are three phases to the myogram. The **latent period** is the very brief interval (about 10 msec) between the application of the stimulus and the onset of contraction. The muscle shortens during the **contraction phase** and

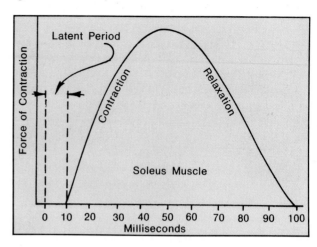

Figure 20.3 A simple twitch of the soleus muscle.

lengthens during the **relaxation phase.** Appendix B shows what sample Physiograph records of muscle contractions look like.

Muscle Preparation

Use a dissecting needle to separate the gastrocnemius muscle from the connective tissue around it. With forceps pull one end of an 18 in. length of thread under the Achilles tendon as shown in Figure 20.4. Tie one end securely to the tendon. Then, use scissors to cut the tendon *distal to the thread* as shown in Figure 20.5.

Pin the dissected leg to the pan or frog board to prevent it from moving. Tie a loop in the other end of the thread and slip it over the hook on the myograph. Adjust the tension of the thread with the fine adjustment knob on the myograph tension adjuster to eliminate any slack in the thread. Compare your set-up with Figure 20.1. *Remember to keep the tissues moist with Ringer's solution throughout the experiment.*

Electrode Placement

You will use only the sleeve electrode in this experiment. Check the electrode to be sure that it is in the correct position. Remove the pin electrodes.

Physiograph Settings

Turn on the **Power** switch. Make sure that the myograph is calibrated so that 100 g of tension is equivalent to 4 cm of pen deflection. See Appendix A for the calibration procedure. *With minimal tension on the myograph,* set the baseline to the centerline with the **position control knob.**

Place the **Record** button in the ON position and gradually increase the tension by moving the myograph upward with the myograph tension adjuster. Tension should be increased until the pen has moved upward 0.5 cm (12.5 g of tension).

Use the **Balance** control to return the pen to the center line. Then, use the **Position** control to move the pen to any desired baseline, usually 2 cm below the center line.

Figure 20.4 Thread is drawn through the space behind the Achilles' tendon by inserting forceps through to grasp the thread.

Figure 20.6 Multiple motor-unit summation.

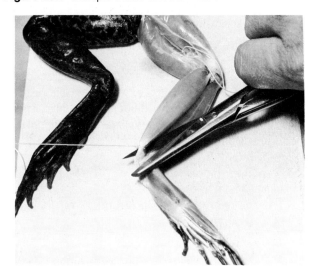

Figure 20.5 After the thread is securely tied to the Achilles' tendon, the tendon is severed between the thread and joint.

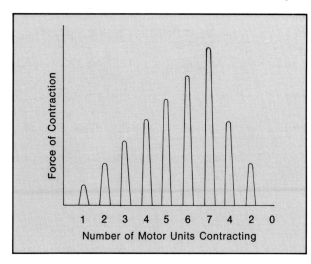

3. Change the chart speed to 2.5 cm/sec and administer enough stimuli so that each member of your team has a recording to attach to the laboratory report.
4. Calculate and record the duration of each phase of the contraction cycle.

Summation

Muscles exert different degrees of pull primarily by **summation**. There are two types of summation: multiple motor unit summation and wave summation.

Multiple Motor Unit Summation

A **motor unit** consists of a group of muscle fibers innervated by a single motor neuron. There are few fibers per motor unit in muscles involved in precise movements, but in postural muscles there may be several hundred.

Multiple motor unit summation yields an increased force by the **recruitment** of additional motor units as the strength of the stimulus increases. See Figure 20.6.

In this experiment, you will gradually increase the voltage (stimulus) to demonstrate multiple motor unit summation and to determine the **maximal stimulus,** which is the minimal stimulus that yields a **maximal contraction.**

1. Set the paper speed at 0.25 cm/sec and lower the pens to the paper. Do not start the paper moving at this time.
2. Return the **Voltage** control on the stimulator to 0.1 volt. Leave other controls as they were in the previous experiment.
3. Be sure the sleeve electrode is in the correct position.

Stimulator Settings

Set the **Width** control at 1 msec and the **Voltage** at the minimum level that produced extension of the foot in the previous experiment.

Recording

1. Start the recording paper moving at 0.1 cm/sec and lower the pens onto the paper. It is not necessary to turn on the timer.
2. Press the **Mode** switch to SINGLE and observe the tracing on the chart. The pen travel should be about 1.5 cm. Adjust the **sensitivity control** to produce the desired pen travel.

4. With the chart paper not moving, administer a stimulus by pressing the **Mode** switch to SINGLE. Then advance the chart paper about 0.5 cm and stop it.
5. Increase the voltage by 0.2 volt and apply another single stimulus. Continue this procedure of advancing the chart and increasing the voltage by 0.2 volt until a maximum contraction is attained. *Be sure to record the stimulus voltage each time on the chart.*
6. Repeat this process enough times to provide a record for each member of your team.

Wave Summation

If a muscle receives a second stimulus prior to relaxing fully, it will respond to both stimuli, and the second contraction will be greater than the first. This phenomenon is called **wave summation.** Figure 20.7 shows that stimuli at 35/sec results in some wave summation. At 70 stimuli/sec, there is considerable summation. At 200 stimuli/sec, complete, sustained contraction (tetanus) occurs. Proceed as follows to observe wave summation.

1. Set the **Voltage** control at the previously determined maximal stimulus and the frequency at 1 Hz (1 stimulus per second).
2. Lower the pens and start the paper moving at a speed of 0.25 cm/sec.
3. Place the **Mode** switch on CONT. (continuous) and note that the monitor lamp on the stimulator is blinking at the stimulus frequency.
4. Record the tracing for 10 seconds. Then turn the stimulator **Mode** switch to OFF, raise the pens, and stop the paper. Record the stimulus frequency (1 Hz) on the chart. Let the muscle rest for *2 minutes.*

Figure 20.7 Wave summation.

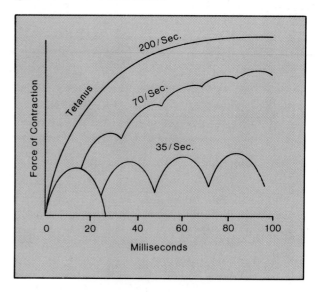

5. Double the frequency to 2 Hz, lower the pens, start the paper, and set the **Mode** switch on CONT. to record again for 10 seconds. Turn the stimulator **Mode** switch to OFF, raise the pens, and stop the paper. Record the stimulus frequency (2 Hz) on the chart. Let the muscle rest for *2 minutes.*
6. In this manner, continue to *double* the frequency, record for 10 seconds, and let the muscle rest for *2 minutes* between recordings until the muscle goes into complete tetanus. Record the stimulus frequency on the chart for each record.
7. After letting the muscle rest for a few minutes, repeat the process enough times to provide a record for each team member.
8. Attach the records to the laboratory report.
9. Complete the laboratory report.

The Spinal Cord and Reflex Arcs

Objectives

1. Identify the spinal cord, its protective coverings, and spinal nerves on charts or models.
2. Identify parts of the spinal cord when viewed microscopically in cross section.
3. Describe the pathways of somatic and visceral reflex arcs.
4. Describe the reflexes performed in this exercise.
5. Define all terms in bold print.

Materials

Model of spinal cord in cross section
Prepared slides of spinal cord, x.s.
Reflex hammers

Automatic stereotyped responses to stimuli enable rapid adjustment to environmental conditions. These responses are called **reflexes,** and they are categorized into two types. **Somatic reflexes** involve skeletal muscle responses. **Visceral reflexes** involve responses by smooth muscle, cardiac muscle, or glands. The neural pathway involved in a spinal reflex is known as a **reflex arc.** It involves receptors, spinal nerves, the spinal cord, and effectors.

The Spinal Cord

The spinal cord extends downward from the brain, within the vertebral canal. In infants, the spinal cord extends the full length of the vertebral canal, but in adults it ends near the second lumbar vertebra. This results from the differences in growth rates of the spinal cord and the axial skeleton. Posterior portions of the axial skeleton have been removed in Figure 21.1 to reveal the spinal cord and spinal nerves.

Extending downward from the end of the spinal cord are the proximal ends of spinal nerves serving body regions below the cord. This aggregation of nerves is called the **cauda equina,** or horse's tail. The **dural sac** (dura mater) ends within the sacrum; only a portion of it is shown (label 12) for better observation of the nerves.

The Spinal Nerves

The spinal nerves emerge from the spinal cord through the intervertebral foramina on each side of the vertebral column. There are eight cervical, twelve thoracic, five lumbar, and five sacral pairs. These thirty pairs of nerves, plus one pair of coccygeal nerves, make a total of thirty-one pairs of spinal nerves.

As shown in the cross sectional view in Figure 21.1, each spinal nerve is joined to the spinal cord by **anterior** and **posterior roots.** The enlargement on the posterior root is a **spinal ganglion** that contains cell bodies of sensory neurons. In the spinal cord, note the outer **white matter** (myelinated fibers) and inner **gray matter** (unmyelinated fibers and neuron cell bodies).

About 2 cm distal from the union of the anterior and posterior roots, each spinal nerve forms two branches (rami) that contain both sensory and motor fibers. The smaller **posterior branch** innervates the skin and muscles of the back. The larger **anterior branch** supplies the lateral and anterior portions of the trunk and limbs.

With the exception of the thoracic spinal nerves, the anterior branches of spinal nerves combine to form complex networks called **plexuses** on each side of the spinal cord. In a plexus, the fibers of the anterior branches are rearranged and combined so that fibers innervating a particular body part occur in the same nerve, although the fibers may originate in different spinal nerves.

Figure 21.1 The spinal cord and spinal nerves.

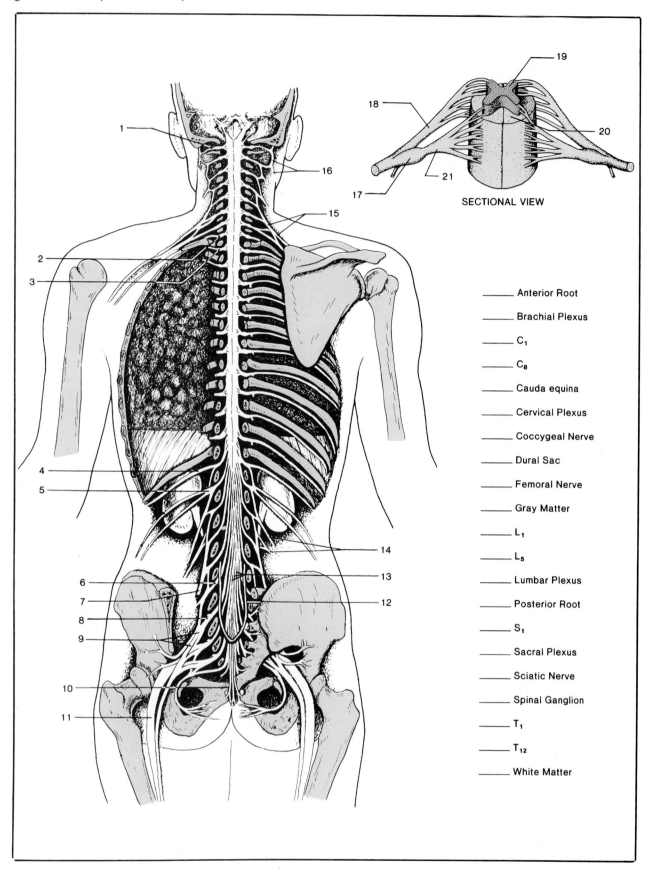

19

18

20

17

21

SECTIONAL VIEW

1

16

15

2

3

4

5

14

13

6

12

7

8

9

10

11

_____ Anterior Root

_____ Brachial Plexus

_____ C_1

_____ C_8

_____ Cauda equina

_____ Cervical Plexus

_____ Coccygeal Nerve

_____ Dural Sac

_____ Femoral Nerve

_____ Gray Matter

_____ L_1

_____ L_5

_____ Lumbar Plexus

_____ Posterior Root

_____ S_1

_____ Sacral Plexus

_____ Sciatic Nerve

_____ Spinal Ganglion

_____ T_1

_____ T_{12}

_____ White Matter

Figure 21.2 The spinal cord and meninges.

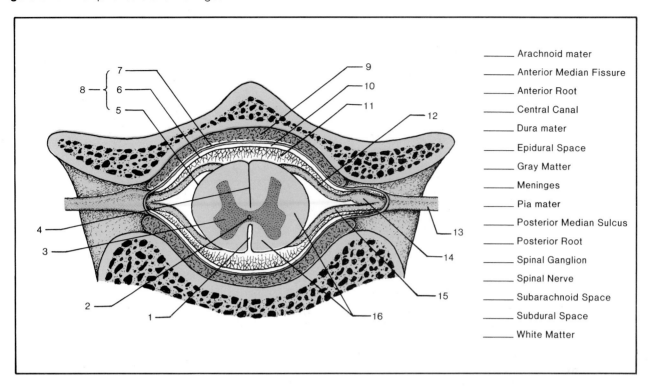

_____ Arachnoid mater
_____ Anterior Median Fissure
_____ Anterior Root
_____ Central Canal
_____ Dura mater
_____ Epidural Space
_____ Gray Matter
_____ Meninges
_____ Pia mater
_____ Posterior Median Sulcus
_____ Posterior Root
_____ Spinal Ganglion
_____ Spinal Nerve
_____ Subarachnoid Space
_____ Subdural Space
_____ White Matter

Cervical Nerves

There are eight pairs of cervical nerves. The first pair (C_1) exit the spinal cord between the skull and the axis. The eighth pair (C_8) exit between the seventh cervical vertebra and the first thoracic vertebra. The first four cervical nerves unite to form the **cervical plexus** (label 16) that serves the muscles and skin of the neck. Cervical nerves five through eight unite with the first thoracic nerve (T_1) to form the **brachial plexus** that is located in the shoulder region. Nerves serving the arm and hand arise from this plexus.

Thoracic Nerves

There are twelve pairs of thoracic nerves, T_1 through T_{12}. Each nerve emerges from the spinal cord just below its corresponding vertebra.

Lumbar Nerves

There are five pairs of lumbar nerves. Lumbar nerves L_1 (label 5) through L_4 unite to form the **lumbar plexus** (label 14) from which the large **femoral nerve** (label 7) emanates to supply the thigh and leg.

Sacral and Coccygeal Nerves

There are five pairs of sacral nerves and one pair of coccygeal nerves (label 10). The **sacral plexus** (label 9) is formed by lumbar nerves L_4 and L_5, and sacral nerves S_1 through S_3. The large nerve emanating from this plexus is the **sciatic nerve** that serves the thigh, leg, and foot.

The Spinal Cord in Cross Section

The cross sectional structure of the spinal cord within the vertebral canal is shown in Figure 21.2. The H-shaped pattern of the **gray matter** is surrounded by **white matter** that is composed mostly of myelinated neuron fibers. Two fissures occur on the midline. The **anterior median fissure** is wider than the **posterior median sulcus.** The small **central canal** lies on the median line in the gray matter. It extends the length of the spinal cord and is continuous with the ventricles of the brain.

The Meninges

Surrounding the spinal cord (and also the brain) are three membranes, the **meninges** (meninx, singular). The outermost membrane, the **dura mater,** also covers the spinal nerve roots. It consists of fibrous connective tissue and is the toughest of the meninges. Between the dura mater and the vertebrae is an **epidural space** that contains areolar connective tissue, adipose tissue and blood vessels.

The innermost membrane is the **pia mater** (label 5), a thin membrane tightly adhered to the spinal cord. The third membrane, the **arachnoid mater,** lies between the pia mater and the dura mater. There is a minute **subdural space** between the dura mater and arachnoid mater. The delicate fibrous structure of the inner arachnoid surface provides support for the spinal cord. Between the arachnoid and pia mater is the **subarachnoid space** that is filled with **cerebrospinal fluid.** This fluid serves as a protective shock absorber around the spinal cord.

Figure 21.3 The somatic reflex arc.

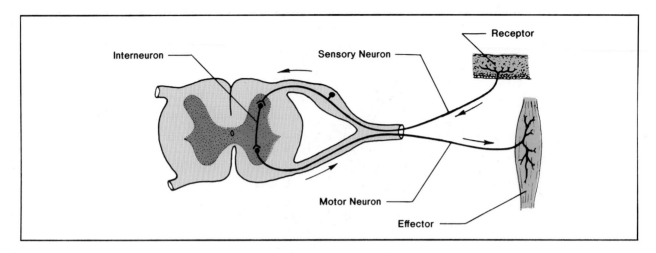

Cerebrospinal fluid may be removed for examination by a **spinal puncture** or **tap** in which a needle is inserted between the third and fourth or fourth and fifth lumbar vertebrae and into the subarachnoid space. The vertebrae may be separated slightly for insertion of the needle by having the patient lie on his side, and draw his knees and chest together.

Assignment

1. Label Figures 21.1 and 21.2.
2. Examine a prepared slide of spinal cord x.s. and locate the parts shown in Figure 21.2. Compare your observations with Figure HA-13. Use high power to locate neuron cell bodies in the gray matter.

The Somatic Reflex Arc

The components of a simple reflex arc are shown in Figure 21.3. Arrows indicate the path of impulse transmission. These components are: (1) a **receptor** that receives stimuli and forms impulses, such as a touch receptor in the skin; (2) a **sensory neuron** that carries impulses into the spinal cord via a spinal nerve and its posterior root; (3) an **interneuron** that transmits impulses to a motor neuron within the gray matter of the spinal cord; (4) a **motor neuron** that transmits impulses to an effector via the anterior root of a spinal nerve; and (5) an **effector,** a muscle or gland, that contracts or secretes when activated by the impulses.

The Visceral Reflex Arc

Reflexes of the viscera are controlled by the **autonomic nervous system** and follow a different pathway than somatic reflexes. Figure 21.4 shows the components of a visceral reflex arc. The major anatomical difference is that visceral reflexes utilize two efferent neurons instead of one. These two neurons synapse outside the central nervous system in an autonomic ganglion.

There are two types of autonomic ganglia. **Vertebral autonomic ganglia** (label 5) are united to form two chains; one chain lies along each side of the vertebral column. **Collateral ganglia** (label 2) are located closer to the visceral organs.

Stimulation of a receptor in a visceral organ initiates impulses that are carried by a **visceral afferent** (sensory) **neuron.** This afferent neuron passes through a vertebral autonomic ganglion and enters the spinal cord via the posterior root of a spinal nerve. It synapses with a **preganglionic efferent neuron** (label 4) within the gray matter of the spinal cord. This first efferent neuron exits the spinal cord via the anterior root of a spinal nerve and synapses with a **postganglionic neuron** at either a vertebral autonomic ganglion or a collateral ganglion. This second efferent neuron carries impulses to the visceral organ where the response occurs.

Recall that the autonomic nervous system consists of two divisions. The **sympathetic division** is involved with spinal nerves of the thoracic and lumbar regions. The **parasympathetic division** involves the cranial and sacral nerves. Most viscera are innervated by nerves from both

Figure 21.4 The visceral reflex arc.

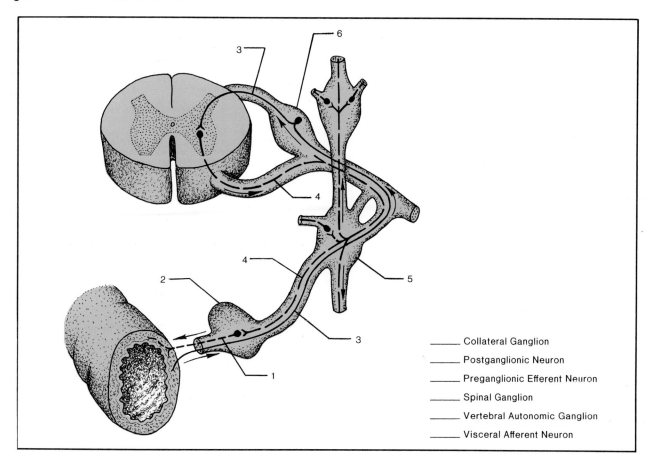

_____ Collateral Ganglion
_____ Postganglionic Neuron
_____ Preganglionic Efferent Neuron
_____ Spinal Ganglion
_____ Vertebral Autonomic Ganglion
_____ Visceral Afferent Neuron

divisions, which enables a dynamic functional balance to be maintained through stimulation by one division and inhibition by the other.

Assignment

1. Label Figure 21.4.
2. Complete Sections A through D on the laboratory report.

Diagnostic Somatic Reflexes

Reflex tests are standard clinical procedures used by physicians in evaluating neural functions. Abnormal reflex responses may indicate pathology of the involved portion of the nervous system. Interpretation of the results requires considerable experience. Our purpose in performing three common tests is not to diagnose but to observe the reflexes and to understand why they are used.

Reflex Responses

Abnormal responses may be either **hypoflexia,** a diminished response, or **hyperflexia,** an exaggerated response. Hypoflexia may result from several causes such as malnutrition, neuronal lesions, aging, or deliberate control. Hyperflexia may be due to a loss of inhibition in motor control areas. Some reflexes occur only if pathological conditions exist.

Figure 21.5 shows the procedures to be used in performing the reflex tests. Each involves striking a tendon to stimulate a muscle's stretch receptors, which normally causes a quick contraction in response. Use the following scale in interpreting the responses of your partner for each reflex. Record your results on the laboratory report.

+ + +   stronger than normal; may be hyperative
  + +   normal; average
    +   somewhat diminished
    0   no response

Figure 21.5 Methods used for producing somatic reflexes.

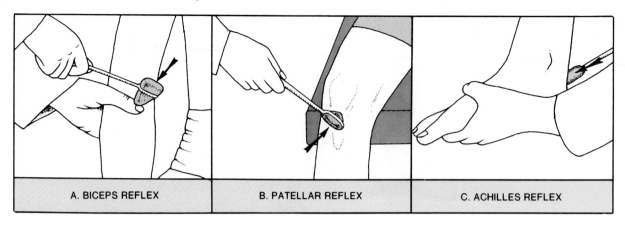

| A. BICEPS REFLEX | B. PATELLAR REFLEX | C. ACHILLES REFLEX |

Biceps Reflex This reflex causes flexion of the forearm. It is elicited by holding the subject's elbow, with the thumb placed on the tendon of the biceps brachii, and striking a sharp blow to the first digit of your thumb with the reflex hammer. Reinforce the response by asking the subject to squeeze his or her thigh with the other hand during the test. Test both arms. This reflex involves the C_5 and C_6 spinal nerves.

Patellar Reflex The subject must be seated on the edge of a table with the leg suspended over the edge. To elicit a response, strike the patellar tendon just below the knee cap. Reinforce the response by having the subject interlock the fingers of both hands and pull each hand against the other isometrically. Test both legs. This reflex involves spinal nerves L_2, L_3, and L_4.

Sometimes a series of jerky contractions occurs following a normal response. This usually indicates damage within the central nervous system.

Achilles Reflex This reflex produces plantar flexion. To elicit the response, hold the foot with one hand in a slightly dorsiflexed position and strike the Achilles tendon with the reflex hammer. Reinforce the response as in the patellar reflex. This reflex involves spinal nerves S_1 and S_2. A diminished response may indicate hypothyroidism.

Assignment

Complete the laboratory report.

Brain Anatomy: External

Objectives

After completing this exercise, you should be able to:

1. Identify the external structures of sheep and human brains on preserved specimens, charts, or models.
2. Describe the function of the cranial nerves.
3. Identify the major functional areas of the human brain.
4. Define all terms in bold print.

Materials

Colored pencils
Human brain models
Sheep brains, preserved
Dissecting instruments and trays

The Meninges

Both the brain and spinal cord are covered by three protective membranes, the **meninges.** Figure 22.1 shows the relationships of the meninges in a frontal section through the upper portion of the cranium and brain.

The **pia mater** is the thin, innermost meninx adhered to the brain surface. The **arachnoid mater** is the intermediate meninx with delicate fibers extending from its inner surface through the **subarachnoid space** to the pia mater. This space contains **cerebrospinal fluid** that serves as a protective shock-absorber for the brain. The outermost meninx is the tough, fibrous **dura mater** that is attached to the inner surface of the cranial bones. Within the dura mater at the superior midline is a **sagittal dural sinus** that receives venous blood from the brain as it returns to the heart. The cerebrospinal fluid diffuses into the venous blood from the **arachnoid granulations** that project into the dural sinus. Note that the meninges extend into the longitudinal fissure that separates the cerebral hemispheres.

Figure 22.1 The meninges.

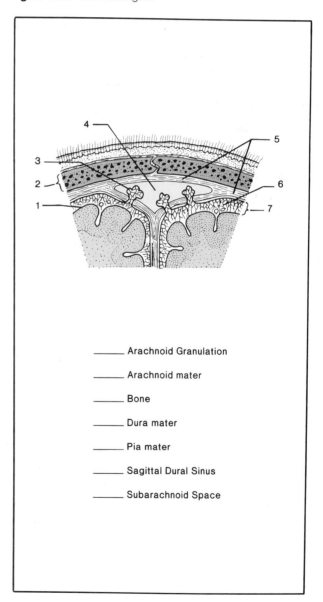

_____ Arachnoid Granulation

_____ Arachnoid mater

_____ Bone

_____ Dura mater

_____ Pia mater

_____ Sagittal Dural Sinus

_____ Subarachnoid Space

Figure 22.2 Lateral aspects of sheep and human brains.

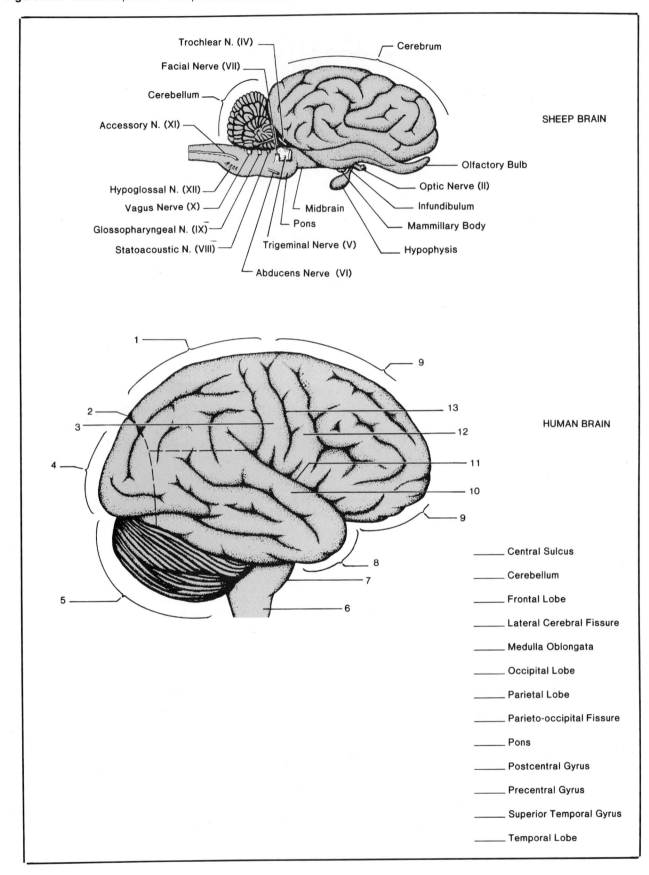

Trochlear N. (IV)
Facial Nerve (VII)
Cerebellum
Accessory N. (XI)
Hypoglossal N. (XII)
Vagus Nerve (X)
Glossopharyngeal N. (IX)
Statoacoustic N. (VIII)
Abducens Nerve (VI)
Midbrain
Pons
Trigeminal Nerve (V)
Cerebrum
Olfactory Bulb
Optic Nerve (II)
Infundibulum
Mammillary Body
Hypophysis

SHEEP BRAIN

HUMAN BRAIN

_____ Central Sulcus

_____ Cerebellum

_____ Frontal Lobe

_____ Lateral Cerebral Fissure

_____ Medulla Oblongata

_____ Occipital Lobe

_____ Parietal Lobe

_____ Parieto-occipital Fissure

_____ Pons

_____ Postcentral Gyrus

_____ Precentral Gyrus

_____ Superior Temporal Gyrus

_____ Temporal Lobe

Figure 22.3 Ventral aspect of sheep brain.

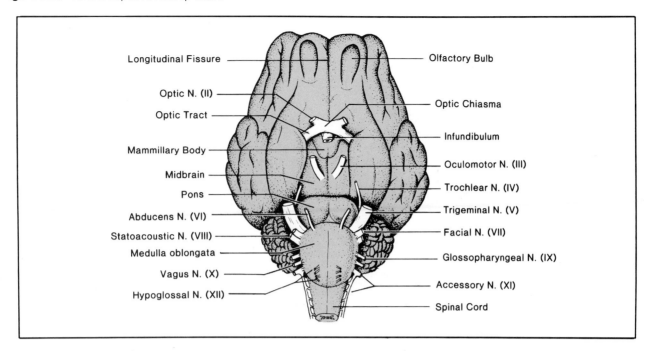

Longitudinal Fissure

Optic N. (II)

Optic Tract

Mammillary Body

Midbrain

Pons

Abducens N. (VI)

Statoacoustic N. (VIII)

Medulla oblongata

Vagus N. (X)

Hypoglossal N. (XII)

Olfactory Bulb

Optic Chiasma

Infundibulum

Oculomotor N. (III)

Trochlear N. (IV)

Trigeminal N. (V)

Facial N. (VII)

Glossopharyngeal N. (IX)

Accessory N. (XI)

Spinal Cord

The inflammation of the meninges, **meningitis,** is a serious condition because of the close association of the meninges and the brain. It is often diagnosed by detecting bacteria or viruses in cerebrospinal fluid removed by a spinal tap.

Major Parts of the Brain

Use the illustrations of the sheep brain in Figures 22.2 and 22.3 and the descriptions that follow to locate and label the external structures on illustrations of the human brain in Figures 22.2 and 22.4.

Cerebrum

The surface of the cerebrum has numerous ridges or convolutions called **gyri** that increase the surface area of the gray matter. They are bordered by shallow furrows called **sulci.** Deeper furrows are called **fissures.** The cerebrum is divided into two **cerebral hemispheres** by a **longitudinal cerebral fissure** (label 20, Figure 22.4) that extends along the median line.

The major fissures that divide each cerebral hemisphere into lobes are shown in Figure 22.2. The **central sulcus** (label 13) separates the **frontal lobe** from the **parietal lobe.** The **lateral cerebral fissure** (label 11) separates the **temporal lobe** from the frontal and parietal lobes.

The **parieto-occipital fissure** (label 2) occurs more extensively on the median side of the hemisphere and is indicated by a dotted line. It separates the **occipital lobe** from the parietal and temporal lobes. Note that the lobes of the cerebrum are named after the cranial bones that cover them.

Cerebral Functions The cerebrum carries out the higher brain functions. The specific functions of some portions of the cerebrum are known.

The **primary motor area** occupies the precentral gyrus of the frontal lobe. It controls voluntary muscular contractions. The **motor speech area** usually is located in the left hemisphere just anterior to the primary motor area on the lower part of the frontal lobe just above the lateral fissure. It controls the production of speech.

The **primary sensory area** occupies the postcentral gyrus of the parietal lobe. It interprets impulses from receptors in the skin, muscles, and viscera as sensations. The **visual area** is in the posterior portion of the occipital lobe and interprets impulses from light receptors in the eyes as visual sensations. Impulses from the inner ear are interpreted as sound sensations by the **auditory area** located on a part of the superior temporal gyrus below the primary sensory area. **Association areas** occupy most of the frontal lobe anterior to the motor areas and much of the other three lobes. They are concerned with memory, reasoning, will, judgment, intelligence, emotions, and personality traits.

Figure 22.4 Ventral aspect of human brain.

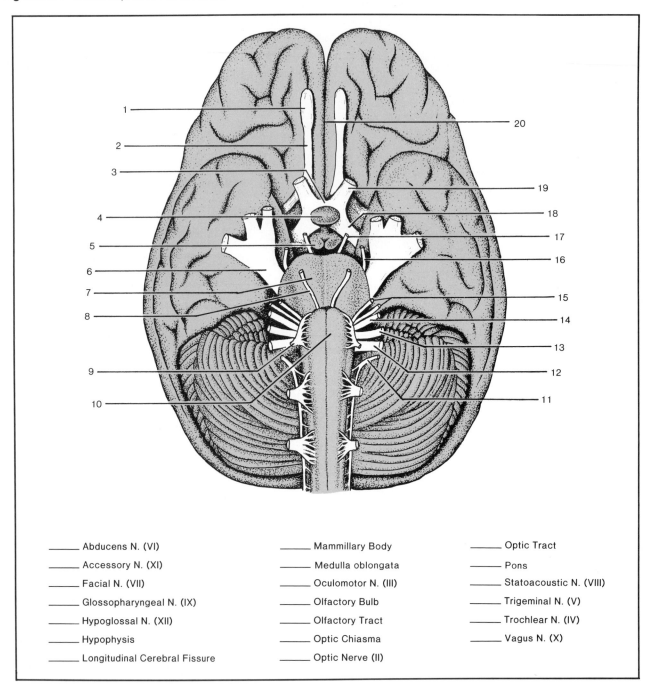

_____ Abducens N. (VI)	_____ Mammillary Body	_____ Optic Tract
_____ Accessory N. (XI)	_____ Medulla oblongata	_____ Pons
_____ Facial N. (VII)	_____ Oculomotor N. (III)	_____ Statoacoustic N. (VIII)
_____ Glossopharyngeal N. (IX)	_____ Olfactory Bulb	_____ Trigeminal N. (V)
_____ Hypoglossal N. (XII)	_____ Olfactory Tract	_____ Trochlear N. (IV)
_____ Hypophysis	_____ Optic Chiasma	_____ Vagus N. (X)
_____ Longitudinal Cerebral Fissure	_____ Optic Nerve (II)	

Damage to part of the motor area may result in loss of purposeful movement. For example, damage to the motor speech area may cause **aphasia,** the inability to speak. Damage to sensory association areas may result in loss of the ability to correctly interpret sensory information.

Cerebellum

The cerebellum is located below the occipital and temporal lobes of the cerebrum posterior to the brain stem. It is divided into two hemispheres by a medial constriction. It provides subconscious control of muscular contractions enabling muscular coordination, balance, and posture.

Midbrain

The midbrain is a short section of the brain stem superior to the pons and is not visible externally in the human brain. It is a pathway for fibers between the cerebrum and lower parts of the brain, and a reflex center for movements in response to visual and auditory stimuli.

Pons

This rounded bulge on the anterior portion of the brain stem contains fibers that connect the cerebellum and medulla with higher parts of the brain.

Medulla Oblongata

This is the lowest portion of the brain stem and is continuous with the spinal cord. The medulla contains centers that control the heart rate, breathing rate and depth, and the diameter of blood vessels.

The Hypothalamus

The ventral portion of the brain that lies between the cerebral hemispheres and the midbrain is the **hypothalamus.** It provides autonomic regulation of body functions, such as temperature, water and electrolyte balance, sleep, and hunger.

The **hypophysis** or **pituitary gland** hangs from the ventral surface of the hypothalamus by a short stalk, the **infundibulum.** See Figure 22.4.

Another external feature is the pair of rounded eminences, the **mammillary bodies,** that lie just posterior to the base of the infundibulum.

Assignment

1. Label Figures 22.1 and 22.2.
2. In Figure 22.2, color the four lobes of the cerebral hemisphere in the human brain for easy recognition: frontal—red; parietal—yellow; temporal—blue; occipital—green.
3. In Figure 22.2, use a pen to outline and label the functional areas of the human cerebrum.

The Cranial Nerves

There are twelve pairs of cranial nerves that arise from various parts of the brain and pass through foramina to innervate parts of the head and trunk. Each is designated by name and number. Most cranial nerves are **mixed nerves** in that they carry both sensory and motor neuron processes. A few carry only sensory fibers.

As you study this section, locate the cranial nerves first on the sheep brain in Figures 22.2 and 22.3 and then on the human brain in Figure 22.4.

I Olfactory   This cranial nerve contains only sensory fibers for the sense of smell. It is not shown on the illustrations since it passes from the olfactory epithelium to the **olfactory bulb** of the brain. It synapses with neurons in the olfactory bulb that carry the impulses on to the olfactory areas of the cerebrum. In humans, an **olfactory tract** connects the bulb with the brain.

II Optic   This sensory nerve functions in vision and contains axons of ganglion cells in the retina. Half of the axons cross over to the other side of the brain as they pass through the **optic chiasma.** They continue through the optic tracts to the thalamus and on to the visual centers of the cerebrum.

III Oculomotor   This nerve emerges from the midbrain and innervates the levator palpebrae muscle that raises the eyelid and four extrinsic muscles of the eyeball: the superior, medial, and inferior rectus muscles and the inferior oblique. Its autonomic fibers control the iris of the eye and the focusing of the lens.

IV Trochlear   This nerve arises from the midbrain and innervates the superior oblique, an extrinsic muscle of the eyeball.

V Trigeminal   This is the largest cranial nerve, and it innervates much of the mouth and face. It emerges from the anterior part of the pons and subsequently divides into ophthalmic, maxillary, and mandibular branches.
The **ophthalmic division** innervates the skin of the scalp, forehead, eyelid, and nose, the lacrimal (tear) gland, and parts of the eye. The **maxillary division** carries sensory impulses from the upper teeth, gums, and lip. The **mandibular division** is a mixed nerve innervating the lower teeth, gums and lip, chewing muscles, lower part of the tongue, and lower part of the face.

VI Abducens   This small nerve originates from the posterior part of the pons and innervates the lateral rectus, an extrinsic muscle of the eyeball.

VII Facial   This nerve (label 15) arises from the lower part of the pons. It innervates facial muscles, salivary glands, and the anterior two-thirds of the tongue, including taste buds.

VIII Statoacoustic or **Vestibulocochlear** This sensory nerve arises from the medulla and consists of two branches innervating the inner ear. The **vestibular branch** carries impulses from the balance receptors. The **cochlear branch** carries impulses from the sound receptors.

IX Glossopharyngeal This mixed nerve emerges from the medulla posterior to the statoacoustic. It innervates much of the mouth lining, posterior part of the tongue, and pharynx. It controls swallowing reflexes and salivary secretion.

X Vagus This nerve (label 12) arises from the medulla and descends into the neck, thorax, and abdomen to innervate the pharynx, larynx, and the thoracic and abdominal viscera. It is a major nerve of the autonomic nervous system.

XI Accessory This nerve arises from both the medulla and spinal cord. The **cranial branch** innervates muscles of the pharynx and larynx. The **spinal branch** innervates the trapezius and sternocleidomastoideus muscles.

XII Hypoglossal This nerve arises from three origins on the medulla and innervates muscles of the tongue. In Figure 22.4 each origin appears as a tree stump with the roots entering the medulla.

Figure 22.5 Exposing the midbrain.

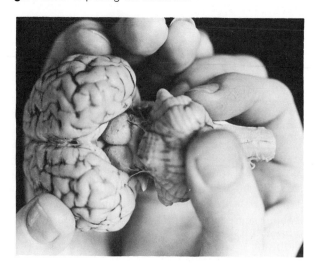

Assignment

1. Label Figure 22.4.
2. Obtain a preserved sheep brain and place it on a dissecting tray for study. Locate the external features as shown in Figures 22.2 and 22.3. It is necessary to depress the cerebellum as shown in Figure 22.5 in order to observe the midbrain. The two pairs of rounded eminences exposed are the **corpora quadrigemina.**
3. Locate the external features studied in this exercise on the model of a human brain.
4. Complete the laboratory report.

Brain Anatomy: Internal

Objectives

After completing this exercise, you should be able to:

1. Identify the internal structures of human and sheep brains on preserved specimens, charts, or models.
2. Describe the circulation or cerebrospinal fluid.
3. Define all terms in bold print.

Materials

Colored pencils
Model of human brain, sagittal section
Sheep brains, preserved
Dissecting instruments and tray
Long, sharp knife

Use the labeled illustration of the sectioned sheep brain, Figure 23.1, and the following descriptions to locate and label the structures of the human brain shown in Figures 23.2 and 23.3.

The Cerebrum

Note in the figures that the convolutions on the medial surface of the cerebral hemisphere have not been disturbed by the midsagittal section. The only part of the cerebrum that has been cut is the **corpus callosum** (label 19). This large, curved band of white matter consists of neuron fibers connecting the two cerebral hemispheres. Locate the **sagittal sinus** in the dura mater along the midline and the **parieto-occipital fissure.**

The Cerebellum

Observe the pattern of gray and white matter in the cut surface of the cerebellum. The functions of the cerebellum are at the subconscious level. It receives impulses from motor and visual centers in the brain, balance receptors in the inner ear, and muscle of the body. This information is integrated and impulses are sent to muscles to maintain posture, balance, and muscle coordination.

Figure 23.1 Midsagittal section of sheep brain.

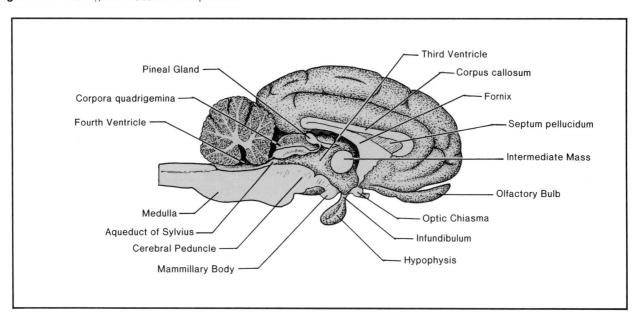

Figure 23.2 Midsagittal section of human brain.

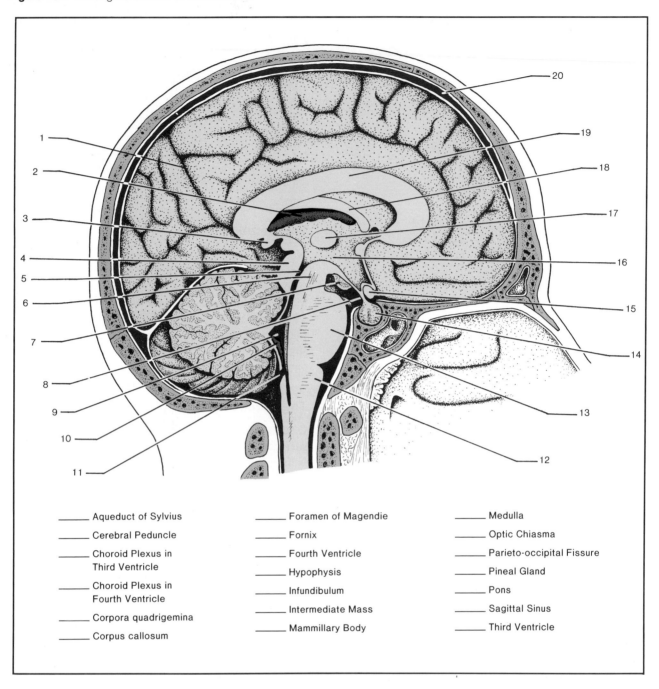

_____ Aqueduct of Sylvius	_____ Foramen of Magendie	_____ Medulla
_____ Cerebral Peduncle	_____ Fornix	_____ Optic Chiasma
_____ Choroid Plexus in Third Ventricle	_____ Fourth Ventricle	_____ Parieto-occipital Fissure
_____ Choroid Plexus in Fourth Ventricle	_____ Hypophysis	_____ Pineal Gland
_____ Corpora quadrigemina	_____ Infundibulum	_____ Pons
_____ Corpus callosum	_____ Intermediate Mass	_____ Sagittal Sinus
	_____ Mammillary Body	_____ Third Ventricle

The Diencephalon

Structures between the cerebrum and the brain stem constitute the **diencephalon.** The thalamus and hypothalamus are the primary components.

The Fornix

The band of white matter inferior to the corpus callosum is the **fornix** (label 18). It contains fibers associated with the sense of smell and is much better developed in sheep than in humans. The small **pineal gland** (label 3) lies on the midline. It secretes a hormone, melatonin, whose function in humans is obscure.

The Thalamus

The **thalamus** consists of two ovoid masses of gray matter, one in each cerebral hemisphere, that are joined by an isthmus of neural tissue called the **intermediate mass of the thalamus** (label 17). The intermediate mass is the only part of the thalamus visible in a sagittal section. The thalamic masses form the lateral walls of the third ventricle.

The thalamus receives sensory impulses coming to the brain and relays them to appropriate sensory centers of the cerebrum. It also provides an uncritical sensory awareness.

The Hypothalamus

This small region extends about 2 cm into the brain from its ventral surface anterior to the brain stem. A short stalk, the **infundibulum** (label 8), hangs from its ventral surface and supports the **hypophysis** that fits into a depression in the sphenoid bone. Posterior to the infundibulum is the left **mammillary body** (label 7). The **optic chiasma,** where half of the fibers of the optic tracts cross, lies just anterior to the infundibulum.

The hypothalamus is closely tied to the autonomic nervous system and controls a variety of functions, such as body temperature, hunger, and water balance.

The Brain Stem

The midbrain, pons, and medulla form the brain stem.

The Midbrain

The short **aqueduct of Sylvius** (label 5) runs longitudinally through the midbrain and connects the third and fourth ventricles. Anterior to this duct is a **cerebral peduncle.** The cerebral peduncles contain the main motor tracts between the cerebrum and lower parts of the brain stem. Posterior to the aqueduct, locate the left portions of the **corpora quadrigemina,** reflex centers for head, eye, and body movements in response to visual and auditory stimuli.

The Pons and Medulla

The **pons** lies between the midbrain and medulla, and is easily recognized by its bulb-like anterior protrusion on the brain stem. It consists primarily of fibrous tracts connecting lower and higher brain centers. The **medulla** is the lowest part of the brain stem and is a connecting link between the brain and spinal cord. It also contains centers that control heart and breathing rates.

Within the brain stem is a network of fibers, the **reticular formation,** that generates impulses that keep the cerebrum awake and alert. Sleep results from its decreased activity, and unconsciousness results if it ceases to function.

The Ventricles and Cerebrospinal Fluid

There are four cavities or ventricles in the brain, which are filled with cerebrospinal fluid. The two **lateral ventricles** are located within the cerebral hemispheres and are separated by a thin membrane. The third and fourth ventricles are on the midline. The **third ventricle** (label 16) lies between the lateral masses of the thalamus; the **fourth ventricle** (label 9) is between the brain stem and the cerebellum.

Each ventricle contains a **choroid plexus,** a mass of specialized capillaries, that secretes the cerebrospinal fluid. The choroid plexuses of the third (label 2) and fourth (label 10) ventricles are shown in Figure 23.2. Locate these plexuses in Figure 23.3, which depicts the circulation of the cerebrospinal fluid.

The cerebrospinal fluid passes from each lateral ventricle into the third ventricle through a **foramen of Monroe** (label 11 in Figure 23.3) that is located anterior to the fornix and the choroid plexus of the third ventricle. From the third ventricle, it flows through the **aqueduct of Sylvius,** a narrow duct that passes longitudinally through the midbrain, and into the fourth ventricle. The fourth ventricle is continuous with the central canal of the spinal cord.

Figure 23.3 Origin and circulation of cerebrospinal fluid.

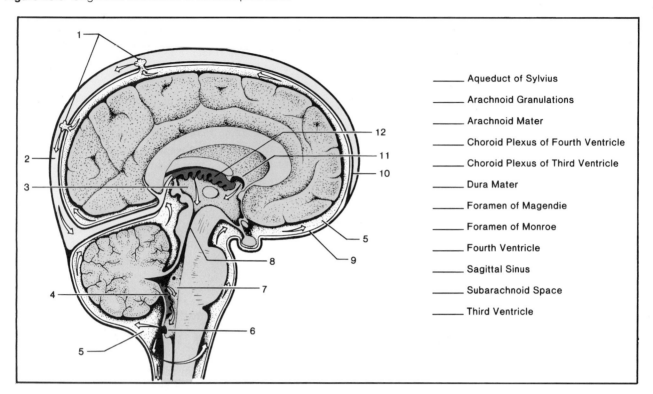

_____ Aqueduct of Sylvius

_____ Arachnoid Granulations

_____ Arachnoid Mater

_____ Choroid Plexus of Fourth Ventricle

_____ Choroid Plexus of Third Ventricle

_____ Dura Mater

_____ Foramen of Magendie

_____ Foramen of Monroe

_____ Fourth Ventricle

_____ Sagittal Sinus

_____ Subarachnoid Space

_____ Third Ventricle

The cerebrospinal fluid then passes from the fourth ventricle through the **foramen of Magendie** and two other small openings into the subarachnoid space below the cerebellum. Some of the fluid moves upwards within the subarachnoid space around the cerebellum and cerebrum, and diffuses into the sagittal sinus through **arachnoid granulations.** The remaining fluid flows downward in the subarachnoid space on the posterior surface of the spinal cord, upward on its anterior surface, continues upward around the cerebrum, and finally diffuses into the sagittal sinus.

Cerebrospinal fluid normally is produced and reabsorbed at equal rates. If an obstruction occurs within the brain, the increased intracranial pressure may cause serious brain damage. In infants, this pressure may cause the brain and cranium to enlarge, a condition known as **hydrocephaly.** Treatment often includes the insertion of a tube to drain excess cerebrospinal fluid into a blood vessel in the neck.

Assignment

1. Label Figures 23.2 and 23.3.
2. In Figure 23.3, color the sagittal sinus blue and the choroid plexuses red.

Sheep Brain Dissection

Obtain a preserved sheep brain and place it on a dissecting tray for study. Review the external features, including the cranial nerves, shown in Figure 22.2 and 22.3.

After you have located the external features make a sagittal section through the brain as follows. Place the brain ventral side down on the dissecting tray. Place a long, sharp knife, not a scalpel, in the longitudinal cerebral fissure and slice through the brain to yield equal right and left halves.

Locate the internal structures shown in Figure 23.1 on the cut surface of the brain. Make a frontal section through the infundibulum of one half of the brain to observe the lateral ventricle. The oval mass of gray matter below the ventricle is one-half of the thalamus. Note the distribution of the gray and white matter. The outer gray matter of the cerebrum composes the **cerebral cortex.** Interior masses of gray matter are called **nuclei.**

When finished with the dissection, dispose of the brain as directed by your instructor. Clean your dissecting instruments, tray, and work station.

Assignment

1. Locate the structures shown in Figure 23.2 on a model of a human brain.
2. Complete the laboratory report.

The Eye

Objectives

After completing this exercise, you should be able to:

1. Identify the parts of the eye and its accessory structures on a preserved eye, chart, or model.
2. Perform and interpret selected visual tests.
3. Define all terms in bold print.

Materials

Beef eye, preferably fresh
Dissecting instruments and tray
Meter stick or tape measure
Ishihara colorblindness test plates
Laboratory lamp
Snellen eye chart
Prepared slide of retina through the fovea centralis

The Lacrimal Apparatus

The eye, lacrimal gland, and extrinsic eye muscles lie protected in the orbital cavity. The eye is protected anteriorly by the **eyelids.** A thin membrane, the **conjunctiva,** lines the inner surface of the eyelids and the anterior surface of the eye. It contains many blood vessels and pain receptors except where it covers the cornea.

The lacrimal apparatus is shown in Figure 24.1. The **lacrimal gland** is located in the upper lateral portion of the eye orbit. It secretes tears that keep the anterior surface of the eye and the conjunctiva moist. They reach the eye surface via 6–12 tiny tubules and flow downward and medially over the anterior surface of the eye. At the medial corner of the eye, tears flow into the **superior** and **inferior lacrimal ducts** through small openings in the upper and lower eyelids, respectively. Then, they enter the **lacrimal sac** and pass through the **nasolacrimal duct** into the nasal cavity.

Figure 24.1 The lacrimal apparatus of the eye.

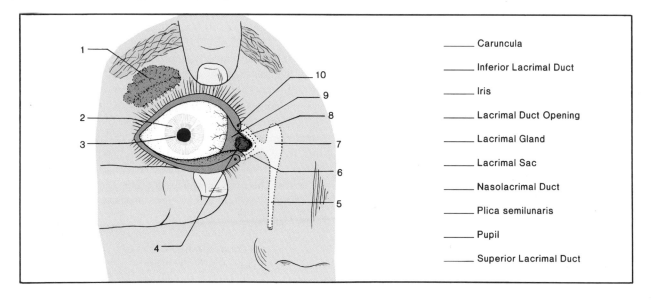

___ Caruncula

___ Inferior Lacrimal Duct

___ Iris

___ Lacrimal Duct Opening

___ Lacrimal Gland

___ Lacrimal Sac

___ Nasolacrimal Duct

___ Plica semilunaris

___ Pupil

___ Superior Lacrimal Duct

Figure 24.2 The extrinsic muscles of the eye.

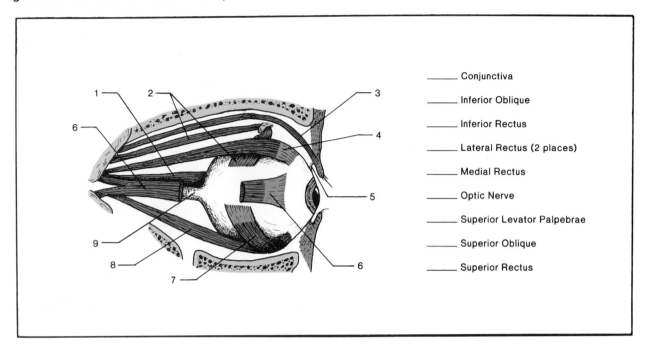

_____ Conjunctiva

_____ Inferior Oblique

_____ Inferior Rectus

_____ Lateral Rectus (2 places)

_____ Medial Rectus

_____ Optic Nerve

_____ Superior Levator Palpebrae

_____ Superior Oblique

_____ Superior Rectus

The small, red, conical body in the medial corner of the eye is the **caruncula.** It secretes a whitish substance that collects in this region. Lateral to the caruncula is a fold of the conjunctiva, the **plica semilunaris,** a remnant of a third eyelid that is well developed in many other vertebrates.

Extrinsic Eye Muscles

The eye orbit contains seven muscles: six extrinsic muscles that move the eye and the **superior levator palpebrae** muscle that raises the upper eyelid. The insertions of these muscles are shown in Figure 24.2. Their origins are on bones at the back of the orbit.

The **lateral rectus** has been partially removed in the illustration. Its opposing muscle is the **medial rectus** (label 2). The **superior rectus** inserts on the upper surface of the eyeball and is opposed by the **inferior rectus** (label 8) inserted on the lower surface. The **superior oblique** passes through a cartilaginous loop allowing it to exert an oblique pull on the eyeball. The **inferior oblique** is oriented in a similar manner, but only a portion of its insertion (label 7) is shown. The interaction of these muscles enables movement of the eye in all directions.

Internal Anatomy

A transverse section of the right eye as seen looking down on it is shown in Figure 24.3. The wall of the eyeball consists of three layers: an outer, fibrous, **scleroid coat;** a middle, pigmented **choroid coat** containing blood vessels; and an inner **retina** containing the light receptors.

The **optic nerve** (label 3) contains fibers leading from the retina to the brain. The point where the optic nerve leaves the retina lacks light receptors and is known as the **optic disc** or **blind spot.** An artery and vein also pass through the optic disc. Note that the optic nerve is wrapped in an extension of the **dura mater.**

The only blood vessels that may be observed directly are those in the eye behind the retina. Physicians use an opthalmoscope to examine them for pathological changes caused by hypertension, diabetes, or atherosclerosis.

Lateral to the optic disc is a small pit, the **fovea centralis,** the area of sharpest vision that is located in the center of a round yellow spot, the **macula lutea.** The macula is 0.5 mm in diameter and contains only cones.

Figure 24.3 Internal anatomy of the eye.

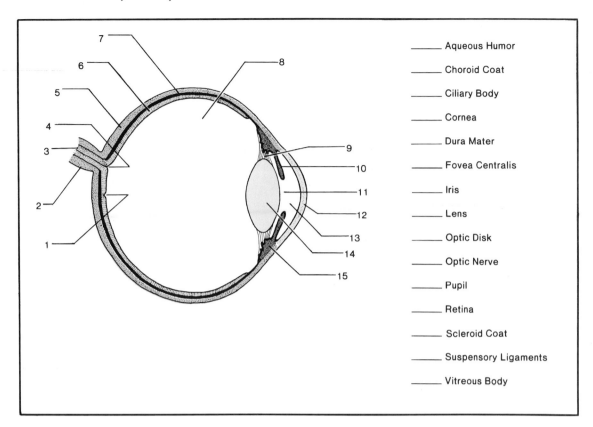

_____	Aqueous Humor
_____	Choroid Coat
_____	Ciliary Body
_____	Cornea
_____	Dura Mater
_____	Fovea Centralis
_____	Iris
_____	Lens
_____	Optic Disk
_____	Optic Nerve
_____	Pupil
_____	Retina
_____	Scleroid Coat
_____	Suspensory Ligaments
_____	Vitreous Body

The **cornea** covers the anterior portion of the eye as an extension of the scleroid coat. It is transparent, and its greater curvature bends the light rays that pass through it. Light rays are precisely focused on the retina by the transparent, crystalline **lens** that is supported by **suspensory ligaments** attached to the **ciliary body.** Contraction or relaxation of smooth muscle in the ciliary body changes the shape of the lens to allow focusing of objects at different distances. Just anterior to the lens is the colored portion of the eye, the circular **iris,** that regulates the amount of light entering the eye through the **pupil.**

Impaired vision in older persons may result from the lens becoming cloudy or opaque, a condition known as **cataract.** The surgical removal of the lens and the insertion of a plastic corrective lens has become commonplace and is usually done on an outpatient basis.

The large cavity between the lens and retina is filled with a transparent, jelly-like material, the **vitreous body.** The cavity in front of the lens is filled with a watery liquid, the **aqueous humor.** Aqueous humor is constantly produced from capillaries near the posterior attachment of the iris to the ciliary body and is absorbed into small veins near the anterior attachment of the iris to the sclera.

Assignment

1. Label Figures 24.1, 24.2, and 24.3.
2. Complete Sections A through D on the laboratory report.
3. Examine a prepared slide of retina sectioned through the fovea centralis. Compare your observations with Figure HA-14 in the Histology Atlas. The photoreceptor cells, rods (black and white vision) and cones (color vision) are located in the outermost layer of the retina next to the choroid coat. Thus, light must pass through neuron processes, the ganglion cell layer, the bipolar cell layer, and the nuclear layer of rods and cones to reach their photosensitive tips. Only cones occur in the fovea centralis.

Figure 24.4 Beef eye dissection.

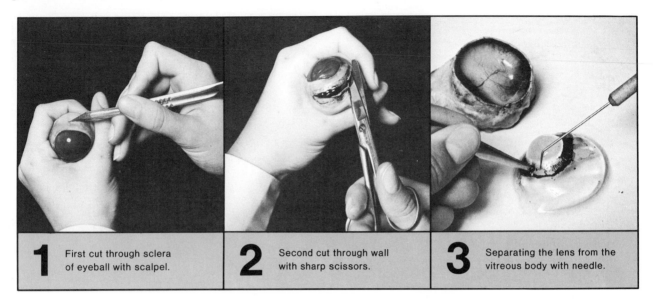

1 First cut through sclera of eyeball with scalpel.

2 Second cut through wall with sharp scissors.

3 Separating the lens from the vitreous body with needle.

Eye Dissection

An understanding of eye anatomy can best be obtained from an eye dissection. The procedures are described below. Figure 24.4 illustrates the major steps to be followed.

1. Examine the external surface of the eyeball. Locate the optic nerve and any remnants of extrinsic muscles. Identify the cornea, iris, and pupil. Is the shape of the pupil the same as in humans?
2. Hold the eyeball as shown in Illustration 1 of Figure 24.4 and make a small incision in the sclera about 0.5 cm from the cornea. Insert scissors into the incision and cut around the eye while holding the cornea upward, as in Illustration 2. Note the watery aqueous humor that exudes.
3. Gently lift off the anterior portion of the eye and place it on a tray with the inner surface upward. The lens usually remains with the vitreous body in fresh eyes, but may remain with the cornea in preserved eyes.
4. In the anterior portion, identify the thickened black ring and the ciliary body, and locate the iris. Are radial and circular muscle fibers evident?
5. Gently pour the vitreous body and lens from the posterior portion of the eye. Use a dissecting needle to carefully separate the lens from the vitreous body as shown in Illustration 3 in Figure 24.4.

6. Hold the lens by the edges and look through it at a distant object. What is unusual about the image? Place it on printed matter. Does it magnify the letters? Compare the consistency of the center and periphery of the lens using a probe or forceps.
7. Observe the thin, pale retina in the posterior part of the eye. It separates easily from the choroid coat and forms wrinkles once the vitreous body is removed. Note the optic disc.
8. Observe the black choroid coat and the iridescent nature of part of it. This reflective surface causes animals' eyes to reflect light, and appears to enhance night vision by reflecting some light back into the retina.
9. Complete Section E on the laboratory report.

Visual Tests

The tests that follow will enable you to understand more about the functioning of the eye. Record the data for your eyes on your laboratory report.

The Blind Spot

Neither rods nor cones are present in the optic disc. Thus, light rays focused on this area produce no image. Determine the presence of the blind spot as follows.

1. Close your left eye and hold Figure 24.5 about 50 cm (20 in) from your face while staring at the cross with the right eye. Note that you can see both the cross and the dot.

Figure 24.5 The blind spot test.

2. While staring at the cross, slowly move the figure closer to the eye and watch for the dot to disappear.
3. At the point where the dot is not visible, have your partner measure and record the distance from your eye to the figure.
4. Repeat the process for the left eye but stare at the dot and watch for the cross to disappear. Record the distance for your left eye.

Near Point Determination

The shortest distance from your eye that an object is in focus is known as the **near point.** To focus on close objects, the contraction of the ciliary body relaxes the suspensory ligaments allowing the lens to take a more spherical shape. The ability for the lens to do this depends upon its elasticity, which deteriorates with age. At age 20, the near point is about 3.5 in.; at age 40, it is 6.8 in.; at age 60, it is 33 in. After age 60, elasticity is minimal, a condition called **presbyopia.** Determine your near point as follows.

1. Close one eye and focus on this letter "T"; gradually move the page closer to your face until the letter is blurred. Then move it away until you get a clear image. At this point, have your partner measure the distance from the page to your eye. Record this distance as the near point.
2. Determine and record the near point for the other eye.

Visual Acuity

The size of the cones and the distance between them in the fovea centralis determine the maximum acuity. Thus, it is possible for a "perfect" human eye to differentiate two points that are only one millimeter apart from a distance of ten meters. Usually poor acuity is due to incorrect focusing of light rays on the retina.

The Snellen eye chart has been developed to measure visual acuity. It is printed with letters of various sizes. When you stand 20 feet from the chart and can read the letters designated to read at 20 feet, you have 20/20 vision.

Figure 24.6 The Snellen eye chart.

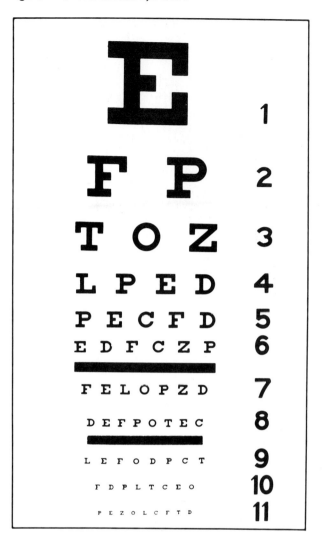

If you can read only the letters designated to be read at 200 feet, you have 20/200 vision. Determine your visual acuity as follows:

1. Stand 20 feet from the Snellen eye chart posted in the laboratory while your partner stands by the chart, indicates the lines to be read, and evaluates your accuracy.
2. Test each eye independently while covering the other eye. If you use corrective lenses, test your acuity with and without them. Record your results.

Astigmatism

An uneven curvature of the lens or cornea prevents some light rays from being sharply focused on the retina. This results in **astigmatism,** a condition in which an image will be blurred in one axis and sharp in other axes. Proceed with the test as follows.

Figure 24.7 Astigmatism test chart.

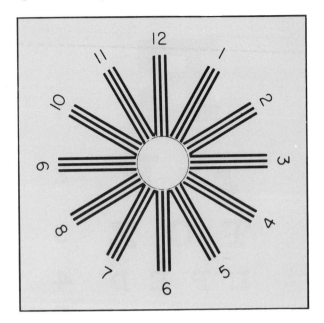

Using one eye at a time, stare at the center of Figure 24.7 and determine if all radiating lines are in focus and have the same intensity of blackness. If so, no astigmatism is present. If not, record the number(s) of the lines that are blurred or less dark.

Colorblindness

There are three types of cones in the retina and each responds maximally to only one kind of monochromatic light: red, green, or blue. The degree of stimulation of each type of cone by a particular wavelength of light determines the color perceived by the brain.

Colorblindness is a sex-linked hereditary trait affecting 8% of the male population and 0.5% of the females. The most common type is red-green colorblindness in which either red-sensitive cones or green-sensitive cones are absent. People with this condition have difficulty distinguishing reds and greens. A totally colorblind person sees only gray.

Test your color vision by viewing the Ishihara colorblindness plates held by your partner in good light. Start with No. 1 and proceed through No. 14, allowing about three seconds for each response. Have your partner record your responses on your laboratory report. Then, reverse roles and repeat the process.

Accommodation to Light Intensity

The diameter of the pupil may vary from 10 mm in darkness to only 1.5 mm in bright light. Sudden exposure of the retina to bright light causes an immediate and proportionate constriction of the pupil. In this reflex, impulses from the retina are carried via the optic nerve to the brain and return to the iris to cause its contraction.

Use a subject with light-colored eyes to test this reflex and perform it in dim light. Hold an unlighted desk lamp about six inches from the left eye. While watching the pupil of the left eye, turn the light on for one second and then turn it off. Did the pupil constrict?

Now have the subject hold the edge of this manual against the forehead and extending down along the nose to exclude light from one side of the face to the other. Repeat the test while watching the pupil of the *right eye.* Record the results. What does this tell you about the pathway of the impulses?

Assignment

Complete the laboratory report.

The Ear

Objectives

After completing this exercise, you should be able to:

1. Identify the parts of the ear on charts and models, and describe their functions.
2. Perform and interpret simple tests of ear function.
3. Define all terms in bold print.

Materials

Tuning fork
Cotton earplugs
Prepared slides of cochlea, x.s.

In this exercise, you will study the anatomy of the human ear and its involvement in both hearing and equilibrium. The ear is anatomically divided into three parts: the external ear, the middle ear, and the inner ear. Figure 25.1 illustrates the basic anatomy of the ear.

External Ear

The external ear consists of two parts: the auricle and the external auditory meatus. The **auricle,** sometimes called the **pinna,** is the outer shell-like structure of skin and elastic cartilage that is attached to the side of the head. The **external auditory meatus** is the canal that extends from the auricle about an inch into the temporal bone and terminates at the **tympanic membrane** or **tympanum.** The skin lining the canal contains some small hairs and modified apocrine sweat glands that produce a waxy secretion called cerumen. The auricle collects sound waves and directs them through the external auditory meatus to the tympanum.

The Middle Ear

The middle ear consists of a small cavity within the temporal bone. This air-filled cavity lies between the tympanum and the inner ear, and it contains three small bones, the ear **ossicles.** These tiny bones form a sensitive lever system that transmits the vibrations of the tympanum to the fluids within the inner ear.

The outermost ossicle, the **malleus,** is a hammer-shaped or club-shaped bone that is attached to the tympanum by the end of its "handle." The middle ossicle is an anvil-shaped bone called the **incus.** The innermost ossicle, the **stapes,** is a stirrup-shaped bone whose "footplate" fits into the **oval window** of the inner ear. Because of the linkage of the ossicles, vibrations of the tympanum cause the stapes to move in and out of the oval window.

The **Eustachian tube** leads downward from the middle ear to the nasopharynx. This slender duct allows the air pressure in the middle ear to be equalized with that of the outside atmosphere. A valve at the nasopharynx end of the tube normally keeps the tube closed. However, yawning or swallowing temporarily opens the valve to allow air pressure equalization.

Otitis media, an acute infection of the middle ear, is a common malady among children. Pathogens usually enter the middle ear via the Eustachian tube. Otitis media is painful and can lead to rupture of the tympanum. It is treated by antibiotics and, in severe cases, by lancing the tympanum and inserting a tube to drain the accumulated fluid.

Figure 25.1 Anatomy of the ear.

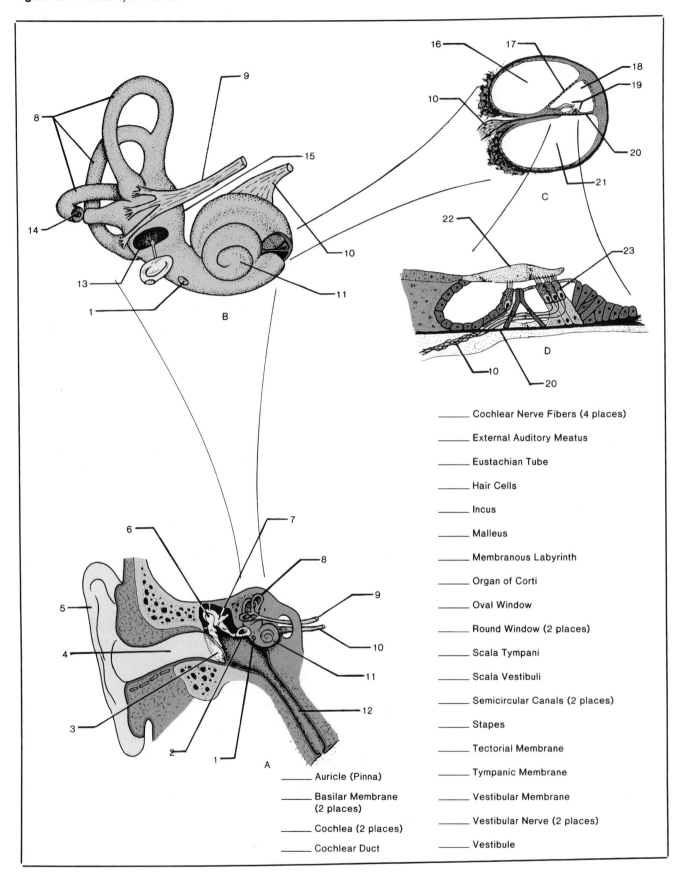

_____ Cochlear Nerve Fibers (4 places)

_____ External Auditory Meatus

_____ Eustachian Tube

_____ Hair Cells

_____ Incus

_____ Malleus

_____ Membranous Labyrinth

_____ Organ of Corti

_____ Oval Window

_____ Round Window (2 places)

_____ Scala Tympani

_____ Scala Vestibuli

_____ Semicircular Canals (2 places)

_____ Stapes

_____ Tectorial Membrane

_____ Tympanic Membrane

_____ Vestibular Membrane

_____ Vestibular Nerve (2 places)

_____ Vestibule

_____ Auricle (Pinna)

_____ Basilar Membrane
(2 places)

_____ Cochlea (2 places)

_____ Cochlear Duct

Figure 25.2 The membranous labyrinth.

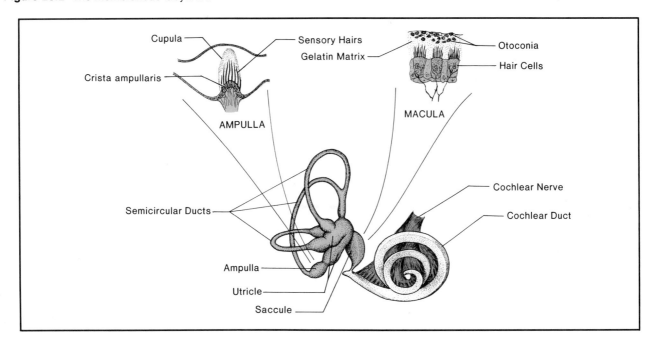

The Inner Ear

The inner ear consists of a **bony labyrinth,** shown in Illustration B, Figure 25.1, that is the hollowed-out portion of the temporal bone. Within the bony labyrinth is a **membranous labyrinth** illustrated in Figure 25.2. The space between the bony and membranous labyrinths is filled with a fluid, the **perilymph.** The fluid within the membranous labyrinth is the **endolymph.** These fluids play important roles in hearing and equilibrium.

Note that the bony labyrinth consists of three semicircular canals, a vestibule, and the cochlea. The **vestibule** is the portion that contains the oval window. The **semicircular canals** branch off one end of the vestibule, and the **cochlea** is the coiled extension at the other end.

Two branches of the statoacoustic nerve, the eighth cranial nerve, carry impulses to the brain from the inner ear. The **vestibular nerve** leads from the vestibule while the **cochlear nerve** emanates from the cochlea.

The Cochlea

The bony labyrinth of the cochlea contains three chambers extending along its full length. See the cross section in the Illustration C, Figure 25.1. The upper chamber, the **scala vestibuli** (label 16), is continuous with the vestibule. The lower chamber, the **scala tympani,** is so-named because it contains the membranous **round window** (label 1). Both of these chambers are filled with **perilymph.** The middle chamber, the **cochlear duct,** is bounded by the **vestibular membrane** above and the **basilar membrane** below.

It is filled with **endolymph.** On the upper surface of the basilar membrane is the **organ of Corti,** which contains **hair cells,** the receptors for sound stimuli.

The detail of the organ of Corti is shown in Illustration D. The hair cells sit on top of the basilar membrane with the tips of the hairs embedded in a gelatinous flap, the **tectorial membrane.** Neuron fibers leading from the hair cells become part of the cochlear nerve.

Vestibular Apparatus

Figure 25.2 shows the membranous labyrinth and the details of the vestibular apparatus that consists of the saccule, utricle, and semicircular canals. The **saccule** and **utricle** are located within the vestibule. Sensory structures, the maculae, are located on the inner walls of these structures. Each **macula** contains hair cells, the receptors for gravitational stimuli. The hairs are embedded in a gelatinous matrix that also contains crystals of calcium carbonate called **otoconia.** Neuron fibers leading from the hair cells become part of the vestibular nerve.

The hair cells that are receptors for dynamic equilibrium occur in the **ampullae,** the dilated basal portions of the semicircular canals. These hair cells are arranged in a crest, the **crista ampullaris,** within each ampulla. The hairs are embedded in a gelatinous mass called the **cupula.** Neuron fibers leading from the hair cells join the vestibular nerve.

Assignment

Label Figure 25.1.

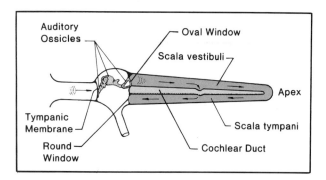

Figure 25.3 Pathway of sound wave transmission in the ear.

Auditory Ossicles
Oval Window
Scala vestibuli
Apex
Tympanic Membrane
Round Window
Scala tympani
Cochlear Duct

The Physiology of Hearing

Hearing is dependent upon: (1) the conduction of sound waves from the tympanum to the inner ear by the ear ossicles, (2) the stimulation of hair cells in the organ of Corti, and (3) the formation of impulses and their transmission to the auditory centers of the brain for interpretation. Figure 25.3 shows the cochlea uncoiled to reveal the relationships of the inner ear components involved in sound wave transmission.

Transmission to the Inner Ear

Sound waves striking the tympanic membrane cause it to vibrate. These vibrations are transmitted to the perilymph of the inner ear by the stapes moving in and out of the oval window.

Activation of the Basilar Membrane

Note in Figure 25.3 that the scala vestibuli and scala tympani are contiguous at the apex of the cochlea. Since the perilymph cannot be compressed, the flexibility of the round window allows the fluid to move back and forth in these chambers in synchrony with the movements of the stapes. The pressure waves created by the sudden movements of the perilymph form bulges in the flexible cochlear duct, which activate the basilar membrane.

The basilar membrane contains about 20,000 transverse fibers of gradually increasing length, which are attached at one end to the bony center of the cochlea. Because they are unattached at the other end, they can vibrate like reeds of a harmonica to specific vibration frequencies. Fibers closest to the oval window are short (0.04 mm); they vibrate when stimulated by high frequency sounds. Those at the apex of the cochlea are long (0.5mm) and vibrate with low frequency sounds. Due to the lengths of these fibers, the human ear can detect sound frequencies from about 30 to 20,000 cycles per second.

Hair Cell Stimulation

When the fibers in a particular part of the basilar membrane vibrate in response to a specific sound frequency, the hairs of the hair cells above them are bent and stressed against the tectorial membrane. This causes the hair cells to form impulses that are carried via the cochlear nerve to the brain, where they are interpreted as sound sensations.

Pitch and Loudness

The hair cells associated with each portion of the basilar membrane send impulses to slightly different portions of the brain. Thus, the detection of pitch is dependent upon the portion of the basilar membrane that is activated by a specific sound frequency and the part of the brain that receives the impulses.

Loudness is determined by the frequency of impulses received by the brain. The louder the sound, the greater the vibration of the basilar membrane, which results in more hair cells sending a greater frequency of impulses to the brain.

Static Equilibrium

When the head is tilted in any direction, the otoconia are pulled by gravity, causing the gelatinous matrix to bend the hairs of the hair cells. Bending of the hairs increases the formation of impulses by the hair cells. The impulses are transmitted via the vestibular nerve to the brain.

Bending the head alone does not give a sense of malequilibrium because proprioceptors and the eyes send impulses to the brain to neutralize the effect of the vestibular impulses. Only when the whole body becomes disoriented are the vestibular impulses not inhibited.

Dynamic Equilibrium

Recall that the semicircular canals are oriented in the three planes of space. When the head moves in any direction, the semicircular canal in the direction of the movement moves in relation to the endolymph within it; i.e., the canal moves but the fluid is stationary. This movement causes the cupula to move in either direction due to the force of the endolymph against it. This bends the hairs of the hair cells and results in the generation of impulses that are carried via the vestibular nerve to the brain. Reception of these impulses by the brain initiates a reflex activating the appropriate muscles to maintain equilibrium.

Assignment

1. Complete Sections A through C on the laboratory report.
2. Examine a prepared slide of cochlea, x.s. and compare your observations with Figures 25.1 and HA-15.

Hearing Tests

There are two basic kinds of deafness. **Nerve deafness** results from damage to the cochlea, organ of Corti, cochlear nerve, or the auditory center. It cannot be corrected. **Conduction deafness** results from damage to the tympanum or ear ossicles. It may be corrected by surgery or hearing aids. Perform the following hearing tests in a quiet room.

A common cause of nerve deafness is exposure to loud noises, such as jet airplanes and amplified rock music, that damage the hair cells. Conduction deafness may be caused by calcium deposits that prevent free movement of the ossicles or by the hardening or rupture of the tympanum.

Rinne Test

This test distinguishes between nerve deafness and conduction deafness.

1. Plug one of your ears with cotton.
2. Set the tuning fork in motion by striking it against the heel of the hand. Hold it about six inches from your unplugged ear with a tine facing the ear as shown in Figure 25.4. Persons with normal hearing and slight hearing loss will hear the sound, but those with a severe hearing loss will not hear the sound or will hear it only briefly.

3. As the sound fades to where it is no longer heard, place the stem of the tuning fork against the mastoid process behind your ear. If the sound reappears, conduction hearing loss is evident. If not, nerve hearing loss exists.
4. Test both ears and record the results.

The Weber Test

When the base of a vibrating tuning fork is placed against the center of the forehead of a person with normal hearing, it is heard with equal loudness in both ears. If conduction deafness exists in one ear, the sound will be louder in the defective ear than in the normal ear. This occurs because the defective ear is more attuned to sound waves conducted through the bone. If nerve deafness exists in one ear, the sound will be louder in the normal ear.

Place the base of a vibrating tuning fork against your forehead and compare the sounds in each ear. Record your results.

Static Balance Test

In addition to the vestibular apparatus, three other factors are important in maintaining equilibrium: (1) visual orientation, (2) proprioceptor sensations, and (3) cutaneous sensations. Impulses from these four sources are subconsciously integrated to provide correct reflex responses in maintaining equilibrium.

Figure 25.4 Tuning fork manipulation in hearing tests.

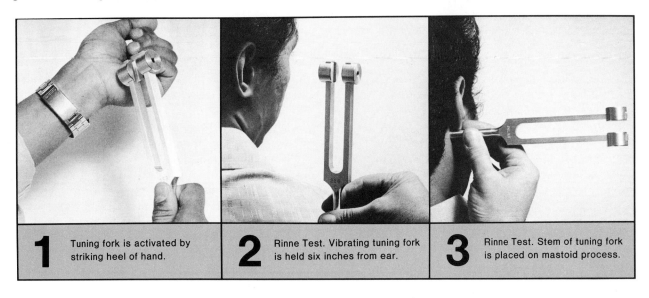

1 Tuning fork is activated by striking heel of hand.

2 Rinne Test. Vibrating tuning fork is held six inches from ear.

3 Rinne Test. Stem of tuning fork is placed on mastoid process.

A simple way to evaluate the effectiveness of the static balance mechanism is to have the subject stand perfectly still with the eyes closed. If there is damage to the system, the subject will waiver and tend to fall.

1. Use a meter stick to draw a number of verticle lines on the chalkboard about 5 cm apart. Cover an area about one meter across. This will help detect body movements.
2. The subject is to remove his/her shoes and stand facing the examiner in front of the lined area.
3. Have the subject stand perfectly still for thirty seconds with arms at the sides and eyes open. Record the degree of swaying movement as small, moderate, or great.
4. Repeat the test with the subject's eyes closed. Record the results.
5. Repeat the test with the eyes closed while the subject stands on only one foot. What happens?

Assignment

Complete the laboratory report.

Blood Tests

Objectives

After completing this exercise, you should be able to:

1. Identify the formed elements of blood when viewed microscopically or on charts.
2. Perform and interpret these tests: differential white cell count, packed red cell volume, coagulation time, and blood typing.
3. Define all terms in bold print.

Materials

1. For Each Test
   Alcohol pads
   Biohazard bag
   Kimwipes
   Paper towels
   Sterile disposable lancets
   Amphyl solution, 0.25%
2. White Cell Differential Count
   Bibulous paper
   Clean microscope slides
   Distilled water in dropping bottle
   Wright's blood stain in dropping bottle
   Wax pencil

3. Volume of Packed Red Cells
   Heparinized capillary tubes
   Micro-hematocrit centrifuge
   Micro-hematocrit tube reader
   Seal-Ease (Clay Adams)
4. Clotting Time
   Capillary tubes, 0.5 mm diameter
   File, 3-cornered
5. Blood Typing
   Clean microscope slide
   Slide warming box with typing slide
   Toothpicks
   Typing sera: anti-A, anti-B, anti-D

Blood is the medium that transports substances to and from body cells. It consists of a fluid **plasma,** which forms 55% of the blood volume, and the **formed elements,** which constitute the remaining 45%. The formed elements consist of **erythrocytes** (red blood cells), **leukocytes** (white blood cells), and **thrombocytes** (platelets). Their relative abundance and functions are shown in Table 26.1. They are illustrated in Figure 26.1 as they appear when stained with Wright's blood stain. Learn the recognition characteristics of each type of formed element.

Table 26.1. Formed elements of the blood

Formed Element	Concentration	Function
Erythrocytes	4,000,000–6,000,000 per cubic millimeter	Transport oxygen and carbon dioxide
Leukocytes	5,000–10,000 per cubic millimeter	Destroy pathogens; neutralize toxins
Granulocytes		
Neutrophils	50–70% of leukocytes	Phagocytosis
Eosinophils	1–3% of leukocytes	Neutralize products of allergic reactions; destroy parasitic worms
Basophils	0.5–1% of leukocytes	Release heparin and histamine; intensify inflammation and allergic reactions
Agranulocytes		
Lymphocytes	20–30% of leukocytes	B-Lymphocytes form antibodies T-Lymphocytes form chemicals that destroy antigens and/or stimulate phagocytosis
Monocytes	2–6% of leukocytes	Phagocytosis
Thrombocytes	250,000–400,000 per cubic millimeter	Initiate clotting process

Figure 26.1 Formed elements of blood.

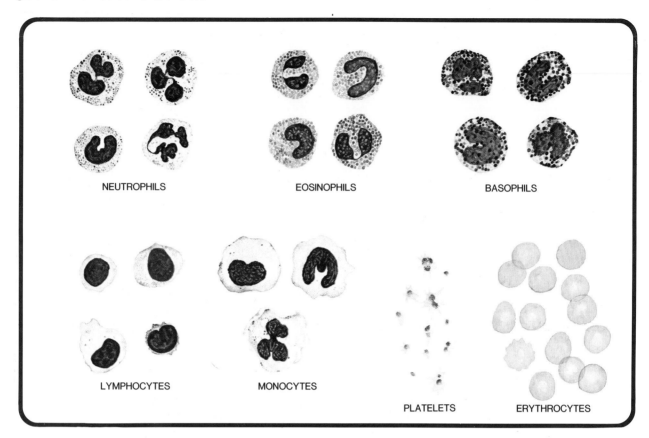

NEUTROPHILS

EOSINOPHILS

BASOPHILS

LYMPHOCYTES

MONOCYTES

PLATELETS

ERYTHROCYTES

In this exercise, you will perform several common blood tests that have diagnostic value. However, our purpose is not to diagnose but to understand the purpose and nature of the tests. Each test involves obtaining a drop of blood by piercing a finger tip with a sterile lancet. *All materials contacting blood are to be placed immediately in a biohazard bag.* Keep your work station clean by washing it with Amphyl solution after each test. Record the results of the tests on the laboratory report.

The Differential White Cell Count

This test is performed to determine the relative percentages of the five types of white blood cells shown in Figure 26.1. Three of them, **neutrophils, eosinophils,** and **basophils,** have cytoplasmic granules and are classified as **granulocytes. Monocytes** and **lymphocytes** lack cytoplasmic granules and are classified as **agranulocytes.**

High or low counts for certain types of white blood cells may be related to pathological conditions. High neutrophil counts usually indicate bacterial infection. High eosinophil counts occur in allergic reactions or parasitic worm infestations. High monocyte counts usually indicate a chronic infection. High lymphocyte counts usually indicate antigen-antibody reactions.

A differential white cell count is performed by preparing a stained blood slide and counting a total of 100 white cells to determine the percentage of each type of cell.

Slide Preparation

Keys to success in preparing a good slide are: (1) placing the right-sized drop of blood on the slide, and (2) spreading the blood evenly over the slide.

1. Clean 3–4 slides with soap and water, and keep them free of fingerprints.
2. Cleanse a finger tip with an alcohol pad and pierce it with a lancet. Place the lancet *immediately* in the biohazard bag. Place a drop of blood 2 cm from one end of a slide. *The drop should be 3–4 mm in diameter.*
3. Spread the blood on the slide using another clean slide as a spreader as shown in Figure 26.2. Note that the blood is *pulled* along by the slide. Hold the spreader slide at an angle of 50° to obtain a good smear. Place the spreader slide directly in the biohazard bag.
4. After the blood has dried, use a wax pencil to draw a line at each end of the smear to confine the stain.

Figure 26.2 Smear preparation technique for differential WBC count.

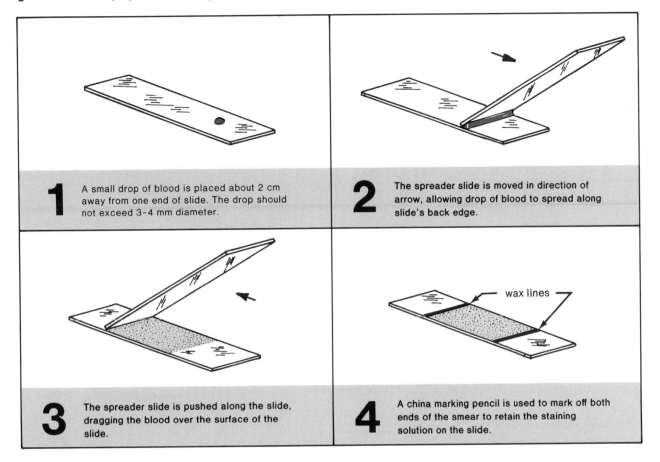

1 A small drop of blood is placed about 2 cm away from one end of slide. The drop should not exceed 3–4 mm diameter.

2 The spreader slide is moved in direction of arrow, allowing drop of blood to spread along slide's back edge.

3 The spreader slide is pushed along the slide, dragging the blood over the surface of the slide.

4 A china marking pencil is used to mark off both ends of the smear to retain the staining solution on the slide.

wax lines

5. Place the slide on a folded paper towel and add Wright's blood stain, *counting the drops* until the smear is covered.

6. Wait *four minutes,* then add an equal number of drops of distilled water. Let it stand for *ten minutes* while blowing on the slide every few minutes to keep the solutions mixed.

7. Gently rinse the slide in slowly running water. Blot the slide dry with bibulous paper.

Making the Count

When the slide is dry, scan it with the low power objective to find an area where the cells are separated from each other. Avoid the more dense areas. Place a drop of immersion oil on the selected area and examine with the oil immersion objective. Note the appearance and relative concentration of erythrocytes and thrombocytes. To make the differential white cell count, move the slide in a path as shown in Figure 26.3. Record on the laboratory report the type of each white cell encountered until you have tabulated 100 cells. When finished, place the slide in the biohazard bag.

Figure 26.3 Examination path for differential count.

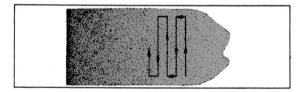

Volume of Packed Red Cells

Hemoglobin, the red blood pigment, is the oxygen-carrying component in erythrocytes. A person with a reduced concentration of hemoglobin is said to be anemic. One of the causes of anemia is a subnormal erythrocyte concentration.

One test used to determine the concentration of erythrocytes in the blood is called the **VPRC** or **volume of packed red cells** (hematocrit). This test determines the volume percentage of the blood occupied by red blood cells. It is performed by using a micro-hematocrit centrifuge and heparinized micro-hematocrit capillary tubes.

Figure 26.4 Determination of the volume of packed red cells.

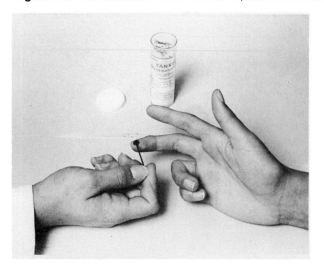

1. Blood is drawn up into heparinized capillary tube for hematocrit determination.

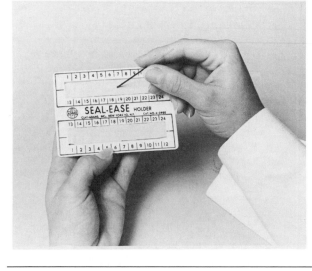

2. The end of the capillary tube is sealed with clay.

3. The capillary tubes are placed in the centrifuge with the sealed end toward the perimeter.

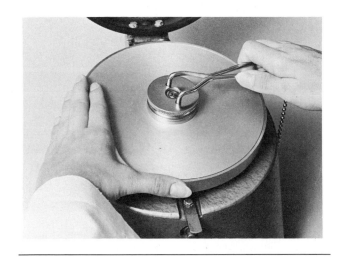

4. The safety lid is tightened with a lock wrench.

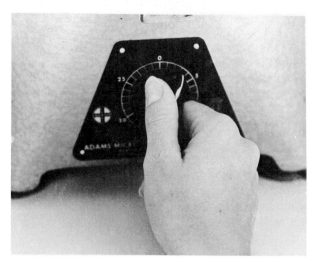

5. The timer is set for four minutes by turning the dial clockwise.

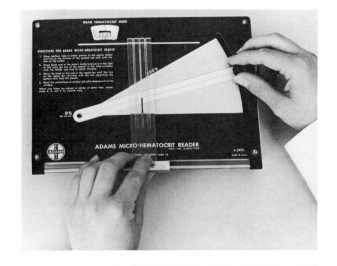

6. The capillary tube is placed in the tube reader to determine the hematocrit.

In males, the normal range is 40–54%; 47% is average. In females, the normal range is 37–47%; 42% is average. Perform the test as described below and shown in Figure 26.4.

1. Cleanse and pierce the finger tip as before and place the lancet in the biohazard bag. Wipe away the first drop of blood with the alcohol pad.
2. Place the red end of the heparinized capillary tube in the second drop of blood and lower the other end so the blood will fill about two-thirds of the tube.
3. Insert the blood end of the tube into Seal-Ease to plug it.
4. Place the tube in the centrifuge with the sealed end against the ring at the perimeter. Record your tube slot number. Load the centrifuge with an even number of tubes that are arranged opposite each other to balance the load.
5. Secure the inside cover and fasten the outside cover.
6. Turn on the centrifuge and set the timer for *four minutes.*
7. Determine the VPRC by using the mechanical tube reader. Instructions are on the instrument. The VPRC may be read on the head of some centrifuges.
8. Place blood-contaminated materials in the biohazard bag.

Clotting Time

Coagulation of the blood is a complex process that culminates in the conversion of a soluble protein, **fibrinogen,** into an insoluble fibrous protein, **fibrin.** Normal clotting time is 3–6 minutes. Figure 26.5 shows the basic procedures. Determine your clotting time as follows.

1. Cleanse and pierce the finger tip and place the lancet in the biohazard bag. *Record the time.* Place one end of a capillary tube into the drop of blood and lower the other end, allowing blood to fill it completely.
2. At *one minute intervals* break off small segments of the capillary tube by scratching the glass first with the file. Place the tube on a paper towel when scoring it with the file. *Separate the broken ends slowly while looking for strands of fibrin between them.* Record the clotting time as the time from the first appearance of blood on your finger to the formation of fibrin strands.
3. Place the tube fragments and paper towel in the biohazard bag.

Figure 26.5 Determination of clotting time.

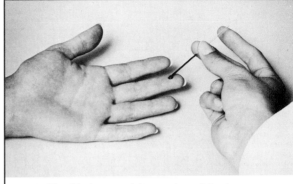

1. Blood is drawn up into a nonheparinized capillary tube.

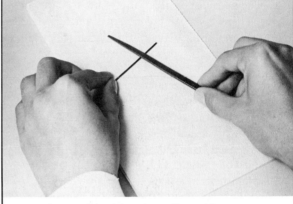

2. At one-minute intervals, sections of the tube are filed and broken off for the test.

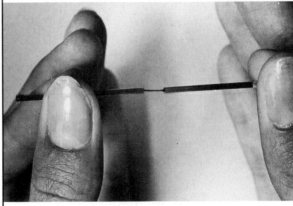

3. When the blood has coagulated, strands of fibrin will extend between broken ends of tube.

Blood Tests   **129**

Table 26.2. Antigen and antibody associations in blood

Blood Type	Antigen	Antibody
O	none	a,b
A	A	b
B	B	a
AB	A,B	none
Rh+	D	none
Rh−	none	d*

*Antibodies are formed by an Rh− person only after Rh+ erythrocytes enter his/her blood.

Blood Typing

Red blood cells possess a variety of antigens on their surfaces. The presence or absence of some of these antigens is used to determine one's blood type, which is genetically controlled and never changes throughout the lifetime of an individual.

An **antigen** is a chemical molecule that stimulates the production of a specific **antibody,** which, in turn, will attach to the antigen. Although there are many antigens on the red blood cells, it is the **A, B,** and **D (Rh)** antigens that are most commonly used in blood typing. In this exercise, we will study blood typing using the A, B, and D antigens to determine your ABO and Rh blood type.

Normally, transfusions involve a donor and a recipient with the same blood type. However, it may be possible for a recipient to receive a transfusion of a different blood type in emergency situations.

All blood types are not compatible, and a transfusion of incompatible blood may be fatal. Thus, typing the blood of both the potential donor and the recipient is essential. Because of differences in concentration, it is the *antigens of the donor* and the *antibodies of the recipient* that must be considered. Table 26.2 shows the relationship of antigens and antibodies in normal blood. By understanding this table and the antigen—antibody reaction, you can predict the compatibility of various blood type combinations.

In the clinical setting, blood is usually diluted with saline when doing ABO blood typing. However, in this exercise, you will use whole blood since you will be determining the ABO group and Rh type simultaneously. Rh typing requires whole blood and a slide warming box to yield a temperature of about 50° C.

You will type your blood by mixing a drop of blood with a drop of **antiserum** that contains specific, known antibodies. If the corresponding antigen is present on the red blood cells, an antigen—antibody reaction occurs causing the erythrocytes to **agglutinate** (clump together). This clumping can be identified visually.

Determine your ABO and Rh blood type as described below.

1. Place a clean microscope slide across the three typing squares on the typing slide of the warming box. Allow 5 minutes for temperature equilibration.
2. Cleanse and pierce a finger tip. Discard the lancet in the biohazard bag. Place a small drop of blood on the slide over each of the three squares.
3. Quickly add one drop of antiserum to the blood in each square: anti-D to the anti-D square, anti-A to the anti-A square, and anti-B to the anti-B square. *Record the time.*
4. Mix the antiserum and blood in each square *using a separate clean toothpick for each one.* Why? Place the toothpicks directly into the biohazard bag.
5. Rock the warming box for *two minutes,* then examine the mixtures for agglutination, starting with the anti-D square. Clumps in the anti-D square must appear within two minutes to be valid. These clumps will be very small and require close scrutiny for detection. Agglutination in any square indicates the presence of the antigen in question. For example, clumping in the anti-A square indicates presence of the A antigen. See Figure 26.7.
6. Place the slide in the biohazard bag. Clean your work station with 0.25% Amphyl solution.
7. Complete the laboratory report.

Figure 26.6 Blood typing with warming box.

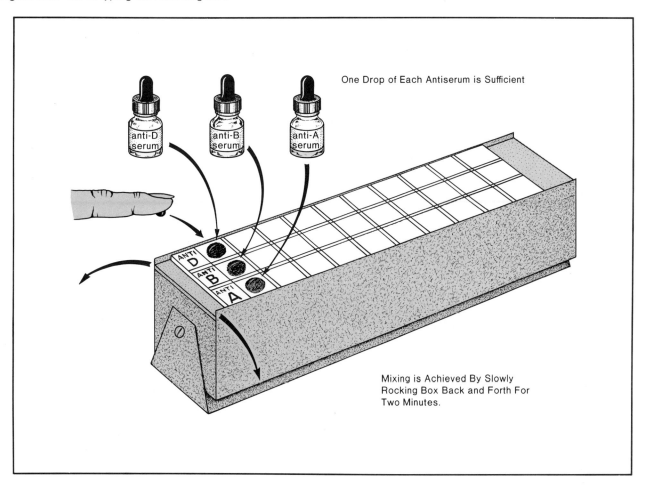

One Drop of Each Antiserum is Sufficient

Mixing is Achieved By Slowly Rocking Box Back and Forth For Two Minutes.

Figure 26.7 Interpretation of agglutination patterns.

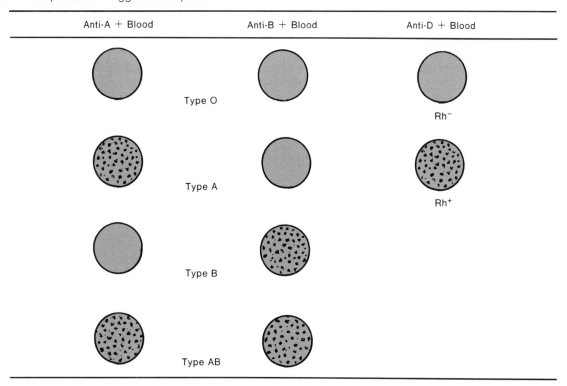

The Heart

Objectives

After completing this exercise, you should be able to:

1. Identify the parts of the heart on preserved specimens, charts, or models.
2. Describe the functions of each part of the heart.
3. Define all terms in bold print.

Materials

Sheep heart, fresh or preserved
Dissecting instruments and tray

Internal Anatomy

Locate the structures described below in Figure 27.1. Vessels colored red carry oxygenated blood; those colored blue carry deoxygenated blood.

Wall of the Heart

The wall of the heart consists of three layers. The thick **myocardium** is composed of cardiac muscle tissue. It is covered by two thin serous membranes attached to the muscle tissue: an inner **endocardium** and an outer **epicardium** or **visceral pericardium.** External to the epicardium is the double-layered **parietal pericardium,** composed of an inner serous membrane and an outer fibrous membrane. The **pericardial cavity** (label 19) lies between the epicardium and parietal pericardium. Fluid secreted by the serous membranes reduces friction, enabling the heart to move freely within the pericardial sac.

Chambers of the Heart

The **left** and **right atria** are the upper, thin-walled chambers, which receive blood returning to the heart. Note that the right atrium is on the left side of the frontal section in the figure. There is no connection between the atria. The lower, thick-walled chambers are the ventricles, which pump blood into the arteries leaving the heart. Note the **ventricular septum** separating the **left ventricle** from the **right ventricle,** and the thicker myocardium of the left ventricle.

The heart has two atrioventricular valves and two semilunar valves. The **tricuspid atrioventricular valve** has three flaps or cusps and occurs between the right atrium and ventricle. The **mitral** or **bicuspid atrioventricular valve** has only two cusps, and separates the left atrium and ventricle. Atrioventricular valves prevent a backflow of blood from the ventricles into the atria during **systole,** the contraction phase of the heart cycle. The cusps have thin cords, the **chordae tendineae,** that anchor them to **papillary muscles** (label 15) on the walls of the ventricles and prevent the valve cusps from being forced into the atria during contraction.

The **pulmonary semilunar valve** and the **aortic semilunar valve** are located at the base of the pulmonary artery and aorta respectively. They each have three cusps and prevent the backflow of blood from the arteries into the ventricles during **diastole,** the relaxation phase of the heart cycle.

Figure 27.1 Internal anatomy of the heart.

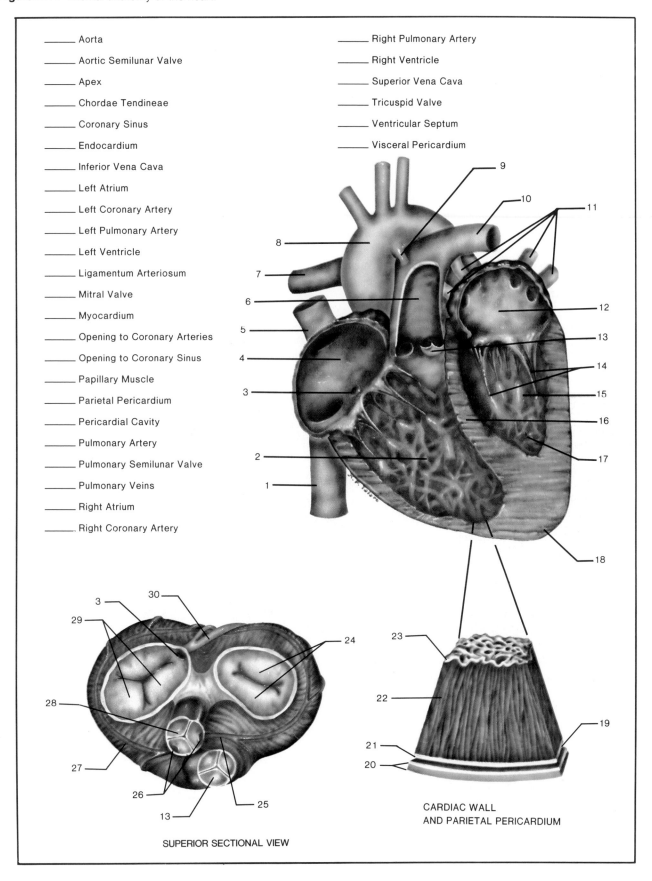

_____ Aorta

_____ Aortic Semilunar Valve

_____ Apex

_____ Chordae Tendineae

_____ Coronary Sinus

_____ Endocardium

_____ Inferior Vena Cava

_____ Left Atrium

_____ Left Coronary Artery

_____ Left Pulmonary Artery

_____ Left Ventricle

_____ Ligamentum Arteriosum

_____ Mitral Valve

_____ Myocardium

_____ Opening to Coronary Arteries

_____ Opening to Coronary Sinus

_____ Papillary Muscle

_____ Parietal Pericardium

_____ Pericardial Cavity

_____ Pulmonary Artery

_____ Pulmonary Semilunar Valve

_____ Pulmonary Veins

_____ Right Atrium

_____ Right Coronary Artery

_____ Right Pulmonary Artery

_____ Right Ventricle

_____ Superior Vena Cava

_____ Tricuspid Valve

_____ Ventricular Septum

_____ Visceral Pericardium

SUPERIOR SECTIONAL VIEW

CARDIAC WALL
AND PARIETAL PERICARDIUM

Vessels of the Heart

The major vessels of the heart are the two venae cavae, four pulmonary veins, a pulmonary artery, and the aorta. During diastole, blood is returned to the atria. Deoxygenated blood is returned to the **right atrium** from body regions above the heart by the **superior vena cava** (label 5) and from body regions below the heart by the **inferior vena cava.** Simultaneously, oxygenated blood is returned to the right atrium by the **left** and **right pulmonary veins.**

During systole, blood is pumped from the ventricles. Deoxygenated blood is carried from the right ventricle to the lungs via the **pulmonary artery,** which branches to form the **left** and **right pulmonary arteries.** Simultaneously, the left ventricle pumps oxygenated blood to all parts of the body except the lungs via the **aorta** (label 8). The emergence of the aorta from the left ventricle is not visible in the figure.

The **ligamentum arteriosum** (label 9) is a nonfunctional remnant of the **ductus arteriosus** that allows blood to flow from the pulmonary artery to the aorta during prenatal life.

Note in the superior sectional view that the **left** and **right coronary arteries** exit from the aorta just above the aortic semilunar valve. These arteries carry blood to nourish the heart itself. Deoxygenated blood is collected by cardiac veins and carried to the **coronary sinus** (label 30), which returns the blood to the right atrium.

Assignment

Label Figure 27.1.

External Anatomy

Locate the external structures of the heart in Figure 27.2 as you study this section. The **parietal pericardium** is intact in the anterior aspect of the figure but has been removed in the posterior aspect.

Anterior Aspect

The heart depicted in this view is oriented in the same position as the interior view of the heart shown in Figure 27.1. This will help you correlate interior and exterior relationships. Note that the tilt of the heart in the **mediastinum** places the **apex** (tip) on the left side of the midline.

Locate the superior vena cava above the right atrium and the inferior vena cava where it projects below the **diaphragm.** Observe that the aorta forms an **aortic arch** over the heart and descends behind the heart to become the **descending aorta.** Find the left and right pulmonary arteries projecting from under the aortic arch, and locate the short ligamentum arteriosum.

The coronary arteries and cardiac veins occur in the **atrioventricular sulci,** shallow grooves between the atria and ventricles, and in the **interventricular sulcus** that lies over the ventricular septum. These vessels are surrounded by fat deposits. It is easy to locate the two ventricles by first identifying the interventricular sulcus.

Posterior Aspect

The four pulmonary veins are clearly visible. The two right pulmonary veins enter the heart near the junction of the superior and inferior venae cavae with the right atrium. The left and right branches of the pulmonary artery are located under the aortic arch.

The coronary vessels lie in sulci on the posterior surface of the heart. All of the cardiac veins empty into the **coronary sinus** (label 1), which returns the blood to the right atrium.

Myocardial infarction, or heart attack, is the death of a portion of the heart muscle. This usually results from a coronary artery being plugged by a thrombus or embolism so that the affected area is deprived of blood. **Angina pectoris** is chest pain resulting from a constriction of a coronary artery, which causes a portion of the heart muscle to receive an insufficient blood supply. Stress, coronary artery spasm, and atherosclerosis may cause angina.

Assignment

1. Label Figure 27.2.
2. Complete Sections A, B, and C on the laboratory report.

Figure 27.2 External anatomy of the heart.

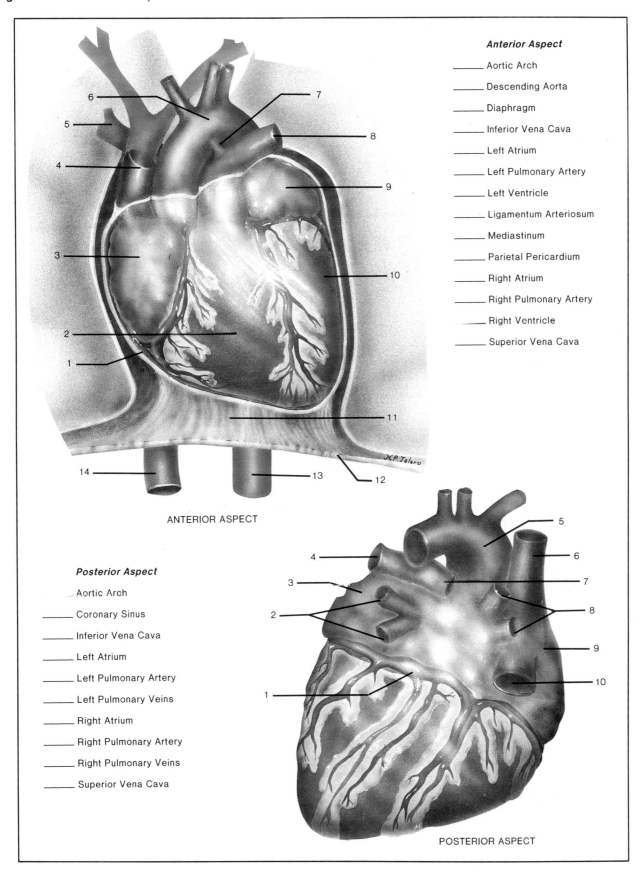

_____ Aortic Arch

_____ Descending Aorta

_____ Diaphragm

_____ Inferior Vena Cava

_____ Left Atrium

_____ Left Pulmonary Artery

_____ Left Ventricle

_____ Ligamentum Arteriosum

_____ Mediastinum

_____ Parietal Pericardium

_____ Right Atrium

_____ Right Pulmonary Artery

_____ Right Ventricle

_____ Superior Vena Cava

ANTERIOR ASPECT

Posterior Aspect

_____ Aortic Arch

_____ Coronary Sinus

_____ Inferior Vena Cava

_____ Left Atrium

_____ Left Pulmonary Artery

_____ Left Pulmonary Veins

_____ Right Atrium

_____ Right Pulmonary Artery

_____ Right Pulmonary Veins

_____ Superior Vena Cava

POSTERIOR ASPECT

Figure 27.3 Sheep heart dissection.

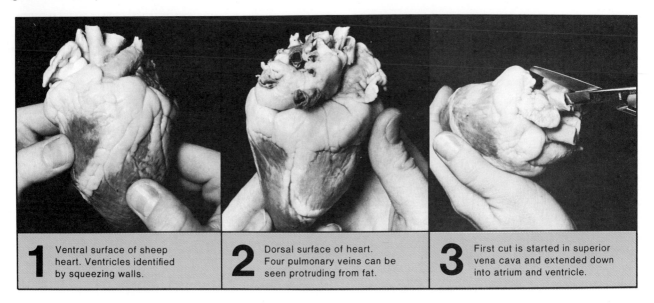

1 Ventral surface of sheep heart. Ventricles identified by squeezing walls.

2 Dorsal surface of heart. Four pulmonary veins can be seen protruding from fat.

3 First cut is started in superior vena cava and extended down into atrium and ventricle.

Sheep Heart Dissection

While dissecting the sheep heart, locate as many of the structures as possible that are shown in Figures 27.1 and 27.2. Follow the directions below and depicted in Figures 27.3 and 27.4. Since sheep walk on four legs while humans are bipedal, the anterior surface of the human heart is comparable to the ventral surface of the sheep heart.

1. Look for remnants of the **parietal pericardium** around the great vessels. It usually has been removed from laboratory specimens.
2. Note that the **epicardium** (visceral pericardium) is firmly attached to the heart.
3. Locate the **interventricular sulcus** and the **left** and **right ventricles** as shown in Illustration 1, Figure 27.3. Remove some fat to expose the **coronary vessels** in this sulcus.

4. Locate the **left** and **right atria.** The right atrium is seen on the left side of Illustration 1, Figure 27.3.
5. Identify the **aorta.** It is the large vessel just to the right of the right atrium in Illustration 1.
6. Locate the **pulmonary artery** between the aorta and left atrium. Trace it to where it branches into the **left** and **right pulmonary arteries.** Try to find the **ligamentum arteriosum.**
7. Examine the dorsal surface of the heart as in Illustration 2, Figure 27.3. Locate the thin-walled **pulmonary veins,** usually embedded in fat. Insert a probe to see that they enter the left atrium.
8. Locate the **superior vena cava** and right atrium. Insert a blade of your scissors into it and cut into the right atrium as shown in Illustration 3, Figure 27.3.

Figure 27.4 Sheep heart dissection.

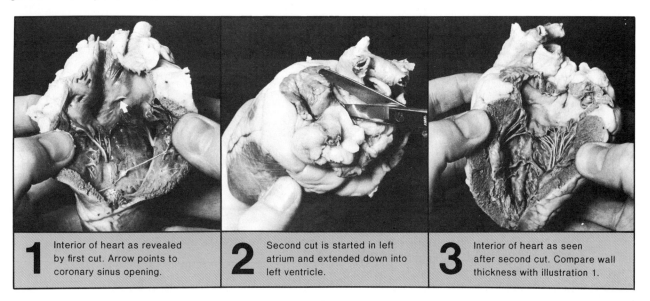

| **1** Interior of heart as revealed by first cut. Arrow points to coronary sinus opening. | **2** Second cut is started in left atrium and extended down into left ventricle. | **3** Interior of heart as seen after second cut. Compare wall thickness with illustration 1. |

9. Observe the **tricuspid valve** between the right atrium and ventricle. Wash under the tap to remove any blood clots present. Fill the right ventricle with water by pouring it through the tricuspid valve. Gently squeeze the right ventricle and observe the closing action of the valve's cusps.

10. Drain the water from the ventricle and continue the cut through the tricuspid valve and into the right ventricle. Spread the heart wall, washing again if necessary. Observe the inner wall of the right atrium and locate the opening of the coronary sinus. It is indicated by an arrow in Illustration 1, Figure 27.4.

11. Observe the inner ventricular surface. Note the tricuspid valve cusps, **chordae tendineae, papillary muscles,** and supporting muscle band extending from the ventricular septum to the ventricular wall.

12. Starting at its lower margin, use scissors to cut upward through the right anterior ventricular wall along the ventricular septum and into the **pulmonary artery** to expose the **pulmonary semilunar valve.** Note the thickness of the cusps.

13. Insert a blade of your scissors into the left atrium, as shown in Illustration 2, Figure 27.4 and cut through the atrium into the left ventricle. Spread the heart wall to observe the **mitral valve.** Compare the thickness of the left and right ventricular walls.

14. With scissors, cut from the left ventricle into the aorta to expose the **aortic semilunar valve.** Locate the openings of the **coronary arteries** in the aortic wall just above the valve. Insert a probe to verify that the openings lead into the coronary arteries.

15. Dispose of the heart as directed by your instructor. Wash and dry your tray and instruments.

16. Complete the laboratory report.

Blood Vessels, Lymphatic System, and Fetal Circulation

Objectives

Upon completion of this exercise, you should be able to:

1. Diagram and describe the basic circulatory plan.
2. Identify the major arteries, veins, and lymphatic vessels on charts or models and describe their locations.
3. Distinguish normal and atherosclerotic arteries and veins when viewed microscopically.
4. Compare circulation in fetus and adult.
5. Define all terms in bold print.

Materials

Colored pencils
Prepared slides of:
Artery and vein, x.s.
Atherosclerotic artery, x.s.

The Circulatory Plan

Figure 28.1 is an incomplete flow diagram of the basic circulatory plan. You are to complete and label this figure as you study this section.

Deoxygenated blood is returned to the right atrium from all parts of the body by the **superior** and **inferior venae cavae.** It then flows into the right ventricle. During contraction, blood is pumped from the right ventricle through the **pulmonary artery** to the lungs. Oxygenated blood is returned from the lungs into the left atrium by the **pulmonary veins** and flows into the left ventricle. During contraction, blood is pumped to all parts of the body *except the lungs* via the **aorta.**

Oxygenated blood is carried to the head and arms by arteries branching from the aorta, but are represented here by a single vessel for simplicity. As the aorta descends (draw it on the right side of the figure), other branches from the aorta carry blood to organs below the heart. The first branch is the **hepatic artery,** carrying blood to the liver.

Figure 28.1 The circulatory plan.

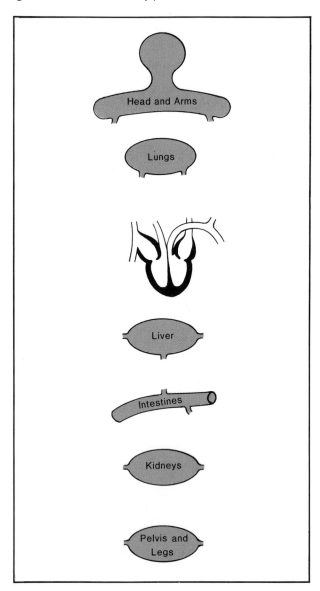

Intestines are supplied by the **superior mesenteric artery.** Kidneys receive blood via **renal arteries.** The pelvis and legs are supplied by several arteries, but only one is shown for simplicity.

Veins from the pelvis and legs combine to form the **inferior vena cava,** which collects blood from regions below the heart and returns it to the right atrium. (Show the inferior vena cava ascending on the left side of the figure.) The **renal** and **hepatic veins** carry blood from the kidneys and liver, respectively, directly to the inferior vena cava. However, blood from the intestines is first carried to the liver by the **hepatic portal vein** before entering the vena cava via the **hepatic vein.**

Assignment

1. Complete and label Figure 28.1.
2. Color vessels carrying oxygenated blood *red,* and those carrying deoxygenated blood *blue.* Indicate the direction of blood flow by arrows.

Major Arteries

The major arteries are shown in Figure 28.2. Blood leaves the left ventricle via the **aorta,** which arches over the heart and descends inferiorly. The **left subclavian** artery, the third artery branching from the aorta, emerges from the **aortic arch** and enters the left shoulder behind the clavicle. In the armpit area, it becomes the left **axillary** artery, (label 13) which, in turn, becomes the left **brachial** artery in the upper arm. The brachial artery branches into the **radial** and **ulnar** arteries, which follow the forearm bones.

The short **brachiocephalic** artery, the first artery to branch from the aorta, emerges from the right side of the aortic arch and quickly branches into: (1) the **right common carotid** artery that goes to the right side of the neck and head, and (2) the **right subclavian** artery that passes behind the clavicle. The right subclavian subsequently becomes the right **axillary** and then the right **brachial** of the upper arm.

The middle artery from the aortic arch is the **left common carotid** that supplies the left side of the neck and head.

Several major arteries branch from the **descending aorta.** The **celiac** artery (label 5) emerges from the aorta just below the diaphragm and supplies the stomach, liver, and spleen. Just below it is the **superior mesenteric** artery that serves most of the small intestine and part of the large intestine. The kidneys receive blood from the **renal** arteries. The **inferior mesenteric** artery emerges below the kidneys and supplies the large intestine and rectum.

In the lumbar region, the aorta divides to form the left and right **common iliac** arteries. Each common iliac then divides into a small **internal iliac** artery and a large **external iliac,** which continues inferiorly to become the **femoral** artery in the thigh. In the upper part of the thigh, the **deep femoral** artery branches from the femoral artery. In the knee region, the femoral becomes the **popliteal** artery, which branches to form the **anterior** and **posterior tibial** arteries.

Major Veins

Refer to Figure 28.2. The **superior vena cava** receives blood from major veins above the heart and returns it to the right atrium. It begins at the base of the neck by the junction of the short left and right **brachiocephalic** veins (label 3) that receive blood from the head and arms.

In the neck are four veins: two medial **internal jugulars** that empty into the brachiocephalic veins and two **external jugulars** that empty into the **subclavian veins.** Each subclavian vein receives blood from a short **axillary** vein that is located near the armpit. Each axillary receives blood from the **cephalic** vein, which drains the lateral surface of the arm, and the **brachial** vein (label 14). The medial **basilic** vein empties into the brachial vein. Note the short **median cubital** vein in the elbow region between the basilic and cephalic veins. It is a common site for venipuncture.

Blood is returned from the legs by both deep and superficial veins. The **posterior tibial** vein (label 22) is a deep vein that collects blood from the foot and lower leg. It becomes the **popliteal** vein in the knee region and continues as the **femoral** vein of the thigh. The surface vein of the leg is the **great saphenous** that originates on the superior surface of the foot and ascends to join with the femoral vein to form the **external iliac** vein. The **internal iliac** (label 10) and external iliac unite to form the **common iliac** vein that empties into the inferior end of the **inferior vena cava.**

Varicose veins, characterized by non-functional valves and dilated veins, often occur in the superficial veins of the lower legs. The tendency for this condition is inherited and is promoted by prolonged standing, lack of exercise, and aging. There are several treatments for this condition, but surgical removal (stripping) may be necessary.

Of the many small veins entering the inferior vena cava, only three are shown in Figure 28.2. The **hepatic** vein returned blood from the liver and is located just below the heart. The two **renal** veins return blood from the kidneys. The branch from the left renal vein is the **left spermatic** (or **ovarian**) vein, which carries blood from the left testis or ovary. The **right spermatic** (or **ovarian**) enters the inferior vena cava just below the right renal vein.

Figure 28.2 Arteries and veins.

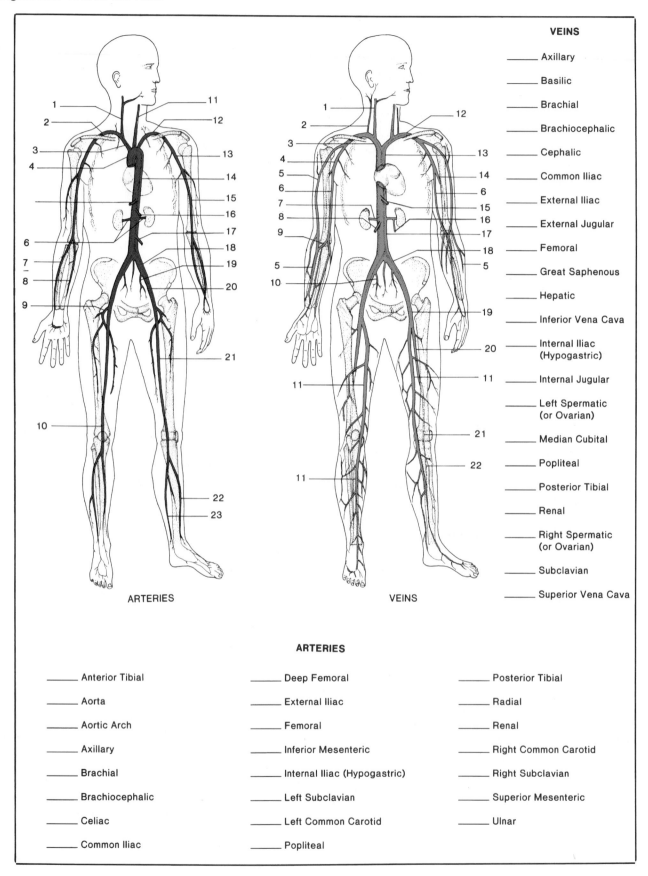

ARTERIES

VEINS

VEINS

_____ Axillary

_____ Basilic

_____ Brachial

_____ Brachiocephalic

_____ Cephalic

_____ Common Iliac

_____ External Iliac

_____ External Jugular

_____ Femoral

_____ Great Saphenous

_____ Hepatic

_____ Inferior Vena Cava

_____ Internal Iliac (Hypogastric)

_____ Internal Jugular

_____ Left Spermatic (or Ovarian)

_____ Median Cubital

_____ Popliteal

_____ Posterior Tibial

_____ Renal

_____ Right Spermatic (or Ovarian)

_____ Subclavian

_____ Superior Vena Cava

ARTERIES

_____ Anterior Tibial

_____ Aorta

_____ Aortic Arch

_____ Axillary

_____ Brachial

_____ Brachiocephalic

_____ Celiac

_____ Common Iliac

_____ Deep Femoral

_____ External Iliac

_____ Femoral

_____ Inferior Mesenteric

_____ Internal Iliac (Hypogastric)

_____ Left Subclavian

_____ Left Common Carotid

_____ Popliteal

_____ Posterior Tibial

_____ Radial

_____ Renal

_____ Right Common Carotid

_____ Right Subclavian

_____ Superior Mesenteric

_____ Ulnar

Figure 28.3 Hepatic portal circulation.

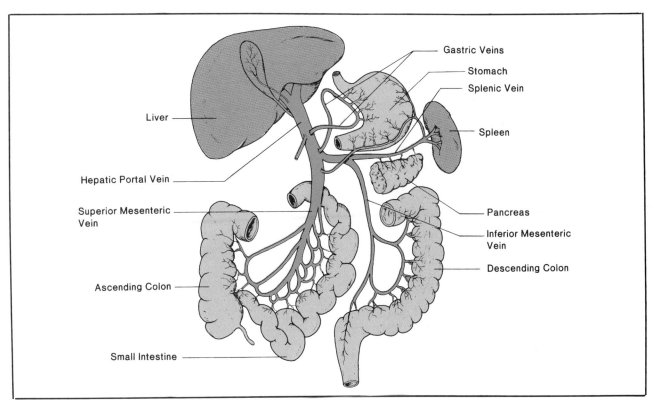

Hepatic Portal Circulation

Figure 28.3 shows that the **hepatic portal** vein receives venous blood from digestive organs and carries it to the liver. This allows the liver to process nutrients absorbed from the digestive tract prior to their input into the general circulation. Blood from the intestines is drained by the **superior mesenteric** and the **inferior mesenteric** veins. The **splenic** vein brings blood from the spleen, pancreas, and part of the stomach to join with the inferior mesenteric. The **gastric** veins serve most of the stomach. Blood from the liver enters the inferior vena cava via the **hepatic vein.**

Assignment

1. Label Figure 28.2.
2. Demonstrate the presence of valves in veins by applying pressure with your fingertip to a prominent vein on the back of your hand near your wrist. Then, continue to apply pressure as you trace the vein toward your knuckles. Note that blood does not flow back to fill the vein until you release the pressure.
3. Examine a prepared slide of artery and vein, x.s. Note the thicker, more muscular wall of the artery.
4. Examine a prepared slide of an atherosclerotic artery, x.s. Locate the fatty plaque that partially plugs the vessel. High levels of blood cholesterol and genetic tendencies lead to the development of atherosclerosis. Make drawings of normal and atherosclerotic arteries, and a normal vein on a separate sheet of paper.
5. Complete sections A and B on the laboratory report.

The Lymphatic System

The lymphatic system collects **interstitial** or **extracellular fluid** from body tissues, returns it to the blood, and cleanses it enroute.

Interstitial fluid enters **lymph capillaries,** tiny blind-ending vessels that are located in the body tissues. After entering a lymph capillary, the fluid is called **lymph.** The lymph capillaries combine to form larger **lymphatic vessels** as shown in Figure 28.4. Lymphatics, like veins, have numerous valves to prevent the backflow of lymph, which is propelled by contractions of skeletal muscles that depress the walls of lymphatic vessels.

Where the lymphatics enter the trunk from the extremities, lymph passes through one or more **lymph nodes.** Lymph nodes are scattered through other parts of the body but are clustered in the following areas: **abdominal, axillary, inguinal, pelvic,** and **cervical.** They vary in size from 1–10 mm across. As it passes through the lymph nodes, lymph is cleansed of bacteria and cellular debris by the phagocytic action of macrophages. Lymph nodes contain massive numbers of lymphocytes concentrated in **germinal centers,** where they replicate. Lymphocytes play a key role in immunity.

Figure 28.4 The lymphatic system.

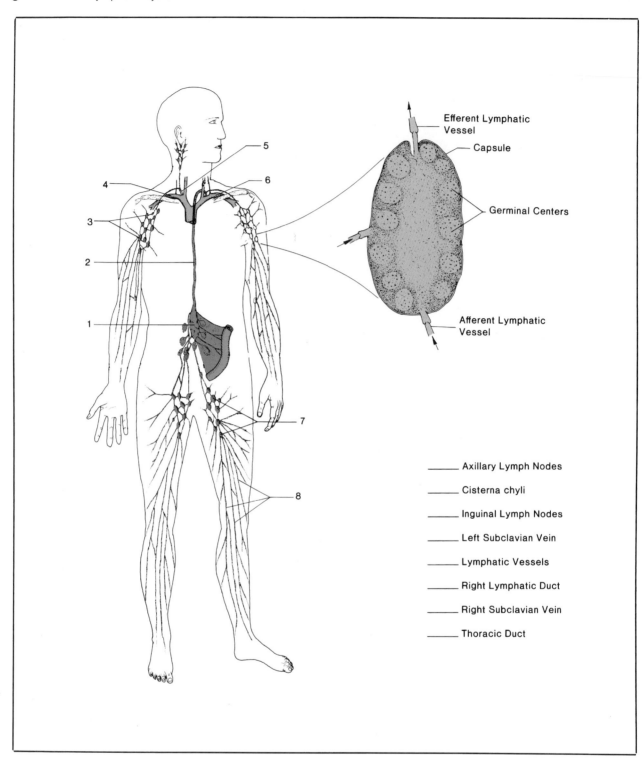

Efferent Lymphatic
Vessel

Capsule

Germinal Centers

Afferent Lymphatic
Vessel

_____ Axillary Lymph Nodes

_____ Cisterna chyli

_____ Inguinal Lymph Nodes

_____ Left Subclavian Vein

_____ Lymphatic Vessels

_____ Right Lymphatic Duct

_____ Right Subclavian Vein

_____ Thoracic Duct

Figure 28.5 Fetal circulation.

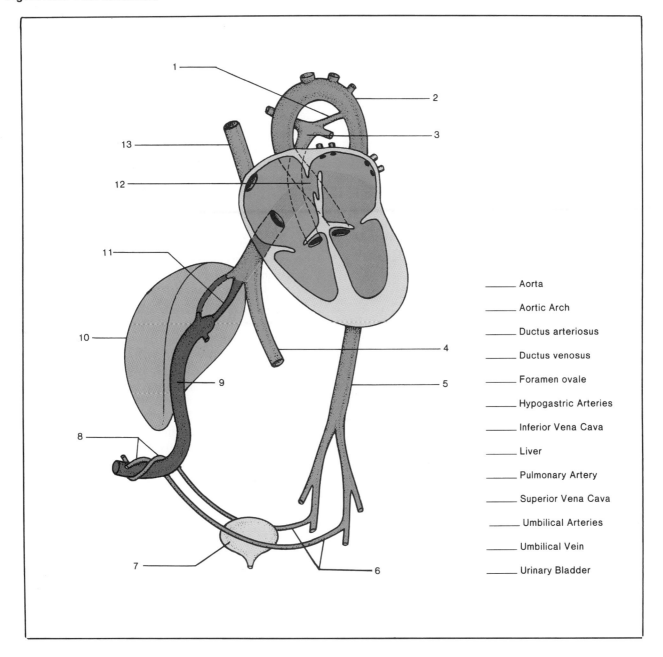

Aorta

Aortic Arch

Ductus arteriosus

Ductus venosus

Foramen ovale

Hypogastric Arteries

Inferior Vena Cava

Liver

Pulmonary Artery

Superior Vena Cava

Umbilical Arteries

Umbilical Vein

Urinary Bladder

The lymphatics from the right arm, right side of the thorax, and right side of the neck join to form the **right lymphatic duct,** which drains into the right subclavian vein.

Lymphatics from the legs, abdomen, left side of the thorax, left arm, and left side of the neck empty into the **thoracic duct.** At its inferior origin is a sac-like enlargement, the **cisterna chyli,** that receives fat-rich lymph from the intestines. The thoracic duct empties into the left subclavian vein.

Cancer cells usually spread by metastasis via the lymphatic vessels. Secondary cancer sites occur in lymph nodes or wherever the cancer cells lodge. A knowledge of the direction of lymph flow in vessels and nodes is important in the diagnosis and prognosis of the spread of cancer.

Assignment

Label Figure 28.4 and add arrows to show the direction of lymph flow.

Fetal Circulation

Circulation during fetal development is necessarily different from that of an adult. During fetal development, the placenta is the site of the exchange of nutrients and oxygen from maternal blood to fetal blood and, simultaneously, of the transfer of wastes and carbon dioxide from fetal blood to maternal blood. Also, the lungs are nonfunctional and receive a reduced supply of blood.

Fetal circulation through the heart is shown in Figure 28.5. Vessels colored purple are carrying a mixture of oxygenated and deoxygenated blood. Locate the red-colored **umbilical vein** that brings oxygenated blood from the placenta to the fetus. Two **umbilical arteries** carrying blood to the placenta are wrapped around the umbilical vein in the umbilical cord. They are extensions of the **hypogastric arteries** (internal iliacs) that pass along each side of the urinary bladder (label 7). Near the liver, the umbilical vein constricts to form the narrow **ductus venosus,** which leads to the inferior vena cava where oxygenated and deoxygenated bloods mix prior to entering the right atrium.

In the fetal heart, there is an opening, the **foramen ovale,** that allows much of the blood to flow from the right atrium into the left atrium and on into the left ventricle. During contraction, blood is pumped from the heart through both the pulmonary artery and the aorta.

Since fetal lungs are nonfunctional, they receive less blood than adult lungs. Much of the blood in the pulmonary artery passes into the aorta via the short **ductus arteriosus** and increases the volume of blood available to the body.

When the fetus is separated from the placenta after birth, circulatory changes occur. The constriction and closure of the ductus venosus helps to decrease the loss of blood via the umbilical vein. It subsequently becomes the **ligamentum venosum** of the liver. The umbilical vein becomes the **round ligament** extending from the umbilicus to the liver. The distal portions of the hypogastric arteries become fibrous cords, and the proximal portions remain functional and supply the urinary bladder.

As soon as the infant begins to breathe, more blood is channeled to the lungs. The decreased use of the ductus arteriosus leads to its closure and conversion into the **ligamentum arteriosum.** The resulting increase in blood volume and pressure in the left atrium closes the flap on the foramen ovale. Subsequently, connective tissue seals this opening to permanently separate the atria.

Failure of the foramen ovale or ductus arteriosus to close after birth results in a circulatory defect that allows continued mixing of oxygenated and deoxygenated bloods. Such defects may deprive body tissues of adequate oxygen, seriously impairing the newborn. These defects are usually repaired surgically.

Assignment

1. Label Figure 28.5 and add arrows to show the direction of blood flow.
2. Complete the laboratory report.

Cardiovascular Phenomena

Objectives

After completion of this exercise, you should be able to:

1. Identify the heart sounds by stethoscope auscultation.
2. Determine the blood pressure of a subject.
3. Describe the control of capillary circulation.
4. Define all terms in bold print.

Materials

Stethoscope
Alcohol pads
Sphygmomanometer
Grass frog
Frog board and wrapping cloth
Dissecting pins
Rubber bands
Epinephrine (1:1,000) in dropping bottles
Histamine (1:10,000) in dropping bottles

In this exercise, you will study heart sounds, blood pressure, and the control of capillary circulation.

Heart Sounds

There are three heart sounds that result from contraction and valvular movements. The **first sound** (lub) results from ventricular contraction and the simultaneous closure of the atrioventricular valves. The **second sound** (dub) results from the simultaneous closure of the aortic and pulmonary semilunar valves. The **third sound** seems to be caused by the vibration of the ventricular walls and the atrioventricular valve cusps during systole. It is difficult to detect unless the subject is lying down.

Abnormal heart sounds are called **murmurs** and usually result from a damaged or defective atrioventricular valve that allows blood to leak back into an atrium. Common causes include damage to the mitral valve by rheumatic fever and **mitral valve prolapse,** an inherited disorder caused by elongation of chordae tendineae, which allows a valve cusp to be pushed too far into the atrium. Many murmurs have no clinical significance.

Listening to heart sounds is called **auscultation.** For best results, you must use the **auscultatory areas** shown in Figure 29.1. These areas do not coincide with the anatomical locations of the valves, because heart sounds are projected to different spots on the rib cage.

Your instructor will demonstrate the heart sounds using an audio monitor. If a sensitive unit is used, such as the Grass AM7, it will not be necessary to bare the subject's chest. After you can distinguish the heart sounds, work in pairs to auscultate each other's heart with a stethoscope.

1. Clean the ear pieces of the stethoscope with the alcohol pad. Fit them into your ears, directing them inward and upward.
2. With the subject in a sitting position, try to maximize the first sound by placing the stethoscope on the tricuspid and mitral auscultatory areas.
3. Now, move the stethoscope to the aortic and pulmonary areas to maximize the second sound.
4. Try to detect a **splitting of the second sound** while the subject is inhaling. This phenomenon is normal and occurs because during inspiration more venous

Figure 29.1 Auscultatory areas.

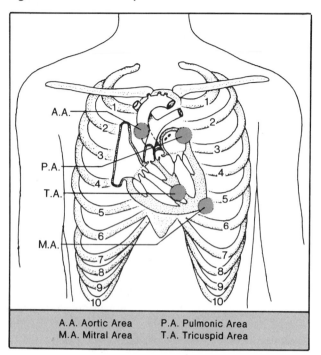

A.A.
P.A.
T.A.
M.A.

1
2
3
4
5
6
7
8
9
10

1
2
3
4
5
6
7
8
9
10

A.A. Aortic Area   P.A. Pulmonic Area
M.A. Mitral Area   T.A. Tricuspid Area

blood is forced into the right side of the heart and causes a delayed closure of the pulmonary semilunar valve. Use hand signals, not verbalization, to inform the subject when to inhale.

5. Now, have the subject recline in a supine position and repeat steps 2 to 4. Try to detect the **third sound** by placing the stethoscope over the apex of the heart.

6. If time permits, compare the sounds before and after exercise.

7. Complete Sections A and B on the laboratory report.

Blood Pressure Determination

The contractions of the ventricles of the heart pump blood into the pulmonary artery and aorta in spurts about 70 times a minute. This causes hydrostatic pressure in the arteries to fluctuate during the heart cycle. Blood pressure is greatest during **systole,** the contraction phase, and least during **diastole,** the relaxation phase.

There are several ways to measure blood pressure, but we will use a standard **sphygmomanometer** and a **stethoscope.** In this method, a stethoscope is used to listen for the **Korotkoff sounds** that are produced by blood rushing through a partially occluded brachial artery.

Figures 29.2 and 29.3 illustrate how to apply the cuff of the sphygmomanometer to the upper arm. When the cuff is properly positioned, the diaphragm of the stethoscope is placed over the brachial artery and the cuff is inflated to shut off the flow of blood there. Then, the pressure within the cuff is slowly decreased by releasing air through a valve on the bulb while listening for a Korotkoff sound.

When the first sound is heard, the pressure on the gauge is noted. This reading is the **systolic pressure.** As air continues to be released, the gauge is watched until the Korotkoff sounds cease. The pressure at this point is the **diastolic pressure.**

Normal systolic pressure in adults is 120 mm Hg; normal diastolic pressure is 80 mm Hg. Such blood pressures are usually designated as 120/80. Systolic and diastolic pressures may vary by ± 10 mm Hg and still be considered normal.

Your instructor will demonstrate the procedure and Korotkoff sounds using a microphone and audio monitor. After you understand the procedure, work with a partner to determine each other's blood pressure. Record your results on the laboratory report.

Figure 29.2 Sphygmomanometer cuff is wrapped around the upper arm, keeping lower margin of cuff above line through the epicondyles of the humerus.

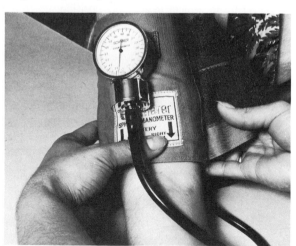

Figure 29.3 The inflated cuff shuts off the flow of blood in the brachial artery. Korotkoff's sounds are listened for as pressure is gradually lowered.

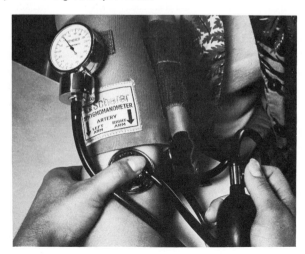

1. Have the subject's arm resting on the table with the palm upward. Wrap the cuff of the spygmomanometer around the upper arm. The method of attachment will vary with the type of instrument. Your instructor will indicate how to secure it.

2. Close the air valve on the neck of the rubber bulb *but not so tight that you can't open it*. Attach the pressure gauge on the cuff so that you can easily read it.

3. Pump air into the cuff by squeezing the bulb. Stop when the pressure gauge reads 180 mm Hg. *Do not leave the cuff inflated for more than one minute.*

4. Slip the diaphragm of the stethoscope under the cuff over the brachial artery—about midway between the epicondyles of the humerus. See Figure 29.3.

5. Loosen the valve control screw *slightly* to slowly release air from the cuff while listening for the Korotkoff sounds. When the first sound is heard, read the pressure on the gauge. This is the systolic pressure.

6. Continue listening to the Korotkoff sounds as air continues to escape from the cuff. At the point where the sounds cease, read the pressure on the gauge. This is the diastolic pressure.

7. Repeat several times until you can get consistent results. Wait several minutes between repetitions.

8. Measure the blood pressure after the subject has done some exercise.

Peripheral Circulation

Arteries and veins are united by an intricate network of capillaries. **Capillaries** are short, thin-walled vessels composed of endothelial cells. They are so numerous that all body cells are within two or three cells of a capillary. The combined diameter of capillaries causes blood to slow down as it passes through them. This slowing allows time for the exchange of materials between the capillary blood and body cells. Interstitial fluid is an intermediate in the exchange.

capillary blood ⇆ interstitial fluid ⇆ body cells

There is insufficient blood to fill all capillaries of the body simultaneously, so many capillaries are intermittently filled and deprived in order to meet the needs of all tissues. Blood flow into a capillary is controlled by a **precapillary sphincter muscle** located at the junction of an arteriole and a capillary. See Figure 29.4.

Figure 29.4 Capillary circulation.

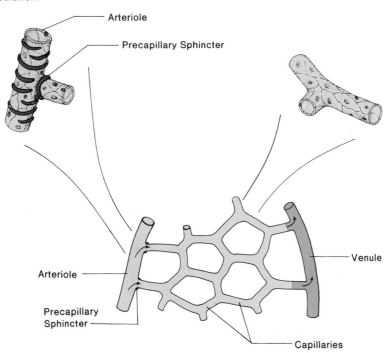

Arteriole

Precapillary Sphincter

Venule

Arteriole

Precapillary Sphincter

Capillaries

Figure 29.5 Blood flow setup.

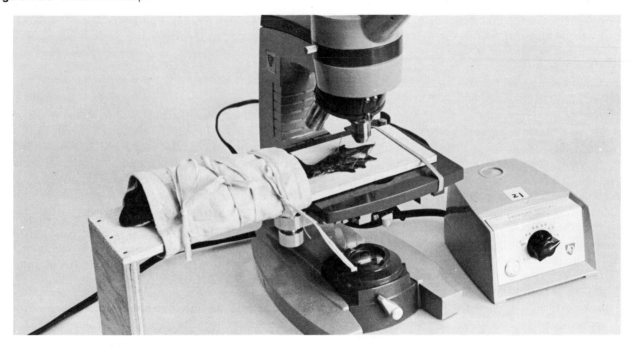

The diameter of arterioles and precapillary sphincters is under the control of the **vasomotor center** in the medulla oblongata. Many factors affect this center, but carbon dioxide concentration of the blood is of major importance. At the localized tissue level, carbon dioxide, epinephrine, and histamine cause a direct influence on capillary blood flow. A high level of carbon dioxide dilates arterioles and precapillary sphincters (vasodilation), increasing capillary blood flow to the affected tissues. Increased levels of epinephrine cause vasodilation in some tissues (heart, lungs and muscles) and vasoconstriction in others. Histamine, a product of allergic reactions, causes vasodilation.

In this section you will study the effect of histamine and epinephrine on capillary blood flow.

1. Obtain and wrap a small frog onto a frog board as shown in Figure 29.5. The cloth should be moist before wrapping.
2. Spread the webbing of a hind foot over the hole in the board and use pins to secure the foot in that position. Attach the board to a microscope stage with rubber bands as shown.

3. Use a 10X objective to observe circulation in the webbing of the foot. Locate an arteriole, with its pulsating, rapid flow, and a venule, with its slower, steady flow.
4. Locate a capillary. Note its small diameter and the blood cells moving slowly in a single file through it. Try to find a site of a precapillary sphincter by observing the irregular flow of blood cells through a capillary. You will not be able to actually see the sphincter.
5. Blot the water from the webbing and add 2 drops of histamine (1:1,000). Observe the effect on capillary flow.
6. Rinse the foot with water and add 2 drops of epinephrine (1:10,000). Observe the effect on capillary flow.
7. Wash the foot and return the frog to the stock table. Clean the microscope.
8. Complete the laboratory report.

The Respiratory Organs

Objectives

After completing this exercise, you should be able to:

1. Identify the components of the respiratory system on charts, models, and a sheep pluck and describe their functions.
2. Trace the flow of air into and out of the lungs.
3. Identify lung and tracheal tissues when viewed microscopically.
4. Define all terms in bold print.

Materials

Model of a midsagittal section of the head
Torso model with removable organs
Corrosion preparation of the bronchial tree
Sheep plucks without livers
Dissecting instruments and trays
Prepared slides of human trachea, normal lung tissue, and emphysematous lung tissue

Body cells must be continuously supplied with oxygen in order to carry out their metabolic activities. The metabolic activities of cells, in turn, produce carbon dioxide that must be removed from the body. The primary function of the respiratory system is to carry out the exchange of these respiratory gases between the body and the atmosphere. Secondary functions include warming and filtering the air that is inhaled.

The exchange of respiratory gases involves three related processes: (1) breathing air into and out of the lungs; (2) the exchange of oxygen and carbon dioxide between the air in the lungs and the blood; and (3) the transport of oxygen and carbon dioxide by the blood.

The respiratory organs include not only the lungs, the primary organs of gas exchange, but also the series of tubes and passageways that enable air to enter the lungs during inhalation and exit the lungs during exhalation. The respiratory organs are often subdivided into upper and lower respiratory tracts. Those respiratory organs located outside the thorax constitute the upper respiratory tract. Respiratory organs within the thorax form the lower respiratory tract

In this exercise, you will study both macroscopic and microscopic aspects of the respiratory organs as well as the basic functions of each organ. The respiratory organs include the nose, pharynx, larynx, trachea, bronchi, and lungs.

The Upper Respiratory Tract

The midsagittal section of the head in Figure 30.1 shows the parts of the upper respiratory system. The **nasal cavity** is divided into left and right chambers by the **nasal septum,** (not shown) which is composed of cartilage and bone. The surface area of the nasal cavity is increased by the superior, middle, and inferior **nasal conchae** that project from the lateral walls. Air enters the nasal cavity via the **nares** or nostrils.

The **palate** separates the nasal cavity from the **oral cavity.** The palate consists of an anterior portion supported by bone, the **hard palate,** and a posterior **soft palate** that terminates in a median finger-like projection, the **uvula.** The small space between the teeth and lips is the **oral vestibule.**

The **pharynx** consists of two parts. The **oropharynx** lies posterior to the oral cavity, and the **nasopharynx** is just above it and posterior to the nasal cavity. The openings of the **Eustachian tubes** (label 14) are visible on the lateral walls of the nasopharynx.

Figure 30.1 Upper respiratory passages.

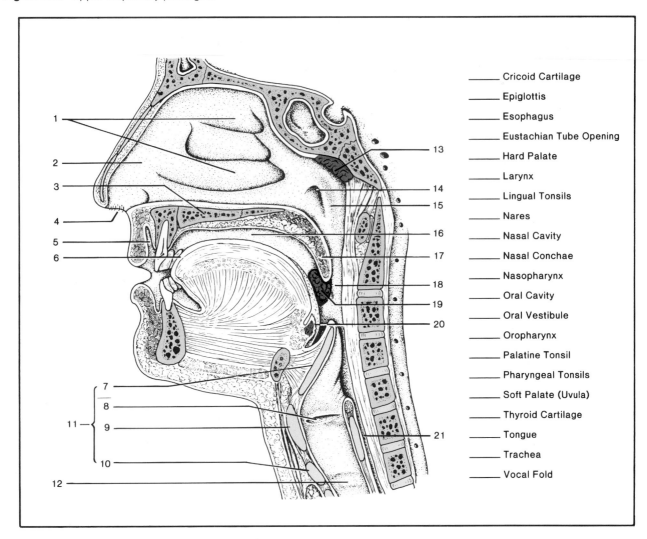

_____ Cricoid Cartilage
_____ Epiglottis
_____ Esophagus
_____ Eustachian Tube Opening
_____ Hard Palate
_____ Larynx
_____ Lingual Tonsils
_____ Nares
_____ Nasal Cavity
_____ Nasal Conchae
_____ Nasopharynx
_____ Oral Cavity
_____ Oral Vestibule
_____ Oropharynx
_____ Palatine Tonsil
_____ Pharyngeal Tonsils
_____ Soft Palate (Uvula)
_____ Thyroid Cartilage
_____ Tongue
_____ Trachea
_____ Vocal Fold

As air passes through the nasal cavity, it is warmed, filtered, and humidified by the mucous membrane lining. Airborne particles are trapped in mucus secreted by goblet cells. The mucus and entrapped particles are moved by beating cilia to the pharynx and swallowed.

From the nasal cavity, air passes through the pharynx, larynx, and trachea on its way to the lungs. The **larynx,** or voice box, has walls formed of an upper **thyroid cartilage** (label 9) and a lower **cricoid cartilage.** The **epiglottis** is a flap-like structure diagonally situated over the **glottis,** the upper opening to the larynx. It prevents food from entering the larynx when swallowing. The opening into the **esophagus,** which carries food to the stomach, lies just posterior to the larynx. Within the larynx are the **vocal folds** (label 8), which vibrate to produce sounds when activated by passing air.

Clumps of lymphoid tissue compose the tonsils. The **pharyngeal tonsils** or **adenoids** (label 13) are located in the roof of the nasopharynx, the **palatine tonsils** occur on each side of the oropharynx, and the **lingual tonsils** are on the posterior inferior portion of the tongue.

The Lower Respiratory Passages

Figure 30.2 illustrates the lower respiratory structures. The **trachea,** or windpipe, extends below the larynx to the center of the thorax where it branches into the **primary bronchi,** which enter the lungs. Within the lungs, the primary bronchi divide to form the **secondary bronchi,** one for each lobe. Continued branching forms the **tertiary bronchi** and **bronchioles,** which end in a cluster of tiny sacs called **alveoli.** The exchange of gases occurs between the

Figure 30.2 Respiratory tract.

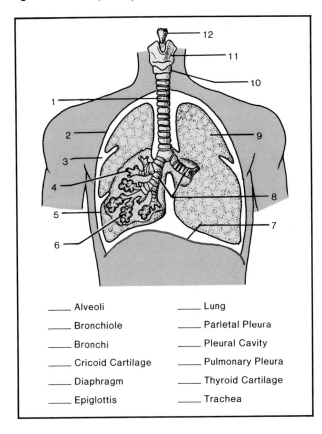

_____ Alveoli	_____ Lung
_____ Bronchiole	_____ Parietal Pleura
_____ Bronchi	_____ Pleural Cavity
_____ Cricoid Cartilage	_____ Pulmonary Pleura
_____ Diaphragm	_____ Thyroid Cartilage
_____ Epiglottis	_____ Trachea

air in the alveoli and blood in capillaries surrounding the alveoli. Cartilaginous "rings" support the walls of both the trachea and bronchi.

Bronchioles lack cartilaginous rings but contain smooth muscle in their walls. In an **asthma** attack, contraction of the muscles and an excessive production of mucus in the air passages restrict the air flow.

Each lung is covered by a tightly adhering membrane, the **pulmonary pleura,** while the **parietal pleura** lines the inner wall of the thoracic cavity. Fluid secreted into the **pleural cavity,** the potential space between these membranes, reduces friction during lung movements. The bottom of the lungs rests against the **diaphragm,** the primary muscle of respiration that separates the thoracic and abdominal cavities.

Assignment

1. Label Figures 30.1 and 30.2.
2. Complete Sections A, B, and C on the laboratory report.
3. Locate the parts of the respiratory system on models available in the laboratory.
4. Examine a corrosion preparation of the bronchial tree and identify the air passages composing it.

Sheep Pluck Dissection

A sheep "pluck" consists of the larynx, trachea, lungs, and heart removed during routine slaughter. The study of a fresh sheep pluck is valuable since its components are similar to those in humans. Although it is as safe to handle as any material from a meat market, use sanitary precautions. Record your observations on the laboratory report.

1. Lay out the pluck on the dissecting tray. Locate the heart, lungs, trachea, larynx, and diaphragm.
2. Identify the epiglottis, and thyroid and cricoid cartilages of the larynx. Try to locate the vocal folds.
3. Use scissors to cut through the trachea below the larynx and examine the cut end. Note the shape of the cartilaginous rings and the smooth, slimy interior of the trachea.
4. Note that the lungs are divided into lobes. Are the same number of lobes found in both sheep and human lungs? Note the smooth surface of the pulmonary pleura covering the lungs.
5. Locate the pulmonary arteries and veins.
6. Trace the trachea down to where it branches into the primary bronchi, removing surrounding tissue as necessary. Note that a bronchus leading to the upper right lobe branches off some distance above the primary bronchi.
7. Use scissors to cut through the trachea at the level of the top of the heart. Then, cut down the dorsal side (opposite the heart) of the trachea to the primary bronchi. Continue the cut along one bronchus into the center of the lung and note the extensive branching of the air passages.
8. Insert a plastic straw into a bronchiole and blow into the lung. Note the expansibility and elasticity of lung tissue.
9. Cut off a piece of a lung and examine the cut surface. Note the sponginess of lung tissue.
10. Dispose of the sheep pluck as directed by your instructor. Wash the tray, instruments, and your hands with soap and water.

Microscopic Study

Compare your observations with Figure HA-16 and make drawings of your observations on the laboratory report.

1. Examine the photomicrographs of the nasal cavity in Figure HA-16. Note the nasal septum, concha, and pseudostratified ciliated epithelium.
2. Examine a prepared slide of trachea. Locate the hyaline cartilage composing cartilaginous rings, and goblet cells and cilia of the epithelial lining.
3. Examine a prepared slide of normal lung tissue. Note the porous nature of the tissue and the thin walls of the alveoli.
4. Examine a prepared slide of emphysematous lung tissue. Note that many walls of the alveoli have been destroyed, which decreases the respiratory surface of the lung. Emphysema also reduces the elasticity of lung tissue so that it is difficult to force air out of the lungs.
5. Complete the laboratory report.

Respiratory Physiology

Objectives

After completing this exercise, you should be able to:

1. Describe the factors that stimulate breathing.
2. Describe the mechanics of breathing.
3. Determine the various lung volumes using a spirometer.
4. Define all terms in bold print.

Materials

Paper bag
Paper cup
Drinking straw
Propper spirometer and disposable, sterile
    mouthpieces
Alcohol pads
Biohazard bag

The Control of Breathing

The rhythmic cycle of respiration is controlled by the **respiratory control center** in the medulla oblongata. The primary stimulus on the respiratory control center for inspiration is an increase in the hydrogen ion concentration of the blood and cerebrospinal fluid. Recall that an increase in carbon dioxide concentration increases the hydrogen ion concentration.

carbon dioxide + water $\leftrightarrows$ carbonic acid

carbonic acid $\leftrightarrows$ hydrogen ion + bicarbonate ion

Both carbon dioxide and hydrogen ions act directly on the respiratory control center. A decreased level of oxygen acts indirectly and is detected by chemoreceptors in the carotid arteries and aortic arch. Impulses from these receptors are then transmitted to the respiratory control center. Usually, oxygen concentration is of little significance in stimulating breathing.

Hyperventilation

Hyperventilation reduces the concentrations of carbon dioxide and hydrogen ions in the blood so that the respiratory center is depressed. This reduces the desire to breathe. Demonstrate this effect as follows. Record your results on the laboratory report.

1. Breathe deeply at the rate of 15 cycles per minute for 1–2 minutes. It will be increasingly difficult to breathe. Stop when you *start* to get dizzy.
2. Place a paper bag over your nose and mouth, and breathe into it for about 3 minutes. Note how much easier it is to breathe into the bag.
3. After breathing normally for 5 minutes, determine how long you can hold your breath.
4. Repeat step one and determine how long you can hold your breath.

Deglutition Apnea

Obviously, breathing and swallowing food or drink cannot occur simultaneously. A reflex called **deglutition apnea** prevents one from wanting or attempting to breathe while swallowing. Demonstrate this reflex and record your results on the laboratory report.

1. Hold your breath until you have a strong need to breathe.
2. Then, sip water through a straw and note whether the desire to breathe decreases.

Figure 31.1 Spirogram of lung capacities.

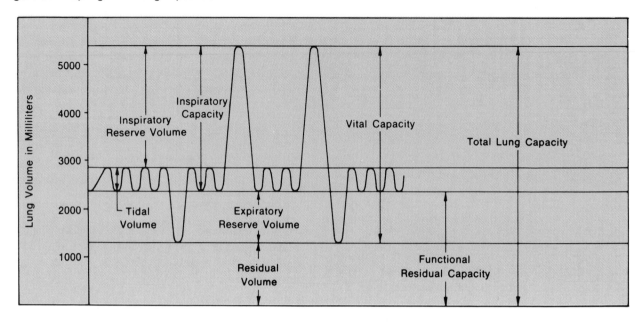

Spirometry and Lung Capacities

A **spirometer** is used to determine the volume of air moving in and out of the lungs during breathing. A recording spirometer may produce a **spirogram** similar to the one shown in Figure 31.1, which indicates the various lung capacities. Note their relationships. Lung capacities vary with sex, age, and weight of the subject and are useful in diagnosing certain pulmonary disorders.

The measurement of lung volumes is essential to determine how well the lungs are functioning and in diagnosing lung disorders. For example, the breakdown of alveoli and loss of lung elasticity in emphysema causes an increase in the residual volume, which decreases the expiratory reserve volume and vital capacity.

In this section, you will use a non-recording spirometer to determine your lung capacities as described below. Record your results on the laboratory report.

Tidal Volume (TV)

The amount of air that moves in and out of the lungs during a normal respiratory cycle is the **tidal volume.** Normal tidal volume averages about 500 ml.

1. Cleanse the stem of a Propper spirometer with an alcohol pad, slip on a sterile mouthpiece, and rotate the dial of the spirometer to zero as shown in Figure 31.2. When using the spirometer always hold the dial upward.
2. Take a normal breath through your nose and exhale through the spirometer. Repeat for a total of three expirations. *Do not inhale or exhale forcibly.*
3. Record the total volume from the spirometer dial. Divide this number by 3 to determine your tidal volume.

Minute Respiratory Volume (MRV)

The amount of tidal air that moves in and out of your lungs in one minute is your **minute respiratory volume.** To determine this, count the number of respirations in one minute and multiply this number by your tidal volume.

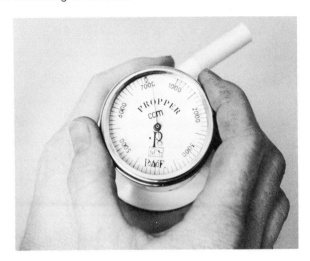

Figure 31.2 Dial face of spirometer is rotated to zero prior to measuring exhalations.

Expiratory Reserve Volume (ERV)

The amount of air that you can expire beyond the tidal volume is your **expiratory reserve volume.** It is usually about 1,100 ml.

1. Set the spirometer dial on 1,000.
2. After 2–3 normal tidal expirations through your nose, exhale forcibly all of the **additional** air that you can through the spirometer.
3. Subtract 1,000 from the dial reading to determine your expiratory reserve volume.

Vital Capacity (VC)

As shown in Figure 31.1, the **vital capacity** of the lungs is the total of the tidal, expiratory reserve, and inspiratory reserve volumes. It averages about 4,500 ml but may vary as much as 20% from established norms and still be considered normal.

1. Set the spirometer dial on zero.
2. After inhaling as deeply as possible and exhaling completely, take a second deep breath and exhale as much air as possible through the spirometer. An even, forced exhalation is best.
3. Repeat three times and record the value for each exhalation. The values should not vary by more than 100 ml. Divide the total by 3 to determine your vital capacity. Compare your vital capacity with the predicted normal values in Appendix C.
4. Place the mouthpiece in the biohazard bag. Wipe the spirometer with an alcohol pad.

Inspiratory Capacity (IC)

The **inspiratory capacity** is the maximum volume of a deep inhalation after expiring the tidal air. It is usually about 3,000 ml. This volume must be calculated since the Propper spirometer cannot measure inspirations. Determine your inspiratory capacity as follows:

$$IC = VC - ERV$$

Inspiratory Reserve Volume (IRV)

This is the volume of air that can be inspired after filling the lungs with tidal air. Calculate your **inspiratory reserve volume** as follows:

$$IRV = IC - TV$$

Residual Volume (RV)

The **residual volume** is air that cannot be expelled from the lungs—about 1,200 ml. It cannot be determined by usual spirometric methods.

Assignment

Complete the laboratory report.

The Digestive Organs

Objectives

Upon completion of this exercise, you should be able to:

1. Identify the parts of the digestive system on charts and models, and describe the function of each.
2. Identify tissues studied microscopically.
3. Define all terms in bold print.

Materials

Torso model with removable organs
Prepared slides of:
    Vallate papillae with taste buds
    Stomach wall, x.s.
    Ileum wall, x.s.
    Jejunum wall, x.s.
    Liver

Refer to Figures 32.1 through 32.5 as you study the descriptions below. The small intestine has been shortened in Figure 32.5 for clarity.

The Oral Cavity

The teeth, tongue, and salivary glands play significant roles in the digestive functions of the oral cavity. When food is taken into the mouth and chewed, it is mixed with saliva secreted by the salivary glands. The food mass is then pushed posteriorly by the tongue into the **oropharynx** where the swallowing reflex is initiated.

The basic structure of the oral cavity is shown in Figure 32.1. Note that the cheeks have been cut and the lips retracted to expose pertinent structures. A mucous membrane, the **mucosa,** lines the lips and cheeks. It forms thickened folds, the **labial frenula,** at the midline on the inner surface of the lips. The **oral vestibule** is the space between the cheeks or lips and the teeth of each jaw.

Figure 32.1 The oral cavity.

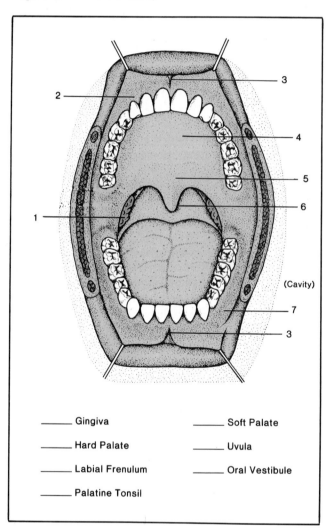

_____ Gingiva   _____ Soft Palate

_____ Hard Palate   _____ Uvula

_____ Labial Frenulum _____ Oral Vestibule

_____ Palatine Tonsil

Figure 32.2 Tooth anatomy.

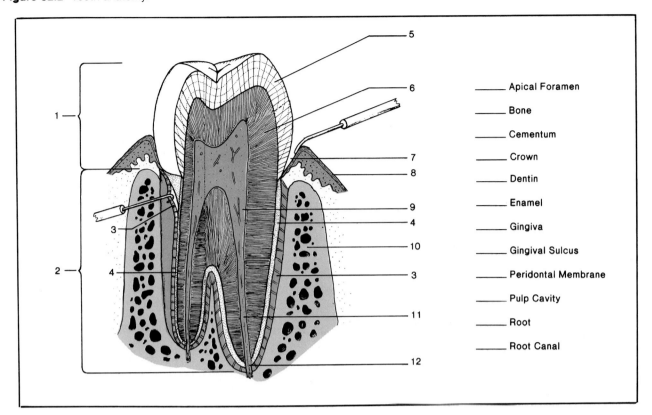

Apical Foramen
Bone
Cementum
Crown
Dentin
Enamel
Gingiva
Gingival Sulcus
Peridontal Membrane
Pulp Cavity
Root
Root Canal

The roof of the mouth is formed by the anterior **hard palate,** which is reinforced by bone, and the posterior **soft palate.** The **uvula** is the finger-like downward projection from the posterior margin of the soft palate. Note that portions of the **palatine tonsils** ae visible on the sides of the oropharynx.

The Teeth

All twenty deciduous teeth usually have erupted by the time a child is two years old. Five teeth are present in each quadrant of the jaws. Starting from the median line, they are: **central incisor, lateral incisor, cuspid, first molar,** and **second molar.**

There ae thirty-two permanent teeth, which begin appearing when a child is about six years of age. The adult dentition is shown in Figure 32.1. Note that there are eight teeth in each quadrant of the jaws. Starting at the median line, they are: **central incisor, lateral incisor, cuspid, first premolar, second premolar, first molar, second molar,** and **third molar.**

Figure 32.2 illustrates the basic anatomy of a tooth. Note the two major subdivisions of a tooth: a **crown,** the portion covered with enamel, and a **root** that is embedded in bone. The **neck** of the tooth is at the junction of the crown and root.

Four substances compose a tooth. **Enamel** is the hardest substance in the body and forms an ideal chewing surface for the crown. **Dentin** forms most of a tooth and lies under the enamel in the crown. It is similar to bone in composition and hardness. **Cementum** (label 4) is modified bone tissue that covers the exterior of the root and attaches the tooth to the **peridontal membrane.** Fibers of the peridontal membrane penetrate into the bone, holding the tooth firmly in place.

The **pulp cavity** occupies the central part of a tooth. It extends down into the roots to form the **root canals.** Pulp consists of loose connective tissues, and contains blood vessels, lymphatic vessels, and nerves that enter via the **apical foramina** of the roots.

Figure 32.3 Tongue anatomy.

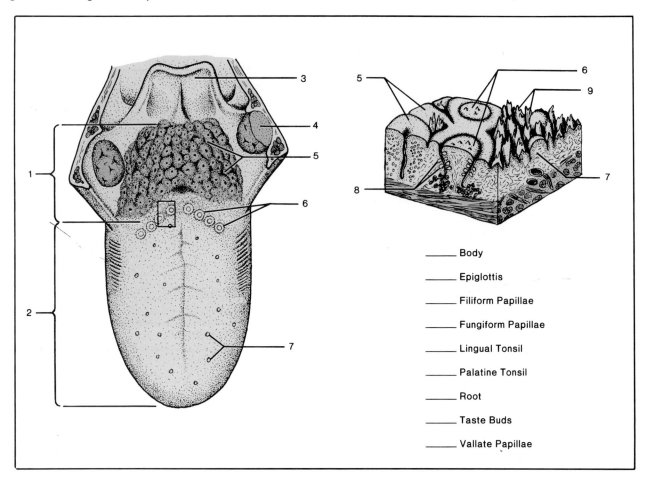

_____ Body

_____ Epiglottis

_____ Filiform Papillae

_____ Fungiform Papillae

_____ Lingual Tonsil

_____ Palatine Tonsil

_____ Root

_____ Taste Buds

_____ Vallate Papillae

The mucosa covering the jaw bones and the neck of each tooth is the **gingiva.** The upper margin of the gingiva is pulled away from the tooth in Figure 32.2 to reveal the **gingival sulcus,** a common site for bacterial putrefaction that may cause peridontal disease.

The Tongue

The tongue consists of skeletal muscle covered by a mucous membrane. The **root** of the tongue is covered by the **lingual tonsils.** Note in Figure 32.3 the **epiglottis** extending upward near the root's posterior margin and the **palatine tonsils** located laterally in the oropharynx. The **body** of the tongue extends from the root to the tip. The **lingual frenulum** (not shown) extends between the lower surface of the tongue and the floor of the oral cavity.

The upper surface of the tongue is covered by tiny projections called **papillae.** The enlarged cut-out in Figure 32.3 shows the different types. **Filiform papillae** have tapered points, are most numerous, and provide a rough texture for handling the food. **Fungiform papillae** (label 7) are low, rounded projections scattered over the surface. Eight to twelve large **vallate papillae** occur in a V-shaped

pattern on the posterior portion of the body. Both fungiform and vallate papillae contain taste buds, which are shown in the circular furrow of a vallate papilla.

Salivary Glands

Saliva is secreted from three pairs of salivary glands as shown in Figure 32.4. The **parotid gland** is located anterior to the ear and exterior to the masseter muscle. Its watery secretion contains salivary amylase, which catalyzes the digestion of starch, and enters the mouth through a duct that opens near the upper second molar.

The **submandibular gland** (submaxillary) lies within the posterior portion of the mandibular arch. Its secretion contains mucin, which aids in holding food together and imparts a slippery nature to the saliva, facilitating swallowing food. The submandibular duct empties into the anterior floor of the mouth.

The **sublingual gland** is located under the tongue in the anterior floor of the oral cavity. Several small ducts carry its secretion into the floor of the mouth under the tongue. Its secretion contains a high level of mucin.

Figure 32.4 The salivary glands.

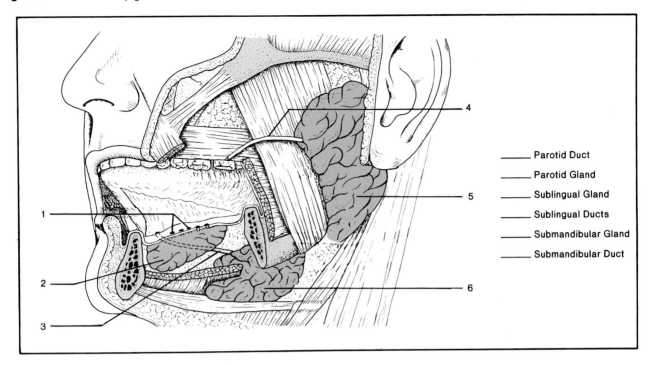

_____ Parotid Duct
_____ Parotid Gland
_____ Sublingual Gland
_____ Sublingual Ducts
_____ Submandibular Gland
_____ Submandibular Duct

Esophagus and Stomach

Refer to Figure 32.5 as you study the discussion of the alimentary canal. When swallowed, food passes into the **esophagus,** a long tube extending to the stomach. Wave-like **peristaltic contractions** begin at the top of the esophagus and carry the food to the **stomach.**

The upper opening of the stomach is guarded by a sphincter muscle, the **cardiac sphincter** or **valve.** It is usually closed but opens to allow the entrance of food. The rounded bulge at the upper end of the stomach is the **fundus** where most of the food to be digested is held. The lower portion, the **pyloric region,** is smaller in diameter and is the primary site of digestion in the stomach. The stomach region between the pylorus and the fundus is called the **body.**

Gastric juice, secreted by the stomach mucosa, is highly acidic and contains enzymes to catalyze the digestion of proteins and milk fats. Food is mixed with gastric juice and is converted into a semiliquid substance called **chyme.** The chyme enters the small intestine through another sphincter, the **pyloric valve** of the stomach.

Sometimes gastric juice is produced excessively or between meals and erodes the mucosa of the stomach or duodenum to produce ulcers. **Gastric ulcers** usually are located in the pyloric region. They are most likely to occur in persons with **gastritis,** an inflammation of the stomach lining. **Duodenal ulcers** often result from chronic stress.

Figure 32.5 The alimentary canal.

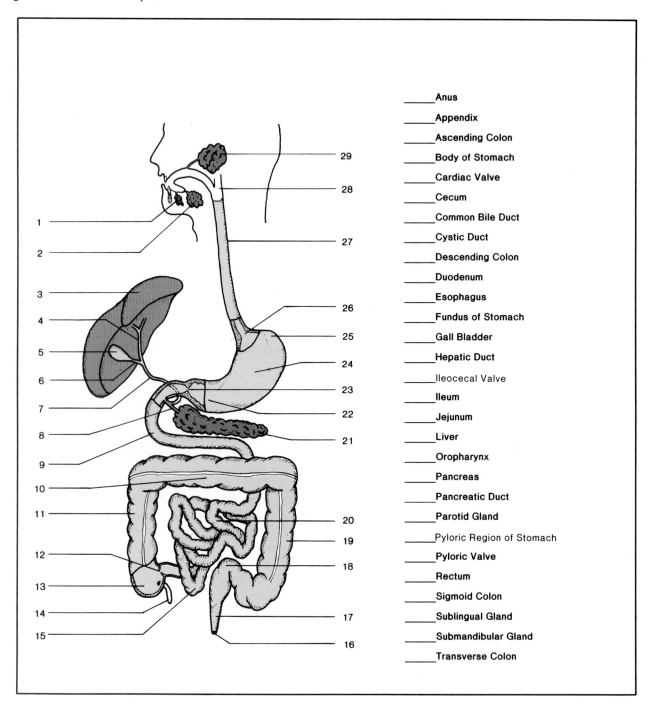

_____Anus	
_____Appendix	
_____Ascending Colon	
_____Body of Stomach	
_____Cardiac Valve	
_____Cecum	
_____Common Bile Duct	
_____Cystic Duct	
_____Descending Colon	
_____Duodenum	
_____Esophagus	
_____Fundus of Stomach	
_____Gall Bladder	
_____Hepatic Duct	
_____Ileocecal Valve	
_____Ileum	
_____Jejunum	
_____Liver	
_____Oropharynx	
_____Pancreas	
_____Pancreatic Duct	
_____Parotid Gland	
_____Pyloric Region of Stomach	
_____Pyloric Valve	
_____Rectum	
_____Sigmoid Colon	
_____Sublingual Gland	
_____Submandibular Gland	
_____Transverse Colon	

The Small Intestine

The small intestine is about 23 ft. long and consists of three parts: the duodenum, jejunum, and ileum. The first 10–12 inches is the **duodenum,** which receives bile from the **liver** and digestive secretions from the **pancreas.**

The **hepatic duct** from the liver joins the **cystic duct** from the gall bladder to form the **common bile duct.** Pancreatic juice is carried from the pancreas by the **pancreatic duct,** which empties into the common bile duct near the duodenum. This union forms a short, dilated duct, the **hepatopancreatic ampulla,** that opens into the duodenum. A sphincter muscle, located at the distal end of the ampulla, controls the passage of bile and pancreatic juice into the duodenum. When no food is in the duodenum, the sphincter constricts so that bile from the liver is forced up the cystic duct to the **gall bladder** for temporary storage. When food is present in the duodenum, the sphincter opens, allowing the passage of bile, and the gall bladder contracts, expelling stored bile into the cystic and common bile ducts.

Gallstones may form in the gall bladder if the bile contains too much cholesterol or is too concentrated. Severe pain and jaundice result if the gallstones block the cystic duct. Often, surgical removal of the gall bladder is necessary.

The middle section of the small intestine is the **jejunum,** which is 7–8 feet in length; the remainder is the **ileum.**

In the small intestine, chyme is mixed with bile and pancreatic juice, which enter via the common bile duct, and with intestinal juice that is secreted by the intestinal mucosa. Digestion of food and absorption of the end products of digestion are completed in the small intestine. Absorption occurs through **villi,** tiny, finger-like projections of the intestinal muscosa. The villi are extremely numerous, and greatly increase the absorptive surface of the small intestine.

The Large Intestine

Indigestible food and water pass from the small intestine into the **large intestine** through the **ileocecal valve.** The large intestine is about 3½–4 feet in length. There are three continuous sections of the large intestine: cecum, colon, and rectum. The ileum opens into the inferior end of the **ascending colon** at its junction with the pouch-like **cecum** from which the narrow **appendix** extends. The ascending colon runs up the right side of the abdominal cavity and is continuous with the **transverse colon** that extends across the abdominal cavity below the diaphragm. The **descending colon** runs downward on the left side of the abdomen and continues as the S-shaped **sigmoid colon.** The final five to six inches is the **rectum,** which terminates with an opening, the **anus.**

The principal functions of the large intestine are the decomposition of undigested food by resident bacteria, absorption of certain vitamins formed by the bacteria, and reabsorption of water. These actions bring about the formation of the feces, which are passed from the body by defecation.

Assignment

1. Label Figures 32.1 through 32.5.
2. Complete the laboratory report.
3. Locate the digestive organs on the human torso model. Note their relative positions.

Microscopic Study

Compare your observations of the following prepared slides with Figures HA-17 and HA-20 of the Histology Atlas. Make drawings of your observations on a separate sheet of paper.

1. Vallate papillae with taste buds. Note the structure of the papillae and the location of the taste buds. Only substances in solution can reach the taste buds.
2. Stomach wall. Locate the columnar epithelium and gastric pits. Try to find chief and parietal cells near the base of a gastric pit. Chief cells produce pepsinogen, a precursor of pepsin, and parietal cells produce hydrochloric acid.
3. Ileum wall. Note how the villi increase the surface area of the mucosa. Try to find the **lacteal,** a small lymph capillary, in a villus. Colloidal fats are absorbed into lacteals.
4. Jejunum wall. Locate the four layers of the intestinal wall: serosa (visceral peritoneum), muscularis, submucosa, and mucosa. Observe the columnar epithelium with goblet cells on the surface of the villi. Nutrients are absorbed in the villi.
5. Liver. Locate a lobule with its **central vein** and **sinusoids** radiating from it between rows of liver cells. Find branches of the **portal vein, hepatic artery,** and **bile duct** at a "corner" of the lobule. On its way to the central vein, blood from the portal vein and hepatic artery flows through the sinusoids, bathing the liver cells. All central veins merge to form the hepatic vein. Bile flows from the liver cells through small bile channels into the bile duct.

Digestion

Objectives

After completion of this exercise, you should be able to:

1. Describe the role of digestive enzymes in the process of digestion.
2. Describe the effect of temperature and pH on the action of salivary amylase.
3. Define all terms in bold print.

Materials

Beaker, 250 ml
Dropping bottles of:
  Iodine (IKI) solution
  0.1% maltose solution
  10% saliva solution
  pH 5 buffer solution
  pH 7 buffer solution
  pH 9 buffer solution
0.1% soluble starch solution in beaker
Medicine droppers
Test tubes, 13 mm diam. $\times$ 100 mm long
Test tube rack, Wasserman type
Electric hot plate
Water baths at 20° C and 37° C
Thermometers, Celsius
Glass marking pen

Organic nutrients are required for the normal metabolic activities of body cells. The processes involved in providing organic nutrients to body cells are: (1) ingestion of food, (2) digestion of food, (3) absorption of nutrients into the blood, and (4) distribution of nutrients to body cells. In this exercise you will investigate the process of digestion using starch as a food and amylase as the enzyme that catalyzes starch digestion.

Digestion and Digestive Enzymes

A normal diet contains carbohydrates, lipids, and proteins plus small quantities of vitamins and minerals. Although vitamins and minerals play a vital role in metabolic processes, it is the carbohydrates, lipids, and proteins that provide the necessary raw materials for tissue repair and growth. In addition, they are the sources of energy required to power body functions.

However, the large molecules of carbohydrates, lipids, and proteins cannot be absorbed by the mucosa of the digestive tract and enter the blood. They must be reduced to absorbable **nutrients** by digestion.

Digestion is a complex process that involves both mechanical and chemical components. Food is mechanically broken down into smaller pieces by mastication (chewing), which increases the surface area of the food. The conversion of food into absorbable nutrients occurs by a chemical process called **enzymatic hydrolysis.** This process may be expressed as:

$$\begin{array}{ccc} \text{large} & & \text{small} \\ \text{nonabsorbable} & \xrightarrow[\text{(+ H}_2\text{O)}]{\text{(digestive enzymes)}} & \text{absorbable} \\ \text{molecules} & & \text{molecules} \end{array}$$

In hydrolysis, water combines with large molecules and splits them into smaller molecules. Hydrolysis without the catalytic action of enzymes is too slow to sustain life. It is the **digestive enzymes** found in saliva and gastric, pancreatic, and intestinal juices that greatly increase the rate of the hydrolysis reactions and enable us to derive nutrients from the food we eat. Relatively few enzymes can catalyze a great number of reactions since the enzymes are not altered in the reaction and are simply recycled. A summary of the major digestive enzymes, their sources, and their actions are shown in Table 33.1.

Table 33.1 A summary of the major digestive enzymes

Enzyme	Digestive Action
Saliva	
amylase	starch to maltose
Gastric Juice	
lipase	milk fat to fatty acids and monoglycerides
pepsin	proteins to peptides
Pancreatic Juice	
amylase	starch to maltose
trypsin	proteins to peptides
chymotrypsin	proteins to peptides
carboxypeptidase	peptides to peptides and amino acids
lipase	lipids to fatty acids and monoglycerides
Intestinal Juice	
maltase	maltose to glucose*
sucrase	sucrose to glucose* and fructose*
lactase	lactose to glucose* and galactose*
lipase	lipids to fatty acids** and monoglycerides**
peptidase	peptides to amino acids***

*End products of carbohydrate digestion.
**End products of lipid digestion.
***End products of protein digestion.

Enzymes are proteins. Each type of enzyme interacts with a specific type of substrate molecule. This specificity results from the three-dimensional shape of the enzyme, which allows it to fit onto a particular type of substrate molecule. The shape of an enzyme is largely determined by weak hydrogen bonds that form between certain amino acids in the molecule. Temperature and pH changes may alter the hydrogen bonding and change the shape of the enzyme so that it is no longer able to fit onto the substrate molecule. When this happens, the enzyme is inactivated. Most enzymes function optimally within rather narrow temperature and pH ranges.

Assignment

Complete Sections A, B, and C on the laboratory report.

Starch Digestion Experiments

The enzyme **amylase** catalyzes the hydrolysis of starch, a polysaccharide, into maltose, a disaccharide. Starch digestion begins in the mouth through the action of salivary amylase and is completed in the small intestine by the action of pancreatic amylase.

In the experiments that follow, you will investigate the effect of temperature and pH on the digestive action of amylase on starch dissolved in an aqueous solution.

Table 33.2 Division of labor

Group	Assignment
A (⅓ of Class)	Controls, 20° C, Boiling
B (⅓ of Class)	Controls, 37° C, Boiling
C (⅓ of Class)	Controls, pH Effects

The experiments are best performed by students working in pairs. If time is limited, it may be necessary to divide the class into three groups with each performing part of the experiments as shown in Table 33.2. Results are to be shared with the entire class.

Solutions will be provided in dropping bottles. You will dispense the solutions with medicine droppers in quantities of "drops" or "droppers." A dropper means *one dropper full of solution,* approximately 1 ml. You must maintain cleanliness and use good laboratory procedures in dispensing the solutions in order to achieve good results. Contamination must be avoided. Do not allow a dropper to touch any solution except the one it is used to dispense. You can avoid problems in performing the experiments if you understand the procedures thoroughly and establish a division of labor among your partners before you start.

Amylase Preparation

Prior to the laboratory session, your instructor prepared either a 10% saliva solution or a 1% commercial amylase solution. The directions that follow assume the use of a 10% saliva solution. Glassware in contact with saliva is to be placed in a 0.25% Amphyl solution after use.

Controls

In the experiments, the activity of the amylase will be determined by using a color test: the iodine (IKI) test for starch. Prepare a set of control tubes as follows to compare with your experimental results:

1. Label two clean test tubes 1 and 2, and add to them the ingredients as follows:
   Tube 1: 1 dropper starch, 2 drops iodine solution
   Tube 2: 1 dropper distilled water, 2 drops iodine solution
2. Your results are interpreted as follows:
   Tube 1: A blue-black coloration is positive for starch.
   Tube 2: An amber coloration is negative for starch.

Figure 33.1 Amylase activity at 20° C and 37° C.

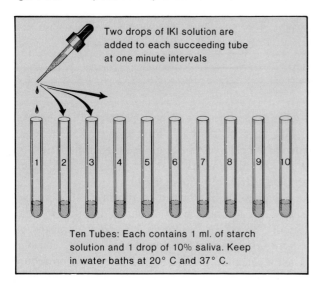

Two drops of IKI solution are added to each succeeding tube at one minute intervals

Ten Tubes: Each contains 1 ml. of starch solution and 1 drop of 10% saliva. Keep in water baths at 20° C and 37° C.

Figure 33.2 Effect of boiling on amylase activity.

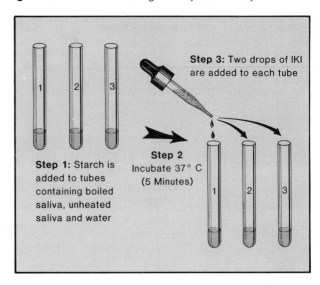

Step 3: Two drops of IKI are added to each tube

Step 1: Starch is added to tubes containing boiled saliva, unheated saliva and water

Step 2 Incubate 37° C (5 Minutes)

Effect of Temperature

In this section, you will compare the activity of amylase at room temperature, 20° C, and 37° C (body temperature). A water bath is provided at each temperature. The general protocol for the experiment is shown in Figure 33.1. Proceed as follows for each temperature to be tested:

1. Label 10 test tubes 1 through 10 and arrange them sequentially in the test tube rack.
2. Add 1 dropper of starch solution to each of the ten test tubes.
3. Place the rack of test tubes and the dropping bottle of saliva in the water bath. Allow 5 minutes for temperature equilibration.
4. *Record the time,* and place 1 drop of saliva solution into each tube starting with Tube 1. Be sure that each drop lands directly in the starch solution and does not run down the side of the tube. Mix by agitation.
5. *Exactly 1 minute* after Tube 1 received the saliva solution, add 2 drops of IKI to Tube 1.
6. *One minute later,* add 2 drops of IKI to Tube 2. Continue to add IKI sequentially to the tubes at 1 minute intervals until all ten tubes have received IKI solution.

7. Remove the tubes from the water bath after adding the iodine solution and compare the color of the solutions with the controls to determine the time required to digest the starch. Record your results on the laboratory report.
8. Discard the tube contents in the sink and place the tubes in the pan containing Amphyl solution.

Effect of Boiling

In this section, you will compare the activity of amylase that has been heated to boiling with unheated amylase. Figure 33.2 depicts the general procedure.

1. Label three test tubes 1, 2, and 3.
2. Add 5 drops of saliva solution to tubes 1 and 2, and 5 drops of distilled water to Tube 3.
3. Place about 100 ml water in a beaker and heat it to boiling on a hot plate. Place Tube 1 in the boiling water bath for 3 minutes.
4. Add 1 dropper of starch solution to each tube and mix by agitation. Place the tubes in a water bath at 37° C for 5 minutes.
5. Remove the tubes from the water bath and add 2 drops of IKI to each tube. record the results on the laboratory report.
6. Discard the tube contents in the sink and place the tubes in the pan of Amphyl solution.

Effect of pH

In this section, you will determine the effect of three different hydrogen ion concentrations on amylase activity. The general procedure is shown in Figure 33.3.

1. Label nine test tubes 1 through 9. Arrange them in sequence in the test tube rack and write the pH designation on each tube as shown in Figure 33.3. Note that there are three tubes at each pH.
2. Add 1 dropper of buffer solution as follows:
   pH 5 buffer to tubes 1, 4, and 7
   pH 7 buffer to tubes 2, 5, and 8
   pH 9 buffer to tubes 3, 6, and 9
3. Add 2 drops of saliva solution to each tube, mix by agitation, and place the rack with the tubes in the water bath at 37° C. Also, place the dropping bottle of starch in the water bath. Allow 5 minutes for temperature equilibration.
4. *Record the time* and, starting with Tube 1, add 1 dropper of starch solution to each tube. Mix by agitation.
5. *Two minutes* after the recorded time, add 4 drops of IKI to tubes 1, 2, and 3. *Four minutes* after the recorded time, add 4 drops of IKI to tubes 4, 5, and 6. *Six minutes* after the recorded time, add 4 drops of IKI to tubes 7, 8, and 9.
6. Remove the tubes from the water bath after adding the iodine solution and compare them with the controls. Record the results on the laboratory report.
7. Discard the contents and place the tubes in the Amphyl solution. Wash your work station with Amphyl solution.
8. Complete the laboratory report.

Figure 33.3 Effect of pH on amylase activity.

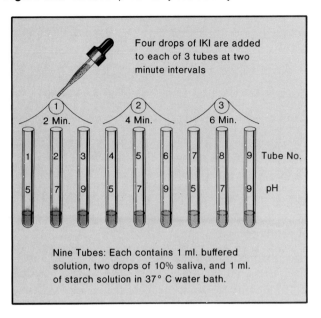

Four drops of IKI are added to each of 3 tubes at two minute intervals

Nine Tubes: Each contains 1 ml. buffered solution, two drops of 10% saliva, and 1 ml. of starch solution in 37° C water bath.

The Urinary Organs

Objectives

After completing this exercise, you should be able to:

1. Identify the components of the urinary system on charts or models and describe their functions.
2. Identify the parts of the kidney on charts, models, or specimens and describe their functions.
3. Identify a glomerulus and Bowman's capsule when viewed microscopically.
4. Define all terms in bold print.

Materials

Models of the urinary system and a kidney
Prepared slides of kidney tissue
Sheep kidneys, fresh or preserved
Sheep kidneys, triple injected
Dissecting kits and trays
Long, sharp knife
Prepared slides of kidney cortex and medulla

Organs of the Urinary System

The components of the urinary system are shown in Figure 34.1: two kidneys, two ureters, the urinary bladder, and the urethra. The **kidneys** are bean-shaped, reddish-brown organs located on either side of the vertebral column and posterior to the parietal peritoneum. Each kidney receives blood from a **renal artery,** which branches from the **abdominal aorta.** Blood leaves each kidney through a **renal vein,** which empties into the **inferior vena cava.**

Urine, formed by the kidneys, is carried from each kidney to the **urinary bladder** through a slender tube, a **ureter.** Peristaltic contractions of the ureter wall propel the urine to the bladder. Each ureter originates as a funnel-like **renal pelvis** in the kidney and descends parallel to the vertebral column and posterior to the peritoneum. The lower end enters the posterior surface of the urinary bladder.

Urine is temporarily stored in the distensible urinary bladder and then voided from the bladder via a short tube, the **urethra.** The male urethra is about 20 cm in length; the female urethra is approximately 4 cm in length.

Figure 34.1 The urinary system.

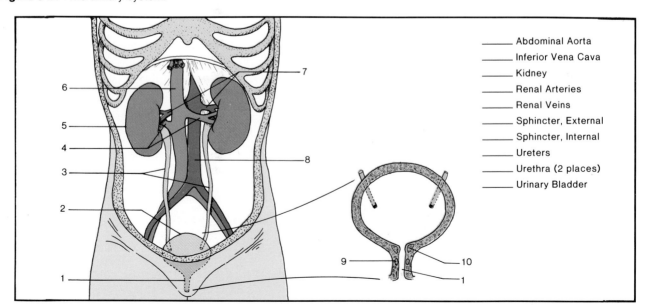

_____ Abdominal Aorta
_____ Inferior Vena Cava
_____ Kidney
_____ Renal Arteries
_____ Renal Veins
_____ Sphincter, External
_____ Sphincter, Internal
_____ Ureters
_____ Urethra (2 places)
_____ Urinary Bladder

Figure 34.2 Anatomy of the kidney.

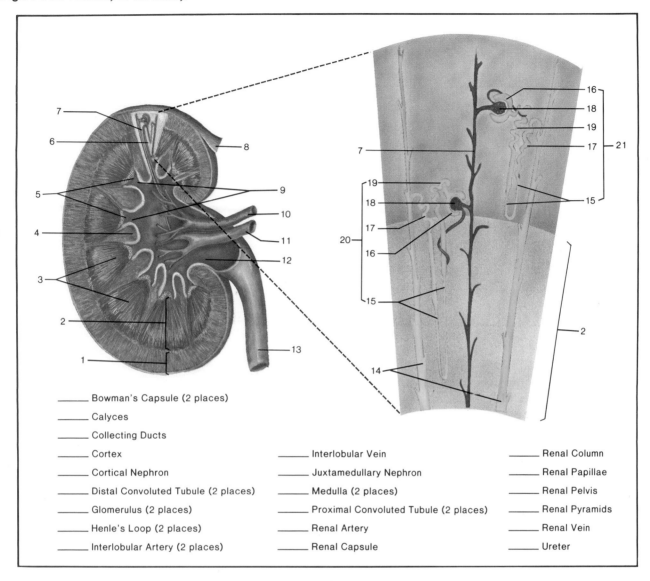

_____ Bowman's Capsule (2 places)

_____ Calyces

_____ Collecting Ducts

_____ Cortex   _____ Interlobular Vein   _____ Renal Column

_____ Cortical Nephron   _____ Juxtamedullary Nephron   _____ Renal Papillae

_____ Distal Convoluted Tubule (2 places)   _____ Medulla (2 places)   _____ Renal Pelvis

_____ Glomerulus (2 places)   _____ Proximal Convoluted Tubule (2 places)   _____ Renal Pyramids

_____ Henle's Loop (2 places)   _____ Renal Artery   _____ Renal Vein

_____ Interlobular Artery (2 places)   _____ Renal Capsule   _____ Ureter

Cystitis, inflammation of the urinary bladder, is more common in females than males since the shorter length of the female urethra provides an easier entrance for pathogens.

The passage of urine from the bladder is called **micturition** and is controlled by two sphincter muscles. The **internal sphincter** is located at the junction of the bladder and urethra. It is formed of smooth muscle and is under parasympathetic control. The **external sphincter** is located in the urethra about 2 cm from the bladder. It consists of skeletal muscle and is under voluntary control.

When about 300 ml of urine has accumulated in the urinary bladder, the stretching of the bladder walls initiates an urge to urinate and a subconscious reflex, which causes the walls to contract. This contraction forces urine past the internal sphincter to the external sphincter, creating a sensation of urgency. When the external sphincter is consciously relaxed, micturition occurs.

Incontinence is the lack of voluntary control of micturition. It is normal in infants until the neurons to the external sphincter develop, and voluntary control is learned through training.

Assignment

1. Label Figure 34.1.
2. Locate the parts of the urinary system on the model and note their relationships to the adjacent structures.

Kidney Anatomy

The basic structure of the kidney is shown in Figure 34.2. The thin outer covering of the kidney is the **renal capsule,** a fibrous membrane. Exterior to this membrane, there is usually a thick protective layer of fatty tissue (not shown).

There are two major parts of the kidney. An outer, reddish-brown **cortex** lies just under the renal capsule. The lighter-colored **medulla** forms the interior of the kidney. The medulla is divided into the cone-shaped **renal pyramids,** which are separated by extensions of the cortex called **renal columns.** The tip of a renal pyramid is the **renal papilla,** which projects into a calyx. The **calyces** are short tubes that receive urine from the renal papillae and empty into the funnel-shaped **renal pelvis.** The renal pelvis is continuous with the enlarged upper end of the ureter.

The Nephron

The functional unit of the kidney is the **nephron.** About one million nephrons are in each kidney. Each nephron consists of two major parts: a renal corpuscle and a renal tubule. A **renal corpuscle** consists of an inner tuft of capillaries, the **glomerulus,** and an outer, double-walled **glomerular** or **Bowman's capsule** that envelopes the glomerulus. A **renal** tubule consists of three sequential segments: (1) the **proximal convoluted tubule** leads from Bowman's capsule, (2) **Henle's loop** is the downward U-shaped portion, and (3) the **distal convoluted tubule** is the terminal segment that empties into a collecting duct. Several tubules empty into a single **collecting duct.**

Most nephrons (80%) are **cortical nephrons** that are entirely located in the cortex; their short Henle's loops rarely enter the medulla. **Juxtamedullary nephrons** have a long Henle's loop that penetrates deeply into the medulla. See Figure 34.2. Figure 34.3 shows a nephron in more detail with associated blood vessels.

Urine Formation   Nephrons form urine by three processes: (1) filtration, (2) reabsorption, and (3) secretion. In this way, water and essential substances in the blood are conserved while the concentrations of surplus substances, including **nitrogenous wastes** (urea, uric acid, and creatinine), are reduced. It is the process of urine formation that maintains the normal concentration of substances in blood plasma.

Blood enters the kidney through the **renal artery** and reaches each nephron via an **interlobular artery** (label 7, Figure 34.2; label 2, Figure 34.3). A short **afferent arteriole** carries blood from the interlobular artery to the glomerulus. An **efferent arteriole** carries blood from the glomerulus to the **peritubular capillaries** that enmesh the nephron tubule. Blood from the capillaries drains into the **interlobular vein** (label 8, Figure 34.2; label 1, Figure 34.3) that empties into the **renal vein.**

Figure 34.3 The nephron.

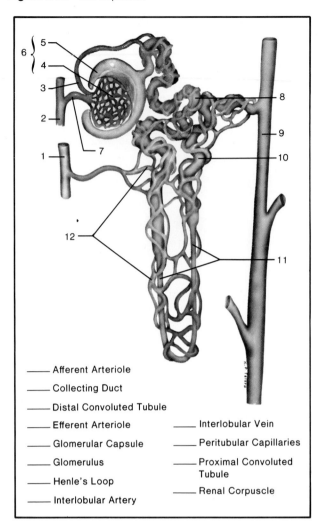

_____ Afferent Arteriole	
_____ Collecting Duct	
_____ Distal Convoluted Tubule	
_____ Efferent Arteriole	_____ Interlobular Vein
_____ Glomerular Capsule	_____ Peritubular Capillaries
_____ Glomerulus	_____ Proximal Convoluted Tubule
_____ Henle's Loop	_____ Renal Corpuscle
_____ Interlobular Artery	

The efferent arteriole has a smaller diameter than the afferent arteriole. This elevates the blood pressure within the glomerulus and forces the **glomerular filtrate,** a dilute fluid derived from blood plasma, into the glomerular capsule. About 600 ml of blood plasma flows through the glomeruli each minute; 860 liters in 24 hours. The kidneys form about 125 ml of filtrate per minute; 180 liters per day. The filtrate consists of all substances present in the blood except the formed elements. However, very few plasma proteins enter the filtrate because their molecules are too large.

As the filtrate passes through the renal tubule, various substances are reabsorbed into the peritubular capillaries. For example, almost all of the proteins that enter the filtrate are reabsorbed by tubule cells. A few substances are secreted from the capillaries into the filtrate.

Table 34.1 Quantity of selected substances in: (1) blood plasma flowing through the kidneys, (2) filtrate passing through Bowman's capsule, and (3) urine formed in a 24-hour period.

Substance	Plasma	Filtrate	Urine
Total Volume	860 l	180 l	1 l
Proteins	7,500 g	10 g	0
Chloride ions	3,180 g	667 g	7 g
Potassium ions	170 g	36 g	3 g
Sodium ions	2,924 g	612 g	5 g
Glucose	860 g	180 g	0
Creatinine	8.6 g	1.5 g	1.5 g
Urea	215 g	45 g	25 g
Uric acid	43 g	9 g	0.8 g

This selective reabsorption and secretion of substances by the renal tubule play a major role in maintaining the constancy of the body fluids.

The remaining filtrate passes into a collecting duct, where it may be further concentrated or diluted by the reabsorption of water or mineral ions. When it reaches the renal pelvis, it is called urine. Table 34.1 compares the quantity of selected substances in blood plasma, tubule filtrate, and urine.

Assignment

1. Label Figures 34.2 and 34.3.
2. Complete Sections A, B, and C on the laboratory report.

3. Examine a sectioned, triple-injected kidney under a demonstration dissecting microscope. Note the many renal corpuscles. Can you distinguish the glomerulus, Bowman's capsule, and tubules?
4. Examine a prepared slide of kidney cortex and medulla. Compare your observations with Figure HA-21 and locate the labeled structures. Note that the walls of Bowman's capsule and the renal tubule are only one cell thick. Make diagrams of your observations on the laboratory report.

Sheep Kidney Dissection

1. Obtain a sheep kidney for study. If it is still encased in fat, remove the fat carefully with your hands. Look for the adrenal gland embedded in the fat near one end of the kidney. Cut the gland in half and note that it has a distinct outer cortex and an inner medulla.
2. Insert a dissecting needle into the kidney to distinguish the tougher renal capsule from the softer underlying tissue.
3. With a long, sharp knife, cut the kidney longitudinally to make a frontal section similar to Figure 34.2. Wash the cut surfaces.
4. Locate all of the macroscopic structures shown in Figure 34.2.
5. Complete the laboratory report.

Urine and Urinalysis

Objectives

After completing this exercise, you should be able to:

1. Identify the normal components and characteristics of urine.
2. List the common abnormal components and characteristics of urine, and identify possible clinical conditions associated with them.
3. Perform a simple urinalysis.
4. Define all terms in bold print.

Materials

Urine "unknowns"
10SG-Multistix reagent strips and analysis charts (Ames)
Urine collection cups, plastic
Test tubes and racks
Amphyl, 0.25%
Biohazard bag

Urine is a complex aqueous solution of organic and inorganic substances. Most of the components are either waste products of cellular metabolism or products formed from certain foods.

The volume, pH, and solute concentration of urine may vary within normal limits in a healthy person as the constancy of the internal environment is maintained. In contrast, the physical and chemical characteristics of urine may change dramatically in certain pathological conditions. Thus, a **urinalysis** is a commonly used clinical test because it provides much information about the condition of the body.

The physical characteristics of normal urine are shown in Table 35.1, and the major organic and inorganic solutes in normal urine are found in Table 35.2. Some variations from these values are normal, but marked variations are considered abnormal.

Table 35.1 Physical characteristics of normal urine

Characteristic	Description*
Volume	0.7 to 2.5 l in 24 hrs. Varies with diet, water intake, concentration of plasma solutes, temperature, physical activity, emotional state, and other factors.
Color	Pale yellow to amber due to normal amounts of urobilinogen. The greater the solute concentration, the darker the color. Certain foods and drugs affect the color.
Turbidity	Clear when fresh but may become turbid after standing.
Odor	Characteristic aromatic odor, but becomes ammonial-like after standing due to bacterial breakdown of urea.
pH	4.8 to 7.5, average of 6.0. Acid in very high protein diets; alkaline in vegetable diets.
Specific Gravity	1.015 to 1.025 in 24-hr specimens. Single specimens from 1.002 to 1.030. The greater the concentration of solutes, the higher the specific gravity.

*Based on 24-hour urine specimen.

Table 35.2 Principal solutes of a normal 24-hour urine specimen

Solute	Grams	Comments
Organic Components		
Urea	25.0	Forms up to 90% of nitrogenous wastes. Produced by liver after deamination of amino acids.
Creatinine	1.5	By-product of muscle metabolism.
Uric Acid	0.8	Product of nucleic acid catabolism. Tends to form crystals; a common constituent of kidney stones (renal calculi).
Inorganic Components		
NaCl	15.0	Most abundant mineral salt. Amount varies with intake.
K^+	3.3	As chloride, phosphate, and sulfate salts.
Sulfates	2.5	As sodium, potassium, magnesium, or calcium compounds.
Phosphates	2.5	As sodium, potassium, magnesium, or calcium compounds.

Abnormal Characteristics of Urine

In this section, you will study the abnormal characteristics of urine, which may have clinical implications.

Physical Characteristics

Volume An excessive production of urine, **polyuria,** occurs in diabetes insipidus and diabetes mellitus, while very low or no urine production, **anuria,** occurs in renal failure.

Color Urobilinogen gives the color to normal urine, but an excess causes a reddish amber coloration. Excess bile pigments give urine a brownish yellow color. Erythrocytes or hemoglobin in urine produce a red or smoky brown color.

Turbidity Abnormal turbidity usually is caused by the presence of phosphates, urates, blood, pus, or bacteria.

Odor Variations in odor may be due to diet or drugs, but only odors related to diseases are considered abnormal. For example, the urine may have a fruity odor in diabetes mellitus.

Hydrogen Ion Concentration Excessive acid reactions occur with fever, **nephritis** (inflammation of the kidney), **glomerulonephritis** (inflammation of the kidney involving glomeruli), and acute **cystitis** (inflammation of the urinary bladder). Excessive alkaline reactions occur in chronic cystitis and prostatic obstruction.

Specific Gravity Specific gravity is a measure of the concentration of solids in the urine. Pure water has a specific gravity of 1.000. A low specific gravity (dilute urine) may result from excessive water intake, diabetes insipidus, or chronic nephritis. A high specific gravity (concentrated urine) may result from low water intake, diabetes mellitus, fever, or acute nephritis.

Constituents

Proteins Proteins are normally absent in the urine since their large size usually prevents their passage into Bowman's capsule. However, under certain conditions, albumin, the smallest and most abundant of the plasma proteins, is present in the urine—a condition known as **albuminuria.** Temporary albuminuria may be caused by excessive physical exertion. Chronic albuminuria results from excessive permeability of the glomerular membrane and may be caused by bacterial toxins and glomerulonephritis.

Traces of mucin, a glycoprotein, may be normally present due to secretion from the urinary epithelium. Excessive mucin suggests irritation of the urinary tract.

Glucose Only trace amounts of glucose are present in normal urine. The presence of larger amounts of glucose in the urine, **glycosuria,** occurs when the glucose level of the blood plasma exceeds normal limits (hyperglycemia), and glucose is not completely reabsorbed from the tubular filtrate. Persistent glycosuria is usually diagnostic of diabetes mellitus. Temporary glycosuria may result from other causes such as the excessive ingestion of carbohydrates.

Ketones When the body uses an excessive amount of fatty acids in cellular respiration, the incomplete metabolism of fatty acids produces ketones, which are released into the blood and excreted in urine. The presence of ketones in urine, **ketonuria,** may result from an excessively low carbohydrate diet or diabetes mellitus. Diabetic ketosis may lead to acidosis, which can culminate in coma, and death.

Hemoglobin The presence of hemoglobin in urine, **hemoglobinuria,** occurs when hemoglobin is released into blood plasma by disintegrating red blood cells. This condition may result from hemolytic anemia, hepatitis, transfusion reactions, burns, malaria, and other causes.

Erythrocytes The presence of red blood cells in urine, **hematuria,** indicates bleeding in the urinary tract, resulting from disease or trauma. It is often the first symptom of malignant kidney tumors.

Leukocytes The presence of leukocytes or other components of pus in the urine, **pyuria,** indicates inflammation in the urinary tract, such as glomerulonephritis, cystitis, or urethritis (inflammation of the urethra).

Bilirubin Bilirubin, a bile pigment, is formed by the breakdown of hemoglobin. Normally, only trace amounts are present in urine. The presence of larger concentrations, **bilirubinuria,** usually indicates liver pathology, such as hepatitis or cirrhosis, in which bilirubin is not removed from the blood by liver cells.

Urobilinogen Bacteria in the large intestine convert bilirubin to urobilinogen. Some of this compound is absorbed into the blood and removed by liver cells in the formation of bile. Excessive amounts in the urine, **urobilinogenuria,** usually indicate an excessive destruction of hemoglobin or liver pathology.

Casts Casts are concentrations of cells or cellular debris that have hardened in the tubules prior to being forced from the tubules by the build-up of filtrate behind them. See Figure 35.1. Casts always indicate kidney pathology, such as glomerulonephritis.

Figure 35.1 Crystals, cells, and casts in urine.

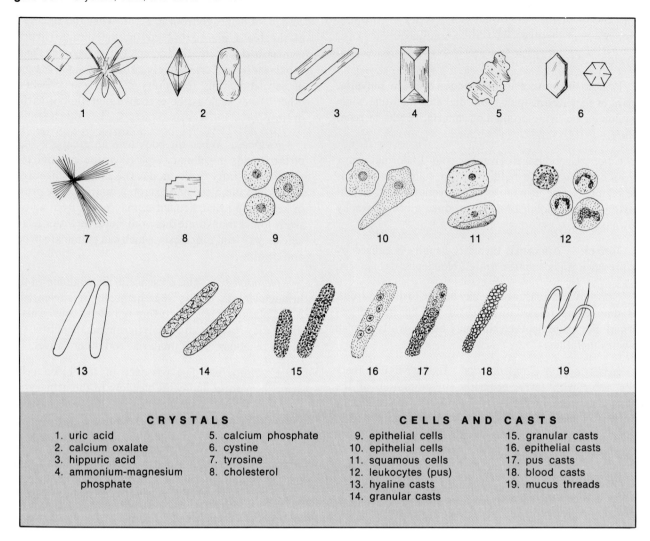

C R Y S T A L S

1. uric acid
2. calcium oxalate
3. hippuric acid
4. ammonium-magnesium phosphate
5. calcium phosphate
6. cystine
7. tyrosine
8. cholesterol

C E L L S   A N D   C A S T S

9. epithelial cells
10. epithelial cells
11. squamous cells
12. leukocytes (pus)
13. hyaline casts
14. granular casts
15. granular casts
16. epithelial casts
17. pus casts
18. blood casts
19. mucus threads

Renal Calculi Calculi (kidney stones) are usually formed of crystals of uric acid, calcium oxalate, or calcium phosphate. They usually form in the pelvis of the kidney. They may lodge in the ureters, bladder, or urethra where they may cause pain, and hematuria and pyuria.

Microbes Urine normally contains low concentrations of bacteria (<1000 bacteria/ml) derived from the urethral flora. An excessive concentration of bacteria in the urine, **bacteriuria,** or the presence of yeasts or protozoans indicates a urinary infection.

Assignment

1. Study Tables 38.1 and 38.2.
2. Complete Sections A and B on the laboratory report.

Urinalysis

A routine urinalysis includes a description of the physical characteristics and a determination of the presence or absence of abnormal components. Microscopic examination of urine sediments and tests for the presence, identification, and concentration of microbes may also be performed; in some cases these tests are the most important parts of the analysis.

In this section, you will perform an analysis of simulated unknown urine specimens as well as a sample of your own urine. The analysis will be done primarily with 10SG-Multistix reagent strips. Figure 35.1 illustrates the types of crystals, cells, and casts that may occur in urine. However, you will not perform a microscopic examination of sediments in this exercise.

Collect a sample of your urine as directed by your instructor. Use the following procedure to analyze your urine specimen and the unknowns. Record your results on the laboratory report.

1. Swirl the specimen in the container to resuspend any sediments. Record the color and turbidity of the specimen.
2. Fill a clean test tube about three-fourths full with a urine sample.
3. Remove a 10SG-Multistix reagent strip from the jar while being careful not to touch the colored reagent squares. The reagent squares from top to bottom test for: leukocytes, nitrite, urobilinogen, protein (albumin), pH, blood (erythrocytes or hemoglobin), specific gravity, ketone, bilirubin, and glucose. Compare the color of the reagent squares with the color chart provided, and note the color of negative and positive tests.

Note that for most tests, the resulting color is to be read only after a minimum number of seconds to obtain accurate results. *Be sure that you know how to read the tests before proceeding.*

4. Dip a 10SG-Multistix reagent strip into the urine sample so that all reagent squares are immersed. Remove the strip, tap off the excess urine on the side of the test tube, and lay it on a paper towel with the reagent squares up. Record the time.
5. Observing the minimum reading times, determine and record the results for each reagent square.
6. Discard the urine as directed by your instructor. Place the reagent strips, urine collecting cup, and paper towels in the biohazard bag.
7. Place the test tubes in the Amphyl solution provided. Wash your work station with Amphyl.
8. Complete the laboratory report.

The Endocrine Glands

Objectives

After completing this exercise, you should be able to:

1. Identify the location of each endocrine gland, the hormones it produces, and their functions.
2. Indicate known effects of hyposecretion and hypersecretion.
3. Identify each endocrine gland when viewing its tissue microscopically.
4. Define all terms in bold print.

Materials

Prepared slides of endocrine glands

The endocrine system consists of a number of glands and tissues that secrete **hormones,** chemical messengers. Hormones are absorbed into the blood and carried throughout the body where they act on specific target tissues. Thus, the endocrine system provides a chemical control of body functions.

As you study this exercise, locate each gland on Figure 36.1 and examine the photomicrographs of its histology.

The Thyroid Gland

The thyroid gland is located just below the larynx. It is formed of two lobes, one on each side of the trachea, that are joined by an isthmus of tissue across the anterior tracheal surface. Histologically, the gland contains numerous secretory **follicles** that secrete the **thyroid hormone.**

The follicles are filled with **thyroglobulin,** a colloidal material that holds excess thyroid hormone until it is released. The parafollicular cells produce a different hormone, **calcitonin.** See Figure 36.2.

The Thyroid Hormone

This hormone consists of several related compounds containing iodine in their molecules. Thyroxine (T_4) and tri-iodothyronine (T_3) are the most important. Thyroid hormone increases the rate of metabolism of most body tissues. For example, it stimulates cellular respiration, heart rate, protein synthesis, mental activity, and growth. Thyrotropin, produced by the anterior lobe of the pituitary gland, stimulates production of thyroid hormone.

Severe hyposecretion (hypothyroidism) in children may result in **cretinism,** a condition characterized by physical and mental retardation. In adults, it may cause **myxedema,** which is characterized by obesity, depression, slow heart rate, and sluggishness.

Severe hypersecretion (hyperthyroidism) results in weight loss, nervousness, sleeplessness, rapid heart rate, and sometimes exophthalmos, the protrusion of the eyeballs.

Calcitonin

This hormone reduces the level of calcium in the body by stimulating bone deposition by osteoblasts and inhibiting calcium removal by osteoclasts. A high level of blood calcium stimulates calcitonin production; a low level inhibits production.

Figure 36.1 The endocrine glands.

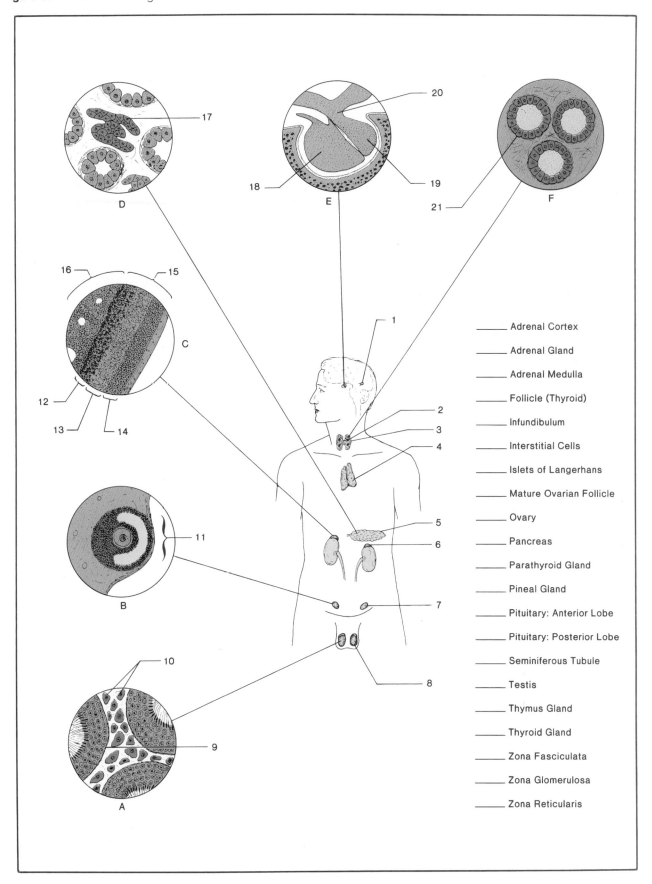

Adrenal Cortex

Adrenal Gland

Adrenal Medulla

Follicle (Thyroid)

Infundibulum

Interstitial Cells

Islets of Langerhans

Mature Ovarian Follicle

Ovary

Pancreas

Parathyroid Gland

Pineal Gland

Pituitary: Anterior Lobe

Pituitary: Posterior Lobe

Seminiferous Tubule

Testis

Thymus Gland

Thyroid Gland

Zona Fasciculata

Zona Glomerulosa

Zona Reticularis

Figure 36.2 The thyroid and parathyroid glands.

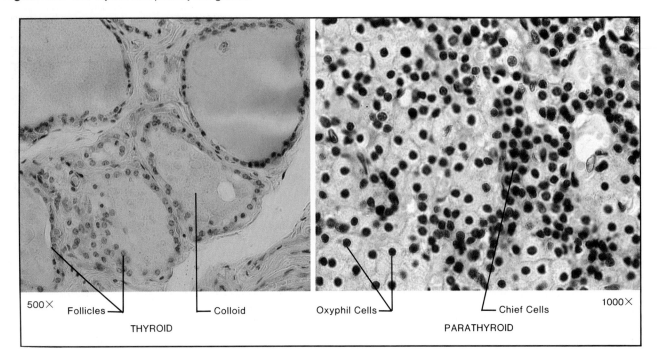

500×   Follicles —   — Colloid   Oxyphil Cells —   — Chief Cells   1000×

THYROID   PARATHYROID

The Parathyroid Glands

These four small glands are embedded on the posterior surface of the thyroid gland. Figure 36.2 illustrates the histology of a parathyroid gland. The small, numerous chief cells produce the parathyroid hormone.

Parathyroid Hormone

This hormone raises the calcium concentration and lowers the phosphorus concentration of the blood. This is done by promoting reabsorption of bone tissue by osteoclasts, and stimulating calcium reabsorption and phosphorus excretion by the kidneys. Parathyroid hormone and calcitonin have opposite effects on the level of blood calcium. A lack of parathyroid hormone results in tremors, tetany, and, without treatment, death.

Production is regulated by negative feedback control. High levels of blood calcium inhibit production of parathyroid hormone, while low levels promote production.

The Pancreas

This pennant-shaped gland has both exocrine and endocrine functions. It is located between the stomach and duodenum. Each of the three pancreatic hormones (insulin, glucagon, and somatostatin) are produced by a different type of cell found in the **islets of Langerhans.** Cells surrounding the islets of Langerhans are acinar cells, which produce pancreatic juice. See Figure 36.3.

Insulin

Insulin decreases blood glucose by: (1) facilitating the passage of glucose through cell membranes into body cells where it can be metabolized, and (2) stimulating the conversion of glucose into glycogen in the liver. A deficiency of insulin results in **diabetes mellitus,** a condition in which **hyperglycemia,** an abnormally high blood glucose level, is present.

The concentration of blood glucose regulates insulin production. A high level stimulates production; a low level inhibits production.

Glucagon

Glucagon increases the blood glucose level by stimulating the liver to: (1) convert glycogen into glucose, and (2) synthesize glucose from non-carbohydrate sources. Thus, insulin and glucagon have opposite effects on the concentration of blood glucose. A deficiency of this hormone results in **hypoglycemia,** a subnormal level of blood glucose.

Regulation is by negative feedback control. A low level of blood glucose promotes glucagon production; a high level decreases glucagon production.

Figure 36.3 Pancreas histology.

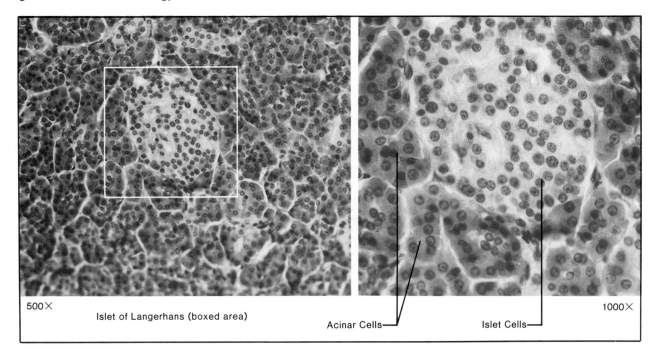

500×

Islet of Langerhans (boxed area)

1000×

Acinar Cells——

Islet Cells——

Somatostatin

This hormone, also known as growth hormone inhibiting factor, is formed in the hypothalamus as well as in the islets of Langerhans. Its function is not well understood, but it inhibits the release of growth hormone from the pituitary gland.

The Adrenal Glands

An adrenal gland sits atop each kidney and consists of two parts: the central **medulla** and the external **cortex.** These two parts arise from different embryological origins and have distinctive functions.

The Adrenal Cortex

The adrenal cortex secretes many different hormones that are collectively called **corticosteroids** (corticoids). These hormones may be subdivided into three different groups. **Mineralocorticoids** control the concentrations of water and electrolytes in the blood, specifically Na^+ and K^+. **Glucocorticoids** control reactions to stress, and the metabolism of glucose and proteins. **Gonadocorticoids,** or sex hormones, occur in such small concentrations that their effect is normally minimal. Of the thirty or more corticoids, the two most important are aldosterone and cortisol.

Figure 36.4 illustrates the histology of the three zones of the adrenal cortex: an outer **zona glomerulosa,** an intermediate **zona fasciculata,** and an inner **zona reticularis.**

Figure 36.4 The adrenal cortex.

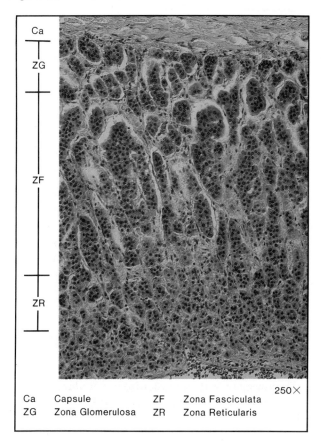

Ca

ZG

ZF

ZR

250×

Ca	Capsule	ZF	Zona Fasciculata
ZG	Zona Glomerulosa	ZR	Zona Reticularis

Aldosterone   This mineralocorticoid is secreted by cells in the zona glomerulosa. It increases the rate of Na^+ reabosrption and K^+ excretion by the kidneys, which, in turn, increases water reabsorption and decreases urine output. In this manner, the concentrations of Na^+ and K^+ in the blood are maintained within normal levels. A deficiency of aldosterone results in an increase of K^+ in body fluids, a decrease of Na^+, and a reduction in water concentration. When severe, these conditions cause reduced cardiac output, shock, and may result in death.

The control of aldosterone secretion is complex. The primary stimulus is an increase in K^+ concentration and a decrease in Na^+ concentration in body fluids.

Cortisol   This glucocorticoid is produced primarily by the zona fasciculata. It serves to maintain normal metabolism and to enhance resistance to stress. Cortisol increases the quantity of amino acids, fatty acids, and glucose available to body cells by causing: (1) a decrease in protein synthesis; (2) the release of fatty acids from adipose tissue; and (3) the formation of glucose from noncarbohydrates. At therapeutic levels, it also tends to block processes that produce inflammation.

A chronic excess of glucocorticoids may result in **Cushing's syndrome,** a condition characterized by protein depletion, muscle and bone weakness, slow healing, hyperglycemia, hydration of tissues, scraggly hair, and possibly masculinization in females.

A chronic deficiency of glucocorticoids may result in **Addison's disease,** a condition characterized by hypoglycemia, dehydration, and electrolyte imbalance, which may be fatal if untreated.

The production of cortisol and other glucocorticoids is stimulated by the secretion of ACTH from the pituitary gland, which is increased in times of stress.

The Adrenal Medulla

Two related hormones are produced by the adrenal medulla: **epinephrine** and **norepinephrine.** The effects of these hormones resemble those of the sympathetic nervous system: increased heart and respiration rates, elevated blood pressure and blood glucose concentration, and decreased digestive action. The action of these hormones and the sympathetic nervous system prepare the body to meet energy-expending emergencies. The production of both hormones is stimulated by sympathetic nerve fibers.

The Thymus Gland

This gland lies in the mediastinum above the heart and consists of two long lobes. It is relatively large in young children but after puberty diminishes in size throughout life. Histologically, both the **cortex** and **medulla** contain large numbers of lymphocytes as shown in Figure 36.5. The thymus gland is involved in the maturation of T-lymphocytes, probably with the aid of the hormone **thymosin,** which it secretes.

The Pineal Gland

The pineal gland is located in the brain under the posterior end of the corpus callosum. At puberty, the gland begins to form small calcium deposits called **pineal sand** whose significance is not known. See Figure 36.5.

Figure 36.5 The thymus and pineal glands.

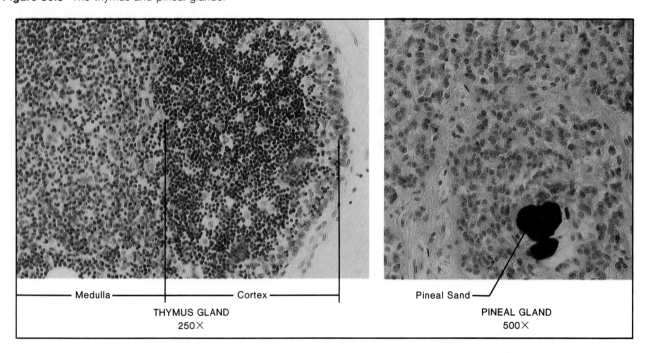

Medulla — Cortex

THYMUS GLAND
250×

Pineal Sand

PINEAL GLAND
500×

Figure 36.6 Testicular tissue.

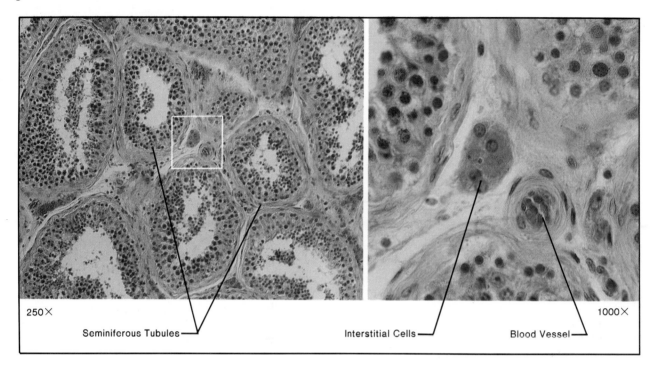

250✕

Seminiferous Tubules

1000✕

Interstitial Cells

Blood Vessel

The function of the pineal gland is not understood. It secretes a hormone called **melatonin,** which possibly helps to regulate the menstrual cycle by inhibiting the release of gonadotropins from the pituitary gland.

The Testes

The male gonads, the **testes,** contain numerous coiled seminiferous tubules that produce spermatozoa. Cells located in spaces between these tubules, the **interstitial cells,** secrete the male hormone, **testosterone.** See Figure 36.6.

Testosterone produced during embryological development is responsible for the development of the male sex organs and the descent of the testes from the body cavity into the scrotum. From birth to puberty, little testosterone is produced, but large amounts are secreted at puberty as sexual maturation takes place.

During puberty, testosterone stimulates the development of the male sex organs and secondary sexual characteristics. The latter includes: (1) the presence of hair on the face, axillae, chest, and pubic areas; (2) the enlargement of the larynx, which deepens the voice; (3) an increase in muscle development and bone thickness; and (4) the development of broad shoulders and narrow hips.

During embryological development, testosterone production is regulated by chorionic gonadotropin, which is produced by the placenta. At puberty, the interstitial cells are stimulated to produce testosterone by the interstitial cell stimulating hormone (ICSH) from the pituitary gland. Testosterone has a negative feedback effect on ICSH production.

The Ovaries

The **germinal epithelium** of the ovary is formed early in embryological development. It produces numerous **primordial follicles** that reside in the ovarian cortex. Some of these follicles migrate inward to become **primary follicles,** which consist of a primary oocyte enclosed in a

Figure 36.7 Ovarian tissue.

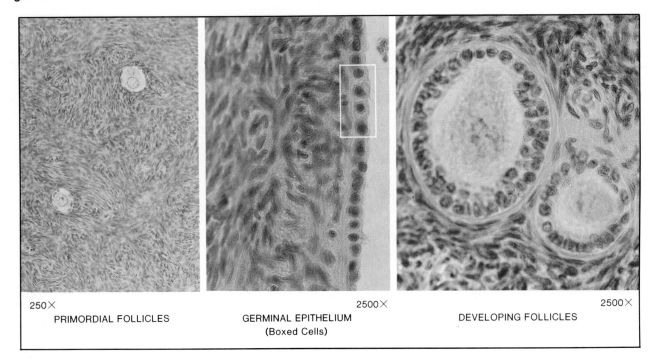

250×
PRIMORDIAL FOLLICLES

2500×
GERMINAL EPITHELIUM
(Boxed Cells)

2500×
DEVELOPING FOLLICLES

sphere of follicular cells. Starting at puberty, follicle stimulating hormone (FSH) from the anterior lobe of the pituitary stimulates some of the primary follicles to develop further. Usually, only one will become a **mature, or Graafian, follicle** containing a secondary oocyte that is released at ovulation. See Figure 36.7. The empty follicle then becomes a **corpus luteum** under the stimulation of luteinizing hormone (LH) from the pituitary.

The ovaries produce the female sex hormones, estrogens and progesterone. FSH stimulates the production of **estrogens** by developing follicles. After ovulation, LH promotes the secretion of **progesterone** by the corpus luteum.

Estrogens

Estrogens consist of several related compounds and are produced primarily by the developing ovarian follicles. But estrogens are also formed by the corpus luteum, by the placenta during pregnancy, and, in minute quantities, by the adrenal cortex.

Estrogens promote growth, and the development of the female reproductive organs and the female secondary sexual characteristics, such as: (1) enlargement and broadening of the pelvic girdle to increase the size of the pelvic opening; (2) development of the breasts; and (3) increased fat deposition under the skin, especially over the hips, buttocks, and thighs.

Progesterone

The primary function of this hormone is to promote changes in the endometrium of the uterus in preparation for the implantation of an early embryo. It also stimulates an increase in the number of secretory cells in the breasts in preparation for milk production.

Without pregnancy, degeneration of the corpus luteum causes a decrease in progesterone production, which results in a breakdown of the endometrium, which results in menstruation. If pregnancy occurs, chorionic gonadotropin stimulates the corpus luteum to form increased amounts of progesterone and some estrogen to maintain the uterine lining until this role is taken over by placental estrogens and progesterone.

Figure 36.8 The pituitary gland.

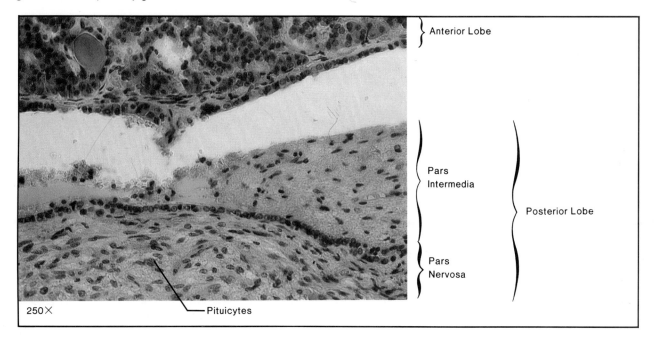

250× ...... Pituicytes

Anterior Lobe

Pars Intermedia

Pars Nervosa

Posterior Lobe

The Pituitary Gland

The pituitary gland, or **hypophysis,** consists of anterior and posterior lobes that have different embryological origins and functions. It is attached to the hypothalamus by a slender stalk, the **infundibulum.** The anterior lobe has secretory cells; the posterior lobe does not. The histological section in Figure 36.8 shows the cleft separating the lobes and the two portions of the posterior lobe.

The Anterior Lobe

Six hormones are produced by various cells in the anterior lobe, or **adenohypophysis.** Except for prolactin and growth hormone, the hormones are **tropic hormones** that stimulate other endocrine glands to produce their hormones. Production is regulated by releasing factors (hormones) formed in the hypothalamus and carried by blood to the anterior lobe via the **hypophyseal portal system.** See Figure 36.9.

Growth Hormone (GH) or somatotropin (STH) This hormone increases the growth rate of body cells by enhancing protein synthesis and utilization of carbohydrates and lipids. Hypersecretion of GH during growing years produces **gigantism** due to excessive growth in the length of long bones. Hypersecretion in an adult causes **acromegaly,** which is characterized by an enlargement of the small bones in the hands and feet, the mandible, and frontal bones. Hyposecretion in the growing years causes **hypopituitary dwarfism.**

The control of GH production is unclear, but two antagonistic hormones from the hypothalamus seem to be involved: **growth hormone releasing factor (GHRF)** and **growth hormone inhibiting factor (GHIF).**

Prolactin (PRL) This hormone stimulates the production of milk after childbirth. Antagonistic hormones from the hypothalamus control its production: **prolactin inhibiting factor (PIF)** is secreted prior to childbirth; **prolactin releasing factor (PRF)** is secreted after childbirth.

Thyrotropin (TSH) This hormone stimulates the production of thyroid hormone. TSH is regulated by the **thyrotropin-releasing factor (TRF)** from the hypothalamus. Thyroid hormone provides a negative feedback control on TRF production.

Adrenocorticotropin (ACTH) This tropic hormone stimulates glucocorticoid production by the adrenal cortex. Its production is promoted by the **corticotropin releasing factor (CRF)** from the hypothalamus. Glucocorticoids exert a negative feedback control on the formation of CRF.

Follicle Stimulating Hormone (FSH) In females, this hormone promotes the development and maturation of ovarian follicles and stimulates the secretion of estrogens. In males, it stimulates the formation of spermatozoa by the testes. FSH production is promoted by the **gonadotropin releasing factor (GnRF)** from the hypothalamus.

Figure 36.9 The pituitary regulatory mechanism.

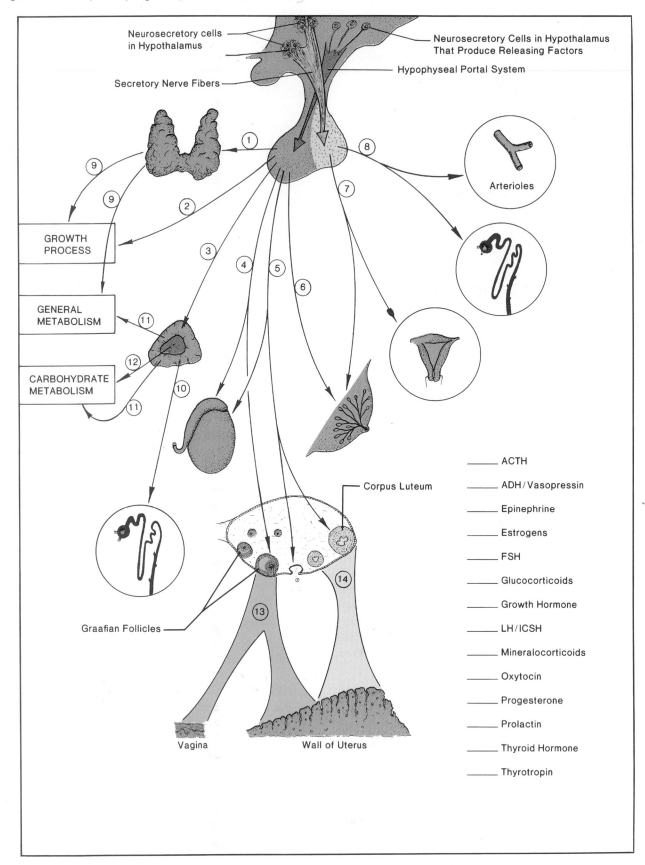

Neurosecretory cells in Hypothalamus

Neurosecretory Cells in Hypothalamus That Produce Releasing Factors

Secretory Nerve Fibers

Hypophyseal Portal System

Arterioles

GROWTH PROCESS

GENERAL METABOLISM

CARBOHYDRATE METABOLISM

Corpus Luteum

Graafian Follicles

Vagina

Wall of Uterus

_____ ACTH

_____ ADH / Vasopressin

_____ Epinephrine

_____ Estrogens

_____ FSH

_____ Glucocorticoids

_____ Growth Hormone

_____ LH / ICSH

_____ Mineralocorticoids

_____ Oxytocin

_____ Progesterone

_____ Prolactin

_____ Thyroid Hormone

_____ Thyrotropin

Luteinizing Hormone (LH) In females, a rapid rise in the level of LH, stimulated by GnRF, causes ovulation and the subsequent development of the corpus luteum. High levels of estrogens and progesterone inhibit GnRF formation, which in turn, inhibits FSH and LH production.

In males, LH is also called **interstitial cell stimulating hormone (ICSH)** since it stimulates the interstitial cells of the testes to secrete testosterone. Testosterone has a negative feedback effect on GnRF production.

The Posterior Lobe

The two hormones from the posterior lobe are formed in the hypothalamus by neurosecretory neurons and released within the posterior lobe, which is also called the **neurohypophysis.** See Figure 36.9. Their production is controlled by the hypothalamus.

Antidiuretic Hormone (ADH) This hormone promotes the reabsorption of water by the kidneys by increasing the permeability of the collecting tubules to water. Thus, ADH controls the osmotic balance of body fluids. Without ADH, excessive water loss occurs by the production of dilute urine. At times of severe blood loss, ADH also constricts arterioles to maintain adequate blood pressure. For this reason, ADH is sometimes called **vasopressin.**

Oxytocin During childbirth, pressure on the cervix initiates a neural reflex that stimulates the formation of oxytocin. In turn, oxytocin increases the strength of uterine contractions to facilitate childbirth.

After childbirth, stimulation of the nipple by a nursing infant initiates a neural reflex that also releases oxytocin. In this case, oxytocin causes contraction of smooth muscles associated with the milk glands, which forces milk into the milk ducts, leading to release of milk from the breast.

Assignment

1. Label Figures 36.1 and 36.9.
2. Complete the laboratory report.
3. Examine prepared slides of endocrine glands. Compare your observations with the photomicrographs. Make labeled drawings on a separate sheet of paper that will help you identify the glands from microscope slides.

The Reproductive Organs

Objectives

After completing this exercise, you should be able to:

1. Identify the parts of the male and female reproductive systems on charts or models, and describe their functions.
2. Describe spermatogenesis and oogenesis.
3. Identify histological structures in prepared slides of ovary and testis.
4. Define all terms in bold print.

Materials

Models of human reproductive organs
Prepared slides of human testis and ovary

The Male Organs

Figure 37.1 illustrates a sagittal section of the male reproductive system. The primary sex organs (gonads) are the **testes** (testicles), which are located in an exterior pouch, the **scrotum.** They produce the spermatozoa and the male sex hormones. In a male fetus, the testes develop within the abdominal cavity and descend into the scrotum 1–2 months before birth. The scrotum provides a temperature of 94–95° F for the testes, which is essential for the production of viable spermatozoa.

A fibrous capsule, the **tunica albuginea,** forms the outer wall of each testis. Testicular septa, formed of connective tissue, divide the testis into several lobules. Each lobule contains **seminiferous tubules** that produce **spermatozoa.** All seminiferous tubules merge to form the **rete testis,** a complex network of tubules. Cilia within the tubules of the rete testis move the spermatozoa through the several **vasa efferentia** and on into the highly-coiled **epididymis** where the spermatozoa mature. Note that the epididymis lies over the superior and posterior portion of the testis. From the epididymis, spermatozoa pass into the **vas deferens** and on toward the urethra. Note the histology of the vas deferens x.s., especially the smooth muscle layers, in Figure HA-22.

Trace a vas deferens from the epididymis and note that it passes over the pubic bone and urinary bladder into the pelvic cavity. The distal end of each vas deferens is enlarged to form an **ampulla.** The ampulla merges with the duct from a **seminal vesicle** to form the **ejaculatory duct.** The ejaculatory duct on each side empties into the **prostatic urethra,** the portion of the urethra that is within the prostate gland. The small **bulbourethral,** or **Cowper's, glands** empty their secretions into the urethra at the base of the penis.

Enlargement of the prostate gland is common in males over age 60. Where the urethra is partially occluded, the obstruction is removed by a surgical procedure performed through the urethral canal. The prostate is a common site of cancer in males.

Semen is formed of spermatozoa and the alkaline secretions of the accessory glands. Two-thirds of the semen is derived from secretions of the seminal vesicles, and about one-third comes from prostatic fluid. Secretions of the bulbourethral glands and spermatozoa account for very little of the semen volume. These alkaline secretions protect the spermatozoa from acid environments, provide nutrients for the sperm, and activate their swimming movements.

Erection of the penis occurs when its three cylinders of erectile tissue fill with blood in response to sexual stimulation. The **corpus spongiosum** is the cylinder of tissue that surrounds the **penile urethra.** Its distal end is enlarged to form the **glans penis.** The two **corpora cavernosa** are located in the dorsal part of the organ and are separated by a medial septum of connective tissue. Observe the histology of the penile urethra and erectile tissue in Figure HA-22.

A sheath of skin, the **prepuce,** begins just behind the glans and extends to cover it. Circumcision is a surgical procedure that removes the prepuce to facilitate sanitation.

Figure 37.1 Male reproductive organs.

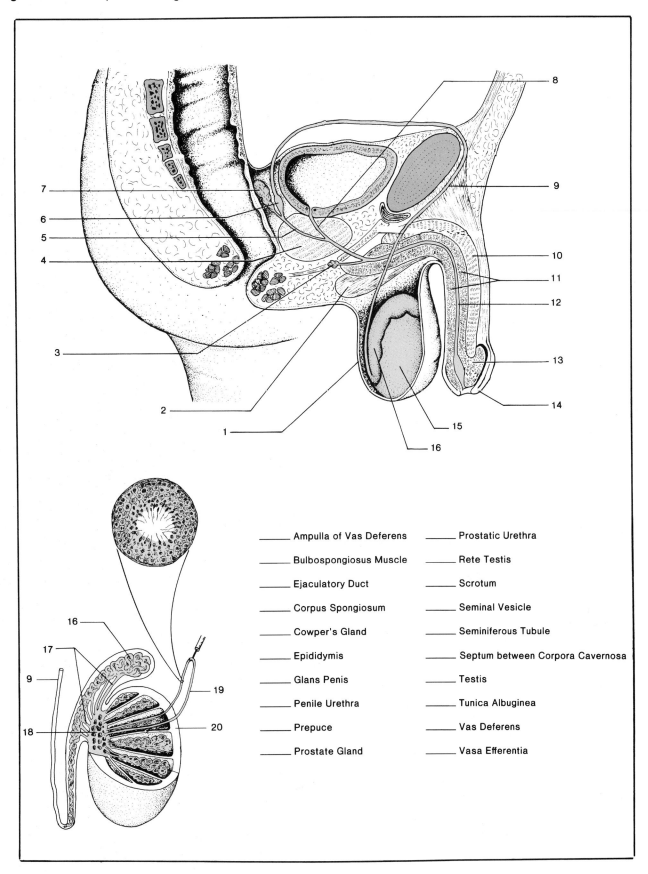

_____ Ampulla of Vas Deferens
_____ Bulbospongiosus Muscle
_____ Ejaculatory Duct
_____ Corpus Spongiosum
_____ Cowper's Gland
_____ Epididymis
_____ Glans Penis
_____ Penile Urethra
_____ Prepuce
_____ Prostate Gland

_____ Prostatic Urethra
_____ Rete Testis
_____ Scrotum
_____ Seminal Vesicle
_____ Seminiferous Tubule
_____ Septum between Corpora Cavernosa
_____ Testis
_____ Tunica Albuginea
_____ Vas Deferens
_____ Vasa Efferentia

Figure 37.2 Spermatogenesis.

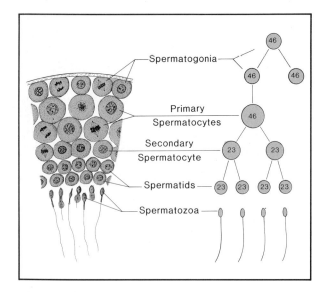

Figure 37.2 Spermatogenesis.

During sexual stimulation, the secretion from the bulbourethral glands provides an alkaline environment within the urethra prior to the passage of the spermatozoa. Peristaltic contractions move the spermatozoa from the epididymis into the urethra. The secretions of the seminal vesicles are mixed with sperm in the ejaculatory ducts, and prostatic fluids are added in the proximal part of the urethra. During ejaculation, rhythmic contractions of the **bulbospongiosus muscle** at the base of the penis propel the semen through the urethra and out of the body.

Spermatogenesis

The process by which spermatozoa are produced in the seminiferous tubules is called **spermatogenesis.** It begins at puberty and continues throughout life. Figure 37.2 illustrates a section of a seminiferous tubule and a diagram of the developmental stages of spermatozoa formation. Both mitosis and meiosis are involved.

All spermatozoa originate from **spermatogonia** that are located at the periphery of a seminiferous tubule. These cells contain 23 pairs of chromosomes, a total of 46 chromosomes, the same as all body cells. Since they contain both members of each chromosome pair, they are said to be diploid. The mitotic division of a spermatogonium forms one replacement spermatogonium and one **primary spermatocyte.**

Each diploid primary spermatocyte divides by **meiotic cell division,** a type of cell division that consists of one chromosome replication and two successive cell divisions. Meiotic division produces four haploid spermatids from the single diploid primary spermatocyte.

The first meiotic division forms two haploid **secondary spermatocytes.** Each secondary spermatocyte contains only 23 chromosomes, one member of each chromosome pair. These chromosomes are already replicated. Each chromosome consists of two chromatids joined at a centromere.

The second meiotic division occurs as each secondary spermatocyte divides to yield two haploid **spermatids,** each with 23 chromosomes. In this division, the chromatids separate to provide a haploid set of chromosomes for each spermatid. The spermatids subsequently mature to become spermatozoa.

Assignment

Label Figure 37.1.

The Female Organs

Refer to Figures 37.3, 37.4, and 37.5 as you study this section.

External Genitalia

Two folds of skin lie on each side of the **vaginal orifice,** the labia majora and labia minora. The larger exterior folds are the **labia majora,** which consist of rounded folds of adipose tissue covered by skin. They merge anteriorly with the **mons pubis,** an elevation of fatty tissue over the symphysis pubis. The outer surfaces of the labia majora possess hair, while their inner surfaces are smooth and moist. The smaller interior folds, the **labia minora,** are devoid of hair and merge posteriorly with the labia majora. Anteriorly, they join to form a hood-like covering, the **prepuce of the clitoris.** The **clitoris** is a small protuberance of erectile tissue located at the anterior junction of the labia minora. It is homologous to the penis in the male and is highly sensitive to sexual stimulation. Collectively, the external female reproductive organs are called the **vulva.**

The area between the labia minora is the **vestibule.** The vagina opens into the posterior portion of the vestibule. The **urethral orifice** is anterior to the vaginal opening and posterior to the clitoris. On either side of the vaginal orifice are the openings of the vestibular glands, which provide a mucous secretion for vaginal lubrication. The **hymen** is a thin mucous membrane that partially covers the vaginal opening. Its condition or absence is not a determiner of virginity.

Internal Organs

The internal female reproductive organs are the ovaries, uterine tubes, uterus, and vagina.

Figure 37.3 Female genitalia.

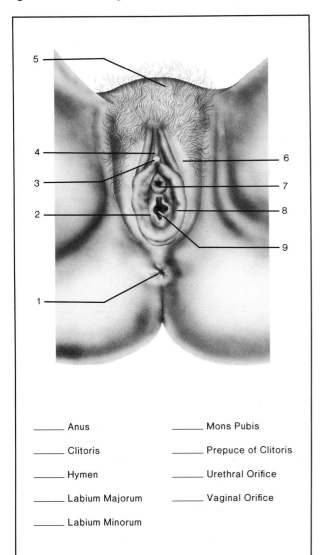

_____ Anus

_____ Clitoris

_____ Hymen

_____ Labium Majorum

_____ Labium Minorum

_____ Mons Pubis

_____ Prepuce of Clitoris

_____ Urethral Orifice

_____ Vaginal Orifice

Pelvic inflammatory disease is a collective term referring to any extensive bacterial infection of the female reproductive organs, but it is most commonly caused by gonorrhea. Inflammation of the uterine tubes may cause scarring, which prevents passage of the ovum and results in infertility.

The Ovaries The primary sex organs (gonads) of the female reproductive system are the **ovaries.** They produce the **ova,** or egg cells, and the female sex hormones. The ovaries are ovoid organs located against either side of the lateral walls of the pelvic cavity.

The Uterine Tubes The uterine tubes (oviducts or Fallopian tubes) extend from the ovaries to the uterus. Near the ovary, each tube is expanded to form a funnel-shaped **infundibulum** that bears a number of finger-like extensions, the **fimbriae.** The fimbriae and infundibula receive the oocytes released from the ovaries. The oocytes are carried toward the uterus by peristaltic contractions of the uterine tubes and the beating cilia of the ciliated columnar cells that line the tubes. Fertilization usually occurs within the upper third of a uterine tube. Observe the histology of a uterine tube in Figure HA-23.

The Uterus The uterus is a hollow, pear-shaped, thick-walled organ within which fetal development takes place. It is located medially over the vagina and is bent anteriorly over the urinary bladder. The upper two-thirds of the uterus is called the **body**; the lower third, the narrow **cervix,** projects into the upper portion of the vagina. The **cervical orifice** is the uterine opening into the vagina.

The uterine wall is composed of three layers. The outer **perimetrium** is a layer of the peritoneum. The thick middle layer, the **myometrium,** is composed of smooth muscle. The inner **edometrium** forms the mucosal lining and consists of two parts. The basal layer is attached to the myometrium. The functional layer is closest to the uterine cavity, is built up and shed during each menstrual cycle and is covered with columnar epithelium on its free surface. Note the histology of the myometrium and endometrium in Figure HA-23.

Figure 37.4 Posterior view of female reproductive organs.

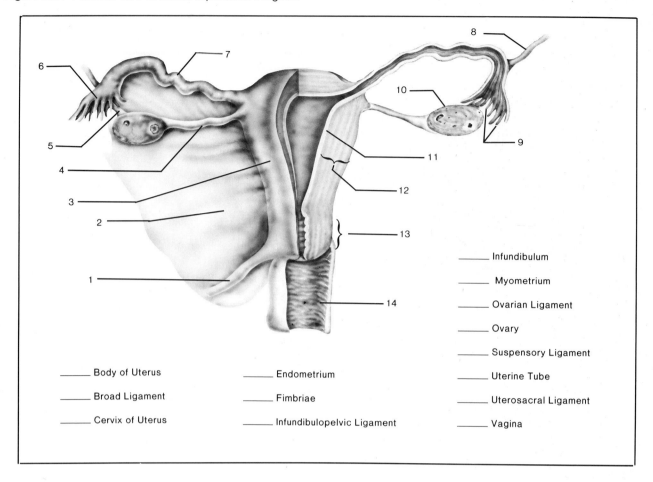

_____ Infundibulum

_____ Myometrium

_____ Ovarian Ligament

_____ Ovary

_____ Suspensory Ligament

_____ Uterine Tube

_____ Uterosacral Ligament

_____ Vagina

_____ Body of Uterus

_____ Broad Ligament

_____ Cervix of Uterus

_____ Endometrium

_____ Fimbriae

_____ Infundibulopelvic Ligament

Ligaments

The ovaries, uterine tubes, and uterus are held in place and supported by several ligaments. The **broad ligament** is a fold of the peritoneum that supports the uterus, uterine tubes, and ovaries. It extends from the lateral surfaces of the uterus to the lateral pelvic walls. Two **uterosacral ligaments** (label 1, Figure 37.4; label 4, Figure 37.5) extend from the cervix to the sacral wall of the pelvic cavity. A **round ligament** (lable 10, Figure 37.5) extends from each side of the uterus to the anterior body wall.

In addition to the broad ligament, each ovary is held in place by two ligaments. An **ovarian ligament** extends from the uterus to the medial surface of the ovary. The **suspensory ligament** (label 5, Figure 37.4) extends from the lateral surface of the ovary to the pelvic wall.

Vagina

The vagina is a fibromuscular canal extending from the uterus to the vestibule. It receives the penis during sexual intercourse and serves as the birth canal.

Figure 37.5 Midsagittal section of female reproductive organs.

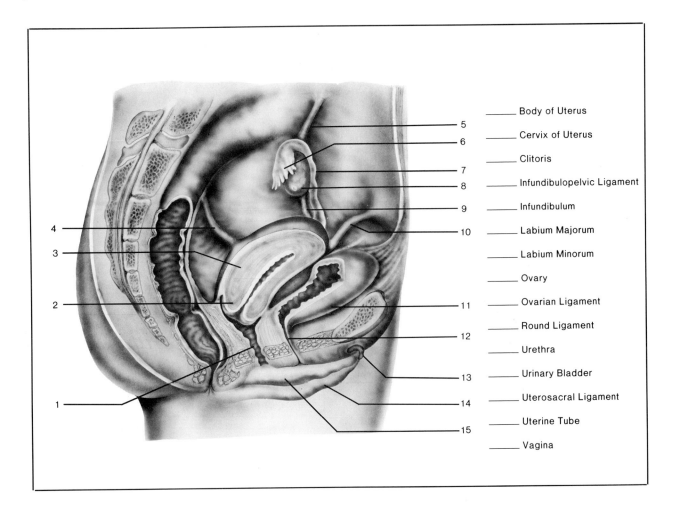

_____ Body of Uterus

_____ Cervix of Uterus

_____ Clitoris

_____ Infundibulopelvic Ligament

_____ Infundibulum

_____ Labium Majorum

_____ Labium Minorum

_____ Ovary

_____ Ovarian Ligament

_____ Round Ligament

_____ Urethra

_____ Urinary Bladder

_____ Uterosacral Ligament

_____ Uterine Tube

_____ Vagina

Oogenesis

The development of the egg cells is called **oogenesis,** and it involves both mitotic and meiotic divisions. See Figure 37.6. The **germinal epithelium** is formed early in the prenatal development of the ovaries. It occurs on the outer surface of the ovaries and consists of as many as 400,000 **oogonia** that form by mitotic division. By the end of the third month of development, mitotic division has ceased and some oogonia have migrated inward to become **primary oocytes.** Each is surrounded by a sphere of cells forming a **primary follicle.** Each primary oocyte is diploid since it contains 23 pairs of chromosomes, or a total of 46 chromosomes.

Starting at puberty, several primary oocytes are stimulated to develop further by FSH. Usually, only one of them will undergo meiotic division. The first meiotic division forms one large **secondary oocyte** and one small, non-functional **polar body.** Both of these cells are haploid since they contain only 23 chromosomes, one member of each chromosome pair. The chromosomes are already replicated and consist of two chromatids joined at the centromere. Each secondary oocyte is located within a

Figure 37.6 Oogenesis.

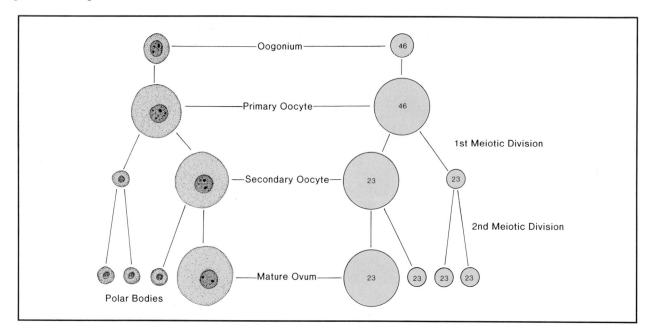

secondary or **developing follicle** that enlarges and fills with fluid to become a **mature,** or **Graafian, follicle.** Ovulation occurs with the rupture of the mature follicle. The extruded secondary oocyte and first polar body enter the infundibulum and are carried toward the uterus by the uterine tube.

If a secondary oocyte is penetrated by a spermatozoan (activation), it undergoes the second meiotic division, which forms the ovum and a polar body. The first polar body may also divide to form two polar bodies. Thus, the meiotic division of a diploid primary oocyte forms four haploid cells: one ovum and three polar bodies. Note that the bulk of the cytoplasm passes first to the secondary oocyte and then to the ovum. After the formation of the ovum, the egg nucleus and the sperm nucleus unite (fertilization) to form a diploid zygote. The polar bodies disintegrate.

Note that the orderly processes of spermatogenesis and oogenesis result in the zygote receiving one member of each chromosome pair from each parent.

Assignment

1. Label Figures 37.3, 37.4, and 37.5.
2. Complete the laboratory report.
3. Examine a prepared slide of human testis. Compare your observations with Figure 36.6 and 37.2, and locate the components labeled in Figure HA-22.
4. Examine a prepared slide of ovary. Compare your observations with Figure 36.7. Locate the germinal epithelium, primary follicles, and mature follicles.

HISTOLOGY ATLAS

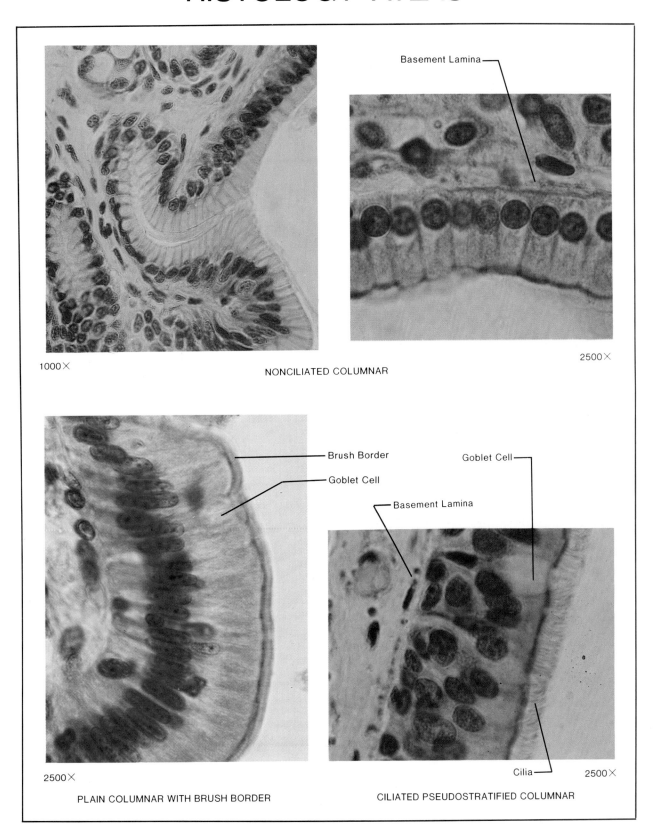

1000× NONCILIATED COLUMNAR 2500×

Basement Lamina

Brush Border
Goblet Cell

Goblet Cell
Basement Lamina

Cilia

2500× PLAIN COLUMNAR WITH BRUSH BORDER CILIATED PSEUDOSTRATIFIED COLUMNAR 2500×

Figure HA-1 Columnar epithelium.

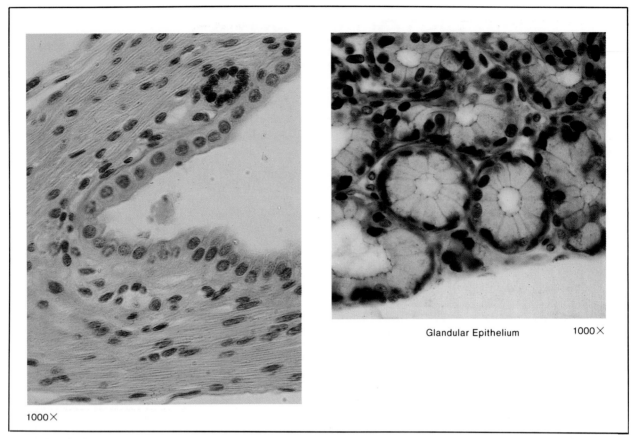

Glandular Epithelium   1000×

1000×

Figure HA-2 Cuboidal epithelium.

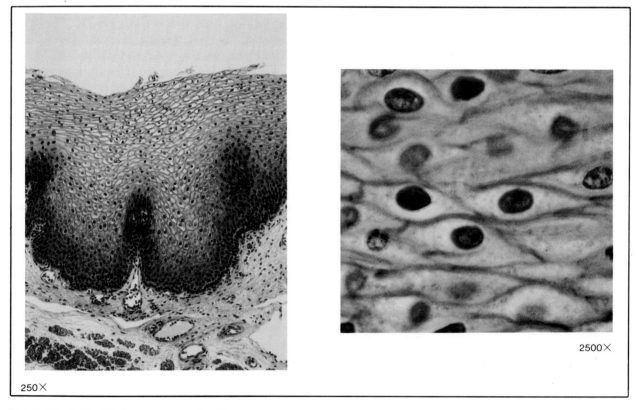

2500×

250×

Figure HA-3 Stratified squamous epithelium.

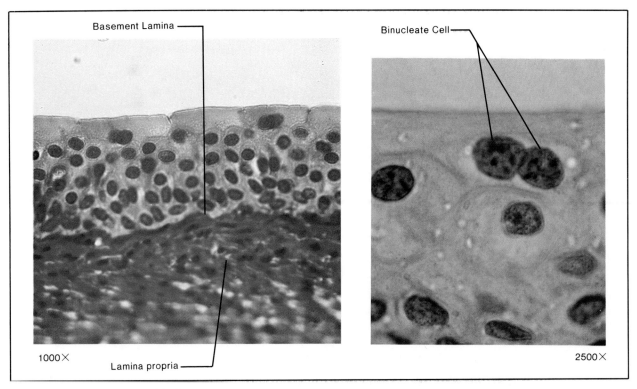

Figure HA-4 Transitional epithelium.

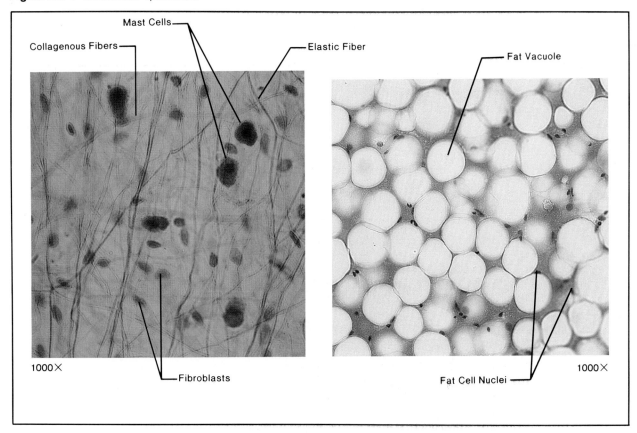

Figure HA-5 Areolar and adipose connective tissues.

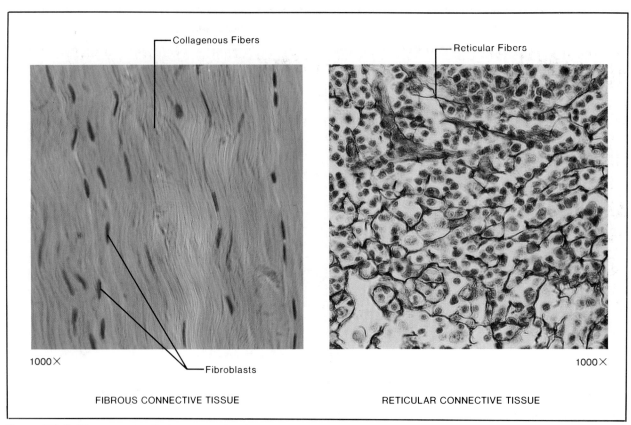

Figure HA-6 Fibrous and reticular connective tissues.

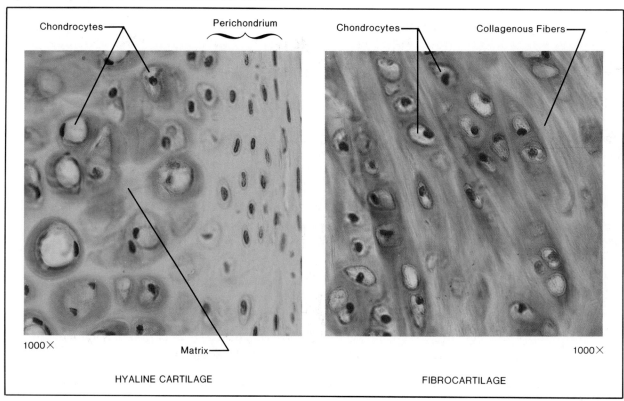

Figure HA-7 Hyaline cartilage and fibrocartilage.

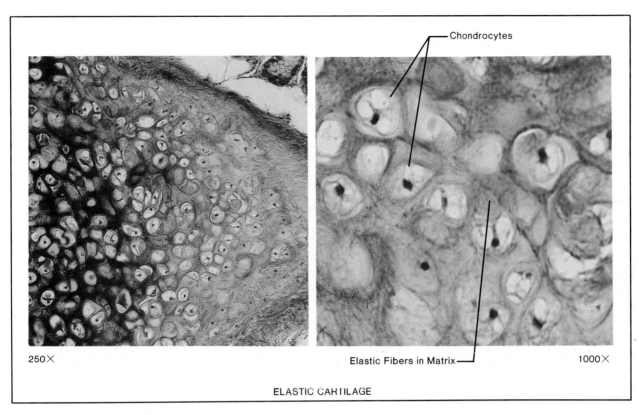

Figure HA-8 Elastic cartilage.

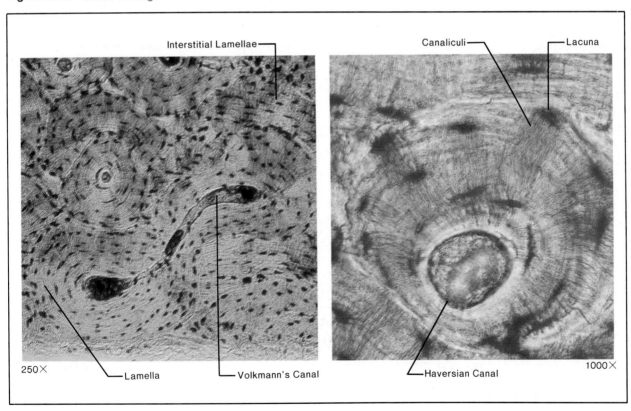

Figure HA-9 Compact bone.

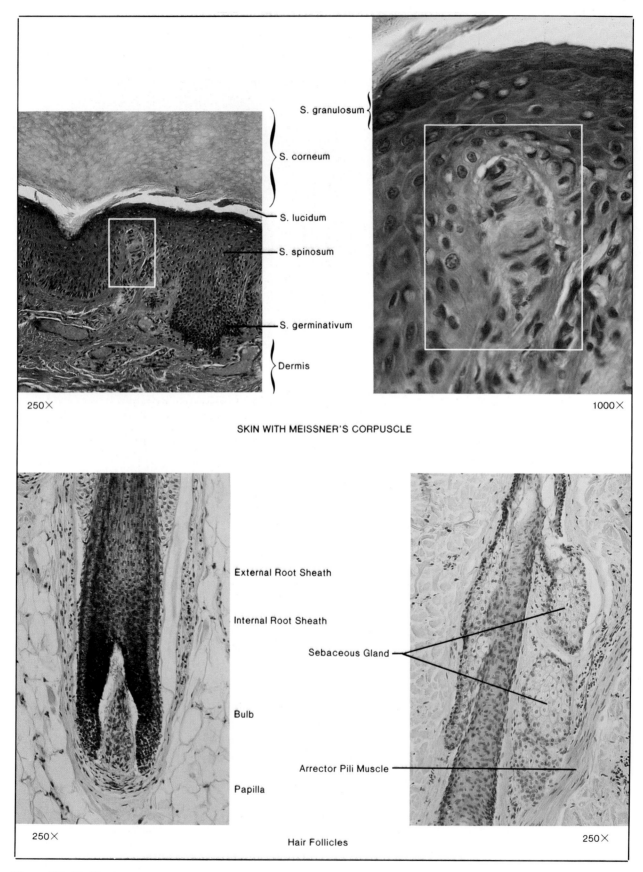

S. granulosum

S. corneum

S. lucidum

S. spinosum

S. germinativum

Dermis

250×

1000×

SKIN WITH MEISSNER'S CORPUSCLE

External Root Sheath

Internal Root Sheath

Sebaceous Gland

Bulb

Papilla

Arrector Pili Muscle

250×

Hair Follicles

250×

Figure HA-10 The integument.

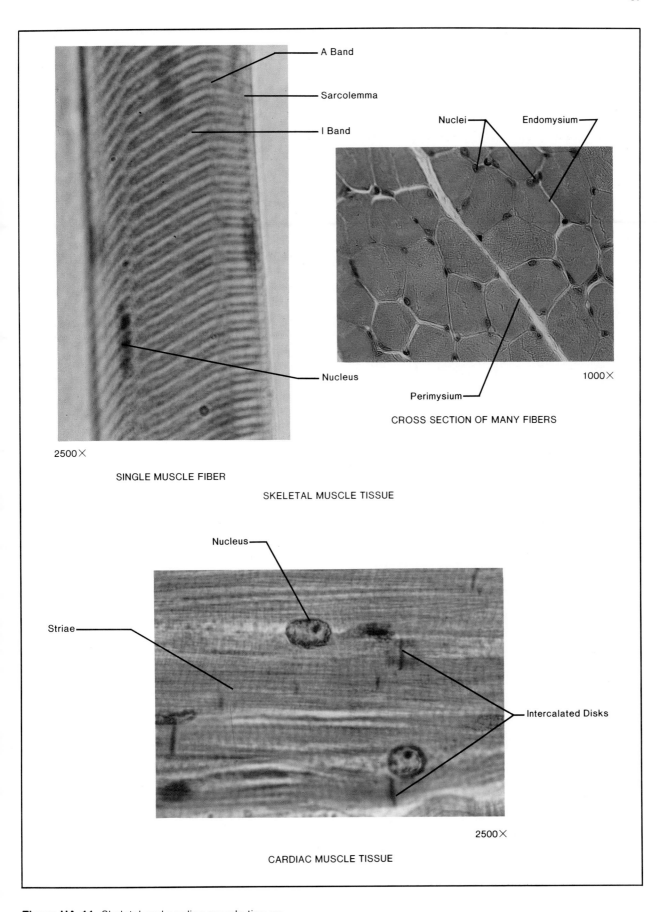

A Band

Sarcolemma

I Band

Nuclei

Endomysium

Nucleus

Perimysium

1000×

CROSS SECTION OF MANY FIBERS

2500×

SINGLE MUSCLE FIBER

SKELETAL MUSCLE TISSUE

Nucleus

Striae

Intercalated Disks

2500×

CARDIAC MUSCLE TISSUE

Figure HA-11 Skeletal and cardiac muscle tissues.

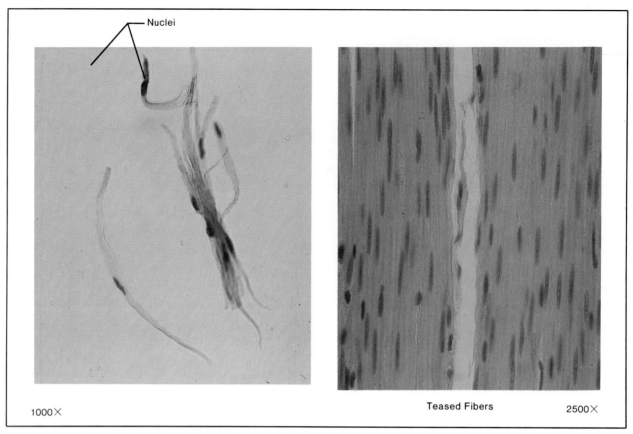

Figure HA-12 Smooth muscle tissue.

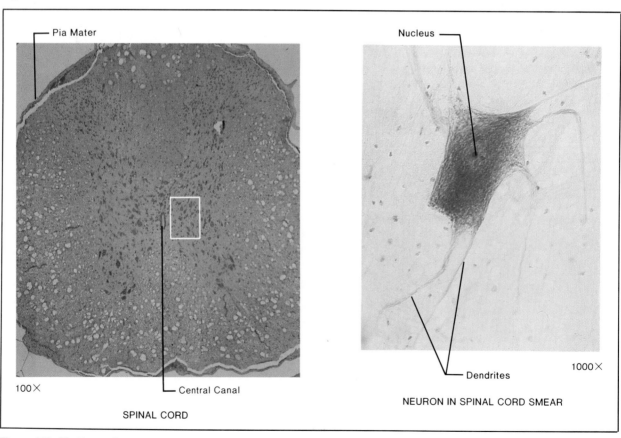

Figure HA-13 Nerve tissue.

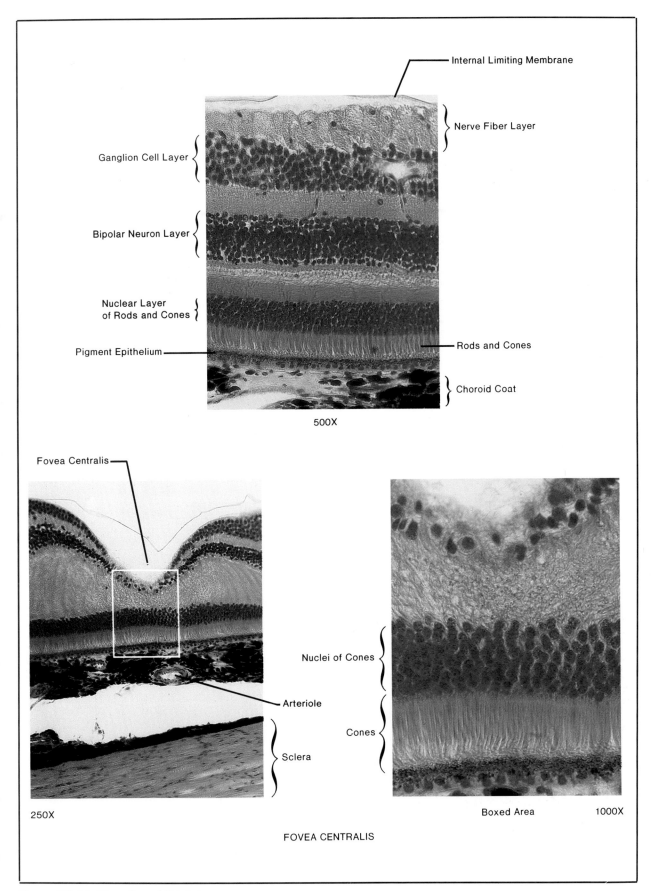

Internal Limiting Membrane

Nerve Fiber Layer

Ganglion Cell Layer

Bipolar Neuron Layer

Nuclear Layer
of Rods and Cones

Pigment Epithelium

Rods and Cones

Choroid Coat

500X

Fovea Centralis

Arteriole

Sclera

250X

Nuclei of Cones

Cones

Boxed Area   1000X

FOVEA CENTRALIS

Figure HA-14 The retina of the eye.

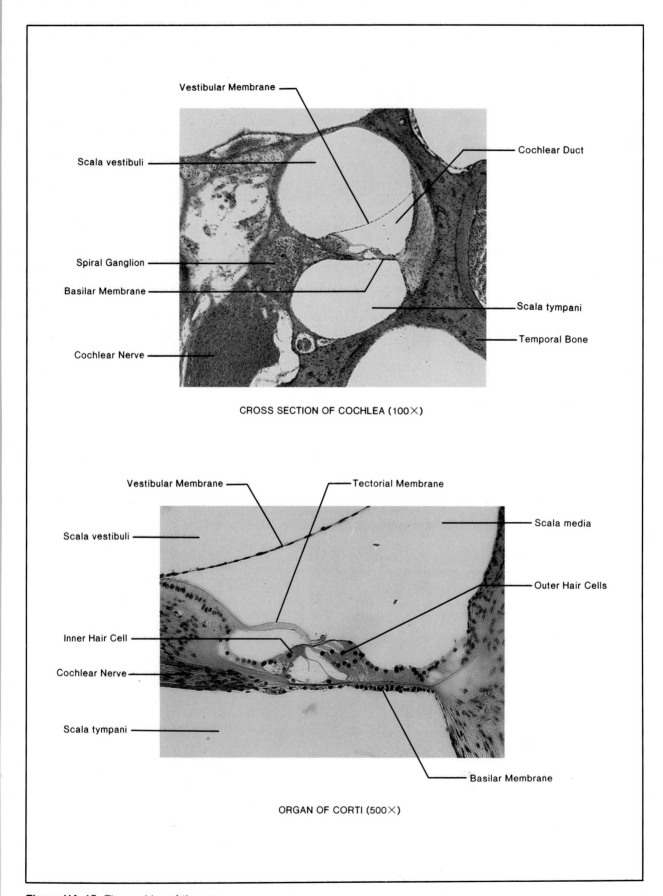

Vestibular Membrane

Scala vestibuli

Cochlear Duct

Spiral Ganglion

Basilar Membrane

Scala tympani

Temporal Bone

Cochlear Nerve

CROSS SECTION OF COCHLEA (100×)

Vestibular Membrane

Tectorial Membrane

Scala media

Scala vestibuli

Outer Hair Cells

Inner Hair Cell

Cochlear Nerve

Scala tympani

Basilar Membrane

ORGAN OF CORTI (500×)

Figure HA-15 The cochlea of the ear.

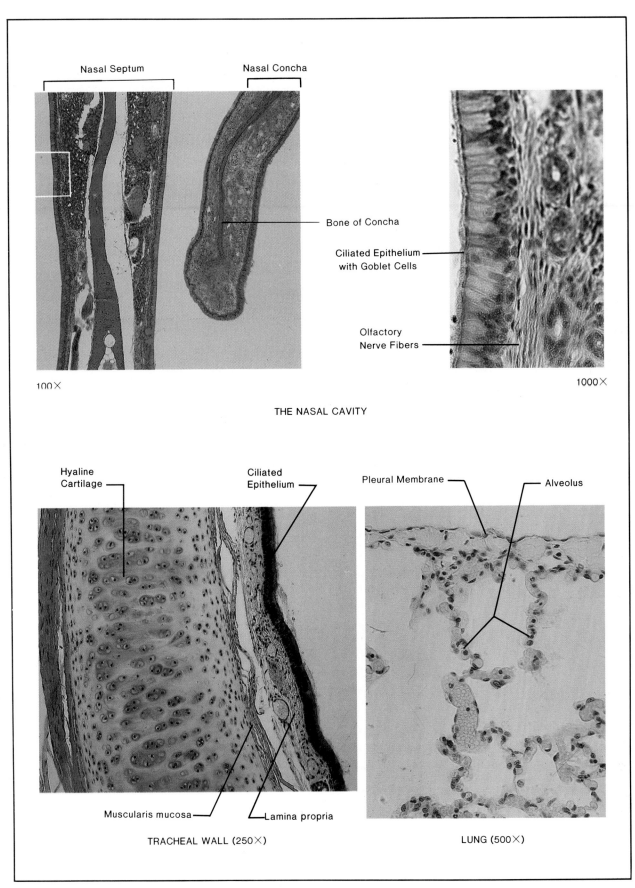

Nasal Septum

Nasal Concha

Bone of Concha

Ciliated Epithelium
with Goblet Cells

Olfactory
Nerve Fibers

100×

1000×

THE NASAL CAVITY

Hyaline
Cartilage

Ciliated
Epithelium

Pleural Membrane

Alveolus

Muscularis mucosa

Lamina propria

TRACHEAL WALL (250×)

LUNG (500×)

Figure HA-16 Respiratory structures.

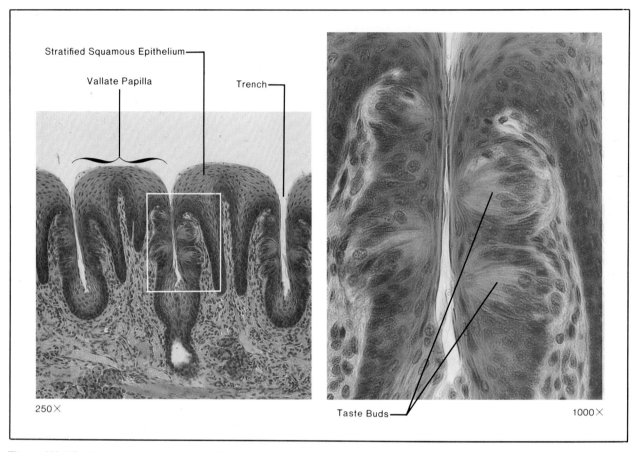

Figure HA-17 Taste buds on vallate papillae.

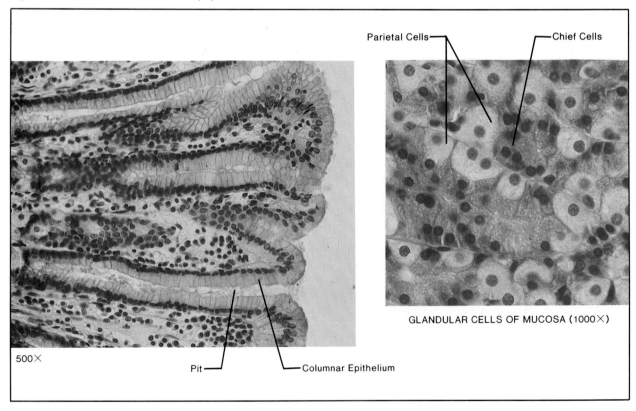

Figure HA-18 Inner surface of stomach fundus.

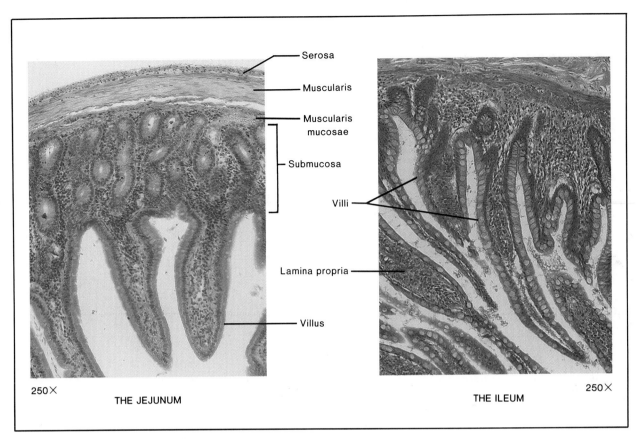

Serosa

Muscularis

Muscularis mucosae

Submucosa

Villi

Lamina propria

Villus

250×

THE JEJUNUM

THE ILEUM

250×

Figure HA-19 The jejunum and ileum.

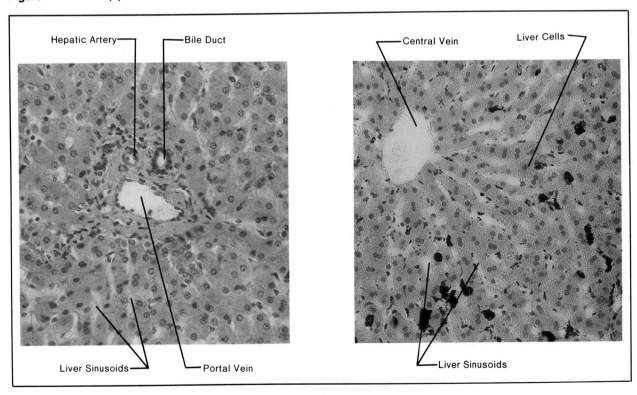

Hepatic Artery

Bile Duct

Central Vein

Liver Cells

Liver Sinusoids

Portal Vein

Liver Sinusoids

Figure HA-20 Liver tissue.

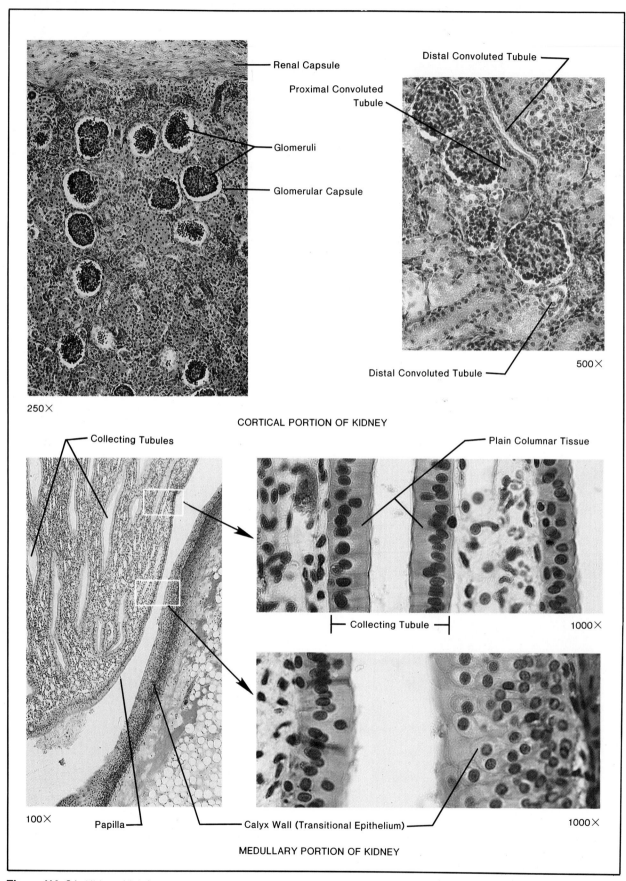

Renal Capsule

Glomeruli

Glomerular Capsule

250×

Distal Convoluted Tubule

Proximal Convoluted Tubule

Distal Convoluted Tubule

500×

CORTICAL PORTION OF KIDNEY

Collecting Tubules

100×

Papilla

Calyx Wall (Transitional Epithelium)

Plain Columnar Tissue

Collecting Tubule

1000×

1000×

MEDULLARY PORTION OF KIDNEY

Figure HA-21 Kidney histology.

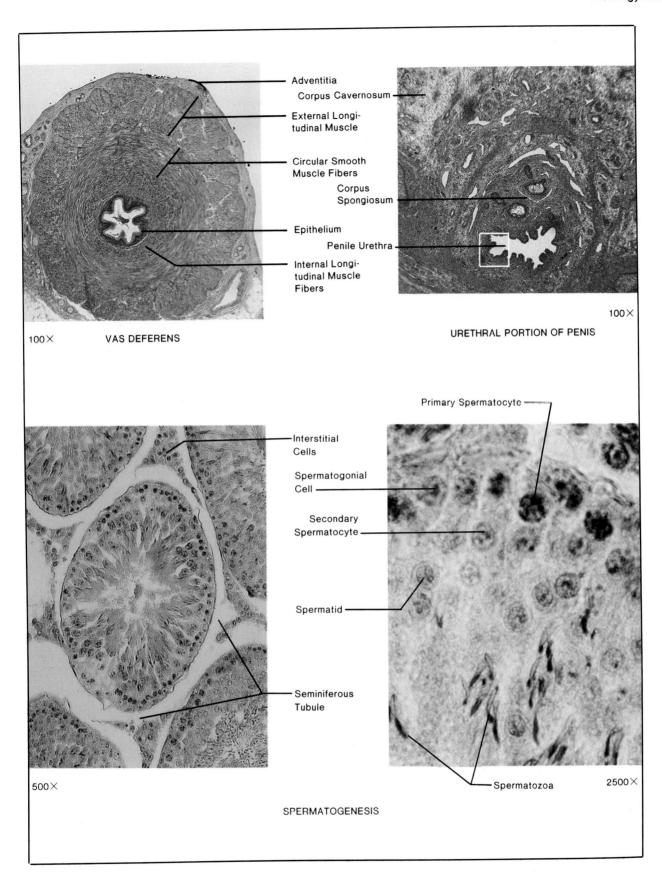

Adventitia
Corpus Cavernosum
External Longi-
tudinal Muscle
Circular Smooth
Muscle Fibers
Corpus
Spongiosum
Epithelium
Penile Urethra
Internal Longi-
tudinal Muscle
Fibers

100×
VAS DEFERENS

100×
URETHRAL PORTION OF PENIS

Primary Spermatocyte

Interstitial
Cells
Spermatogonial
Cell
Secondary
Spermatocyte

Spermatid

Seminiferous
Tubule

Spermatozoa

500×

2500×

SPERMATOGENESIS

Figure HA-22 Male reproductive tissue.

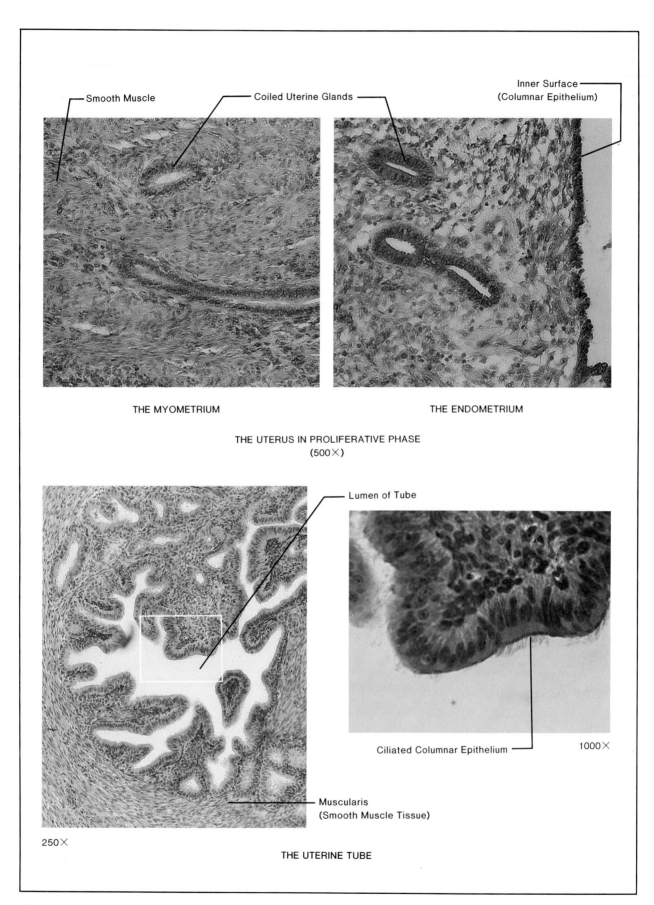

Smooth Muscle — Coiled Uterine Glands — Inner Surface (Columnar Epithelium)

THE MYOMETRIUM

THE ENDOMETRIUM

THE UTERUS IN PROLIFERATIVE PHASE
(500×)

Lumen of Tube

Ciliated Columnar Epithelium

1000×

250×

Muscularis
(Smooth Muscle Tissue)

THE UTERINE TUBE

Figure HA-23 The uterus and uterine tube.

Student _____

Lab Section _____

The Microscope

A. *Figures*

Write the figure labels in the spaces provided.

Figure 1.1

1. _____
2. _____
3. _____
4. _____
5. _____
6. _____
7. _____
8. _____
9. _____
10. _____
11. _____
12. _____
13. _____
14. _____
15. _____
16. _____
17. _____
18. _____
19. _____
20. _____

Figure 1.6

1. _____
2. _____
3. _____
4. _____

Figure 1.2

1. _____
2. _____
3. _____
4. _____
5. _____
6. _____
7. _____
8. _____
9. _____
10. _____
11. _____
12. _____
13. _____
14. _____
15. _____
16. _____

Figure 1.3

A. _____
B. _____
C. _____
D. _____
E. _____
F. _____

Figure 1.4

1. _____
2. _____
3. _____
4. _____
5. _____
6. _____
7. _____
8. _____
9. _____
10. _____
11. _____
12. _____
13. _____
14. _____
15. _____

Figure 1.5

1. _____
2. _____
3. _____
4. _____
5. _____
6. _____
7. _____
8. _____

B. Terminology

Write in the answer column the term from the list below that corresponds to the statements that follow.

Directional Terms		*Sections*	*Membrane*
anterior	lateral	frontal	mesentery
caudad	medial	midsagittal	pericardium, parietal
cephalad	posterior	sagittal	pericardium, visceral
distal	proximal	transverse	peritoneum, parietal
inferior	superior		peritoneum, visceral
			pleura, parietal
			pleura, pulmonary

Directional Terms

1. The hand is _____ to the elbow.
2. The knee is _____ to the ankle.
3. The mouth is _____ to the nose.
4. The ear is _____ to the nose.
5. The teeth are _____ to the cheeks.

Sections

6. Divides the body into equal left and right halves.
7. Divides the body into superior and inferior parts.
8. Divides the body into anterior and posterior parts.
9. Divides the body into unequal left and right parts.

Surfaces

Write the surface on which the following are located.

10. Kneecap
11. Ear
12. Umbilicus (navel)
13. Buttocks
14. Palm of Hand
15. Nose

Membranes

16. Double-layered membrane around the heart.
17. Serous membrane attached to surface of lungs.
18. Serous membrane attached to surface of stomach.
19. Double-layered membrane supporting intestines.
20. Serous membrane lining wall of abdominal cavity.

C. Body Regions and Surface Features

Write the term from the list below that corresponds to the following statements.

antebrachium	cubital	hypochondriac	pectoral
antecubital	epigastric	hypogastric	plantar
axilla	flank	iliac	popliteal
brachium	groin	lumbar	

1. The elbow.
2. The forearm.
3. Sole of the foot.
4. Upper chest region.
5. The posterior surface of the knee.
6. Depression at junction of thigh with abdomen.
7. The armpit.
8. Abdominal region overlying the urinary bladder.
9. Abdominal region superior to the lumbar region.
10. Abdominal region directly above the umbilical region.

Terminology

1. _____
2. _____
3. _____
4. _____
5. _____
6. _____
7. _____
8. _____
9. _____
10. _____
11. _____
12. _____
13. _____
14. _____
15. _____
16. _____
17. _____
18. _____
19. _____
20. _____

Regions and Surfaces

1. _____
2. _____
3. _____
4. _____
5. _____
6. _____
7. _____
8. _____
9. _____
10. _____

Laboratory Report 2

Cell Anatomy

A. Complete the chart

Organ Systems	Functions	Components
Cardiovascular		
Digestive		
Endocrine		
Integumentary		
Lymphatic		
Muscular		
Nervous		
Reproductive		
Respiratory		
Skeletal		
Urinary		

B. Terms

Define these terms and give examples of each.

Organ _____

_____

Organ system _____

_____

C. Matching

Write the name of the organ system(s) that corresponds to the words or phrases that follow.

Cardiovascular   Lymphatic   Respiratory
Digestive   Muscular   Skeletal
Endocrine   Nervous   Urinary
Integumentary   Reproductive

Functions

1. Protective covering of the body.
2. Transports materials throughout the body.
3. Enables gas exchange between blood and atmosphere.
4. Secretes hormones.
5. Digests food to form absorbable nutrients.
6. Provides supporting framework for the body.
7. Contractions enable body movements.
8. Removes wastes from the blood to form urine.
9. Enables rapid responses to environmental changes.
10. Collects tissue fluid from intercellular spaces.
11. Provides chemical control of body functions.
12. Perpetuates the species.
13. Produces blood cells.
14. Shields body from ultraviolet radiation.
15. Cleanses lymph and returns it to the bloodstream.

Components

1. Stomach and intestines.
2. Pituitary, adrenal, and thyroid glands.
3. Trachea, bronchi, and lungs.
4. Arteries, veins, and capillaries.
5. Cartilages and ligaments.
6. Kidneys, ureters, and urethra.
7. Spleen, tonsils, and adenoids.
8. Epidermis and dermis.
9. Uterus, oviducts, and vagina.
10. Liver and pancreas.
11. Nasal cavity, pharynx, and larynx.
12. Brain and spinal cord.
13. Sensory receptors and nerves.
14. Pancreas, ovaries, and testes.
15. Vasa deferentia, urethra, and penis.

Functions

1. _____
2. _____
3. _____
4. _____
5. _____
6. _____
7. _____
8. _____
9. _____
10. _____
11. _____
12. _____
13. _____
14. _____
15. _____

Components

1. _____
2. _____
3. _____
4. _____
5. _____
6. _____
7. _____
8. _____
9. _____
10. _____
11. _____
12. _____
13. _____
14. _____
15. _____

Introduction to Human Anatomy

A. Parts of the Microscope

Record in the answer column the microscope part described by the statements below.

arm   mechanical stage
base   objective
condenser   ocular
diaphragm   revolving nosepiece
focusing knob, coarse   stage
focusing knob, fine   stage aperture
lamp

1. Platform on which slides are placed for viewing.
2. Lens that is closest to the microscope slide.
3. Controls amount of light entering the condenser.
4. Focusing knob used with low power objective only.
5. Concentrates light on object being observed.
6. Light source.
7. Serves as a "handle" in carrying a microscope.
8. Lens you look through to view the image.
9. Part to which objectives are attached.
10. Opening in center of the stage.
11. Enables precise movement of the slide.
12. Bottom of the microscope.
13. Focusing knob used with all objectives.
14. Focusing knob with the larger diameter.
15. May be rotated to bring different objectives into viewing position.

Parts

1. _____
2. _____
3. _____
4. _____
5. _____
6. _____
7. _____
8. _____
9. _____
10. _____
11. _____
12. _____
13. _____
14. _____
15. _____

B. Magnification

Record the magnification of the ocular(s) and objectives on your microscope and calculate the total magnification obtained when using each objective.

Ocular	×	Objective	=	Total Magnification
_____ ×		_____ ×		_____ ×
_____ ×		_____ ×		_____ ×
_____ ×		_____ ×		_____ ×

C. True–False

Record your answer to these statements as true or false.

1. A microscope should be carried with two hands.
2. You should start your observations with the low power objective.
3. Blue light gives better resolution than white light.
4. Malfunctions should be reported to your instructor.
5. If necessary, you should disassemble the ocular to clean it.

True–False

1. _____
2. _____
3. _____
4. _____
5. _____

D. Observations

Diagram the appearance of the letter "e" as observed with the:

unaided eye   low power objective   high-dry objective

                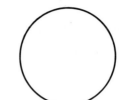

E. Completion

Write your responses to the following statements in the answer column.

1. Tissue used to clean the lenses.
2. Objective with the greatest working distance.
3. Objective with the least working distance.
4. Usually, the condenser should be kept at its _____ position.
5. What effect (increase, no change, or decrease) does an increase in magnification of an objective have on:
   a. diameter of field?
   b. working distance?
   c. light intensity?
6. The contrast of the image is _____ (increased, unchanged, or decreased) when light intensity is reduced.
7. Controls light intensity in microscopes with a constant light source.
8. Circle of light seen while looking into the ocular.
9. When the slide is moved to the left, the image moves to the _____ .
10. Maximum resolution with a light microscope.
11. Three liquids that may be used to clean lenses.
12. When an object is in focus with one objective and is still in focus when another objective is rotated into position, the microscope is said to be _____ .
13. Only focusing knob used with high-dry and oil immersion objectives.

Completion

1. _____
2. _____
3. _____
4. _____
5a. _____
5b. _____
5c. _____
6. _____
7. _____
8. _____
9. _____
10. _____
11. _____
12. _____
13. _____

Laboratory Report 4

Student _____

Lab Section _____

Body Organization

A. Figure 4.1
List the labels for Figure 4.1.

Figure 4.1

1. _____
2. _____
3. _____
4. _____
5. _____

6. _____
7. _____
8. _____
9. _____

10. _____
11. _____
12. _____
13. _____

B. Cell Structure and Organelle Function
Select the structure(s) from the following list that are described by the statements below.

AGER
centrioles
centrosome
chromatin granules
cilia
cytoplasm
flagellum
Golgi apparatus
lysosomes

microvilli
mitochondria
nuclear membrane
nucleolus
nucleus
plasma membrane
RER
ribosomes
secretory vesicles

1. Short, hair-like projections on a cell surface.
2. Short, rod-like bundles of microtubules oriented at right angles to each other.
3. Increase absorptive surface area of certain cells.
4. Provides motility of sperm cells.
5. A stack of membranous sacs continuous with the ER.
6. Packages materials for extracellular transport.
7. Sites of ATP production via cellular respiration.
8. Structures in nucleus composed of DNA and protein.
9. A double membrane surrounding the nucleus.
10. A spherical structure in the nucleus composed of RNA and protein.
11. ER studded with ribosomes.
12. Organelle with an inner membrane having many folds.
13. Controls passage of materials into and out of cells.
14. ER without ribosomes.
15. Sacs of digestive enzymes.
16. Sites of protein synthesis.
17. Control center of the cell.
18. Assembles RNA and protein that will form ribosomes.
19. Sacs containing materials destined for export.
20. Contains the chromosomes of the cell.
21. Parts of chromosomes visible in non-dividing cells.
22. Tiny cytoplasmic organelles composed of RNA and protein.

Structure & Function

1. _____
2. _____
3. _____
4. _____
5. _____
6. _____
7. _____
8. _____
9. _____
10. _____
11. _____
12. _____
13. _____
14. _____
15. _____
16. _____
17. _____
18. _____
19. _____
20. _____
21. _____
22. _____

C. Completion

The plasma membrane consists of a bilayer of (1) molecules in which (2) molecules are embedded. Its (3) permeability regulates the passage of materials into and out of the cell. In a similar manner, the nuclear membrane regulates the passage of materials between the nucleus and the (4).

The nucleus is the control center of the cell because the genetic information encoded in the (5) molecules of the (6) controls cellular functions. The nucleolus assembles (7) and (8), which will move into the cytoplasm to become ribosomes.

The (9) provides passageways for the movement of materials within the cytoplasm. Proteins synthesized on (10) of the RER are carried to the (11), which packages them in (12) for export from the cell.

Energy for cellular work is made available by cellular respiration occurring in the (13), which contain their own RNA and (14), and are able to (15) themselves.

Completion

1. _____
2. _____
3. _____
4. _____
5. _____
6. _____
7. _____
8. _____
9. _____
10. _____
11. _____
12. _____
13. _____
14. _____
15. _____

D. Microscopic Study

1. Make a series of sketches to show the changing shape of *Amoeba* during amoeboid movement. Use arrows to indicate the direction of movement of the *Amoeba* and of the flowing cytoplasm.

2. Draw a few cheek epithelial cells and label the nucleus and cytoplasm.

3. Draw a few human sperm to show the relative size of the head, which consists primarily of the nucleus, and the flagellum.

Mitotic Cell Division

A. *Matching*

Write the name of the correct response in the answer column.

anaphase   metaphase   telophase
interphase   prophase

1. Chromosomes line up at equator of spindle.
2. Chromosomes become visible as rod-like structures.
3. Nuclear membrane is intact.
4. Chromatin granules are present.
5. Each pair of centrioles moves to opposite ends of cell.
6. Cleavage furrow forms.
7. Centrioles replicate.
8. Nuclear membrane disappears.
9. Cytokinesis is completed.
10. Chromosomes are replicated.
11. Sister chromatids separate and move to opposite poles of the spindle.
12. New nuclear membranes are formed.
13. Daughter chromosomes uncoil and become less distinct.
14. Spindle is formed of microtubules.
15. Daughter cells are formed.

Matching

1. _____
2. _____
3. _____
4. _____
5. _____
6. _____
7. _____
8. _____
9. _____
10. _____
11. _____
12. _____
13. _____
14. _____
15. _____

B. *Completion*

1. Briefly describe the process of mitotic cell division. _____

_____

_____

_____

_____

_____

_____

2. There are 23 pairs of chromosomes in human body cells. How many chromosomes are in

a parent cell? _____ each daughter cell? _____

3. How many daughter chromosomes are in a human cell at anaphase? _____

4. Compare the genetic composition of parent cell and daughter cells. _____

_____

Why is this relationship important? _____

_____

C. Microscopic Study

Diagram the appearance of the whitefish cells in each phase of the cell cycle and label the identifiable structures.

Interphase

Prophase

Metaphase

Anaphase

Telophase

Daughter Cells

Laboratory Report **6**

Diffusion and Osmosis

A. Concepts

1. Define diffusion. _____

_____

_____

2. Define osmosis. _____

_____

_____

3. Describe the relationship between solute concentration and the osmotic pressure of an aqueous solution. _____

_____

_____

4. Consider two sodium chloride (NaCl) solutions separated by a semipermeable membrane. Solution A is a 5% solution; solution B is a 10% solution. Both water and NaCl can readily pass through the membrane. Indicate:

the direction of net water movement. _____

the direction of net NaCl movement. _____

when this movement will stop. _____

the hypotonic solution. _____

5. If a 1.5% NaCl solution is separated by a semipermeable membrane from another NaCl solution that is isotonic, indicate the:

salt concentration of the isotonic solution. _____

direction of net water movement. _____

B. Brownian Movement

1. Describe the movement of the ink particles observed at 1000×. _____

_____

2. Explain the cause of Brownian movement. _____

_____

C. Diffusion and Temperature

1. For each beaker, record the temperature of the water and the time required for molecules of potassium permanganate to diffuse throughout the water.

Beaker A: Temp. _____   Diffusion Time _____

Beaker B: Temp. _____   Diffusion Time _____

2. Describe the relationship between the rate of diffusion and temperature. _____

_____

D. Diffusion and Molecular Weight

1. Record the diameter of the colored area around each crystal.

   Potassium permanganate: _____ mm   Methylene blue: _____ mm

2. Describe the relationship between the rate of diffusion and molecular weight. _____

_____

E. Osmosis

1. Record the change (mm) in the height of the water-sucrose columns during a 30-minute interval.

   A: _____ mm   B: _____ mm

2. Which osmometer has the higher concentration of sucrose? _____

   Explain your response. _____

   _____

F. Cell Membrane Integrity

1. Record the degree of transparency and the effect on the red blood cells (lysis, crenation, normal) of the solutions.

Solution	Transparency	Cell Shape
1. 2.0% glucose		
2. 5.0% glucose		
3. 0.3% NaCl		
4. 0.9% NaCl		
5. 2.0% NaCl		

2. Indicate the number of the solution that is:

   isotonic _____   hypertonic _____   hypotonic _____

Laboratory Report 7

Student _____

Lab Section _____

Epithelial and Connective Tissues

A. Epithelial Tissues

Select from the list of tissues those that are described by the following statements.

columnar, simple ciliated
columnar, simple nonciliated
columnar, stratified
columnar, pseudostratified nonciliated
columnar, pseudostratified ciliated

cuboidal
squamous, simple
squamous, stratified
transitional

1. Epidermis of the skin.
2. Lining of the intestine.
3. Lining of the blood vessels.
4. Lining of the urinary bladder.
5. Lining of anal canal.
6. Lining of oviducts.
7. Secretory epithelium in the pancreas.
8. Lining of large ducts in parotid gland.
9. Lining of nasal cavity and trachea.
10. Single layer of elongate cells without cilia.
11. Single layer of cubelike cells.
12. Multilayered; outer layer of thin, flat cells.
13. Ciliated cells appear multilayered, but are not.
14. Multilayered; cells flatten when stretched.
15. Multilayered; outer layer of elongate cells.

B. Connective Tissues

Select from the list of tissues those that are described by the statements that follow.

adipose tissue
areolar tissue
bone
dense fibrous tissue

elastic cartilage
fibrocartilage
hyaline cartilage
reticular tissue

1. Superficial fascia.
2. Framework of lymph nodes.
3. Tendons and ligaments.
4. Supporting rings in trachea.
5. Supports pinna of external ear.
6. Occurs on ends of long bones.
7. Intervertebral discs.
8. Fat storage.
9. Most flexible supporting tissue.
10. Solid matrix of calcium salts.
11. Most rigid supporting tissues.
12. Cartilage with many collagenous fibers.
13. White, glassy appearance.
14. Serves as insulation material.
15. Preforms most bones in fetal skeleton.

Epithelial Tissues

1. _____
2. _____
3. _____
4. _____
5. _____
6. _____
7. _____
8. _____
9. _____
10. _____
11. _____
12. _____
13. _____
14. _____
15. _____

Connective Tissues

1. _____
2. _____
3. _____
4. _____
5. _____
6. _____
7. _____
8. _____
9. _____
10. _____
11. _____
12. _____
13. _____
14. _____
15. _____

C. Terminology

Write the term defined below in the answer column.

1. Protein in white fibers.
2. Fiber-producing cell.
3. Mucus-secreting cell.
4. Lines medullary cavity.
5. Intercellular substance.
6. Produces bone matrix.
7. Cartilage cell.
8. Plates in spongy bone.
9. Rings of compact bone.
10. Space around bone cell.

D. Figure 7.8

Record the labels for Figure 7.8 in the answer column.

E. Microscopic Study

Use the space below to diagram a few cells of the tissues studied. Add labels and make notations that will help you recognize these tissues in a lab practicum. Use extra sheets as necessary.

Terminology

1. _____
2. _____
3. _____
4. _____
5. _____
6. _____
7. _____
8. _____
9. _____
10. _____

Figure 7.8

1. _____
2. _____
3. _____
4. _____
5. _____
6. _____
7. _____
8. _____
9. _____
10. _____
11. _____
12. _____

Student _____

Lab Section _____

The Integument

A. Matching

Select the structure from the list below that is described by the following statements and record it in the answer column.

apocrine sweat gland
arrector pili muscle
dermal papilla
dermis
eccrine sweat gland
epidermis
hypodermis

Meissner's corpuscle
Pacinian corpuscle
sebaceous gland
stratum corneum
stratum germinativum
stratum granulosum
stratum spinosum

1. Layer of epidermis containing melanocytes.
2. Contains blood vessels that nourish hair follicle.
3. Touch receptor in dermal papilla.
4. Sweat gland emptying secretion into hair follicle.
5. Inner layer of skin formed of connective tissue.
6. Outermost layer of epidermis.
7. Epidermal layer just interior to stratum granulosum.
8. Most abundant type of sweat gland.
9. Pressure receptor located deep in dermis.
10. Secretes oily lubricant into hair follicle.
11. Superficial fascia.
12. Activation produces goosebumps.
13. Formed of loose connective and adipose tissues.
14. Four layers of stratified squamous epithelium.
15. Has numerous collagen fibers that give strength to skin.
16. Forms watery secretion containing waste materials.
17. Epidermal layer in which cell division occurs.
18. Secretion is responsible for body odor.
19. Epidermal layer with granule precursors of keratin.
20. Responsible for the pattern of fingerprints.
21. Sweat gland activated primarily by thermal stimuli.
22. Epidermal layer producing cells that form a hair.
23. Contains insulating fatty tissue.
24. Epidermal layer preventing evaporative water loss.
25. Skin layer containing nerves and blood vessels.
26. Epidermal layer of flat, dead, keratinized cells.
27. Consists of papillary and reticular layers.
28. Larger, but less numerous, sweat glands.
29. Secretes sebum.
30. Contains coiled portion of eccrine sweat glands.

Matching

1. _____
2. _____
3. _____
4. _____
5. _____
6. _____
7. _____
8. _____
9. _____
10. _____
11. _____
12. _____
13. _____
14. _____
15. _____
16. _____
17. _____
18. _____
19. _____
20. _____
21. _____
22. _____
23. _____
24. _____
25. _____
26. _____
27. _____
28. _____
29. _____
30. _____

B. Figure 8.1

Record the labels for Figure 8.1 in the answer column.

C. Completion

1. Describe the function of melanin. _____

_____

2. Explain why a "tan" is temporary. _____

_____

_____

_____

3. Distinguish between these skin cancers.

Carcinoma: _____

_____

Melanoma: _____

_____

D. Microscopic Study

Use the space below to make drawings and notations that will help you understand skin structure, and recognize skin and its components on a lab practicum.

Figure 8.1

1. _____

2. _____

3. _____

4. _____

5. _____

6. _____

7. _____

8. _____

9. _____

10. _____

11. _____

12. _____

13. _____

14. _____

15. _____

16. _____

17. _____

18. _____

19. _____

20. _____

21. _____

22. _____

The Skeletal Plan

A. Figures
List the labels for Figures 9.1, 9.2, and 9.3.

Figure 9.1

1. _____
2. _____
3. _____
4. _____
5. _____
6. _____
7. _____
8. _____
9. _____
10. _____

Figure 9.2

1. _____
2. _____
3. _____
4. _____
5. _____
6. _____
7. _____
8. _____

Figure 9.3

1. _____
2. _____
3. _____
4. _____
5. _____
6. _____
7. _____
8. _____
9. _____
10. _____
11. _____
12. _____
13. _____
14. _____
15. _____
16. _____
17. _____
18. _____

Long Bone

1. _____
2. _____
3. _____
4. _____
5. _____
6. _____
7. _____
8. _____
9. _____
10. _____
11. _____
12. _____
13. _____

B. Long Bone Structure
Select the response that matches the statements that follow and write it in the answer column.

articular cartilage   epiphyseal line
cancellous bone   epiphysis
compact bone   medullary cavity
diaphysis   periosteum
endosteum   red marrow
epiphyseal disk   yellow marrow

1. Shaft portion of a long bone.
2. Hollow chamber in the bone shaft.
3. Type of marrow in the medullary cavity.
4. Enlarged end of a long bone.
5. Lining of the medullary cavity.
6. Type of bone tissue forming most of the diaphysis.
7. Fibrous membrane covering the bone.
8. Site of linear growth.
9. Protective surface on ends of a long bone.
10. Type of bone tissue forming interior of epiphyses.
11. Reduces friction in joints formed by long bones.
12. Type of marrow in cancellous bone of the femur.
13. Formed by fusion of epiphysis and diaphysis.

C. Bone Fractures

Select the type of fracture that matches the statements that follow and write it in the answer column.

comminuted   oblique
compacted   segmental
fissured   spiral
greenstick   transverse

1. Results from twisting of the bone.
2. Complete fracture at 90° to bone axis.
3. Break on one side only due to bending of the bone.
4. One part of a bone is forced into another part of the same bone.
5. A linear, incomplete splitting of the bone.
6. Complete fracture *not* at 90° to bone axis.
7. Bone broken into several pieces.
8. A single piece is broken out of the bone.

Write the type of fracture in the answer column that is described by the statements below.

9. Fractured bone breaks through the skin.
10. Fractured bone is not broken clear through.
11. Fracture caused by excessive vertical force.
12. Fractured bone does not break through the skin.

D. Parts of the Skeleton

Select the response that matches the statements that follow and write it in the answer column.

clavicle   os coxa   sternum
femur   patella   symphysis pubis
fibula   radius   tibia
humerus   scapula   ulna
hyoid   skull   vertebrae

1. Shoulder blade.
2. Collar bone.
3. Breastbone.
4. Shinbone.
5. Kneecap.
6. Upper arm bone.
7. Bones of spinal column.
8. Thigh bone.
9. Lateral bone of forearm.
10. Joint between os coxae.
11. Horseshoe-shaped bone under lower jaw.
12. One-half of pelvic girdle.
13. Medial bone of forearm.
14. Thin bone lateral to tibia.
15. Bony framework of the head.
16. Two bones forming the shoulder girdle.

Fractures

1. _____
2. _____
3. _____
4. _____
5. _____
6. _____
7. _____
8. _____
9. _____
10. _____
11. _____
12. _____

Skeleton

1. _____
2. _____
3. _____
4. _____
5. _____
6. _____
7. _____
8. _____
9. _____
10. _____
11. _____
12. _____
13. _____
14. _____
15. _____
16a. _____
16b. _____

Laboratory Report 10

Student _____

Lab Section _____

The Skull

A. *Figures*

List the labels for the figures in the answer columns.

Figure 10.1

1. _____
2. _____
3. _____
4. _____
5. _____
6. _____
7. _____
8. _____
9. _____
10. _____
11. _____
12. _____
13. _____
14. _____
15. _____
16. _____
17. _____
18. _____
19. _____
20. _____

Figure 10.6

1. _____
2. _____
3. _____
4. _____

Figure 10.7

1. _____
2. _____
3. _____
4. _____
5. _____
6. _____
7. _____
8. _____

Figure 10.2

1. _____
2. _____
3. _____
4. _____
5. _____
6. _____
7. _____
8. _____
9. _____
10. _____
11. _____
12. _____
13. _____
14. _____
15. _____
16. _____

Figure 10.4

1. _____
2. _____
3. _____
4. _____
5. _____
6. _____
7. _____
8. _____
9. _____
10. _____
11. _____
12. _____
13. _____
14. _____
15. _____
16. _____
17. _____

Figure 10.3

1. _____
2. _____
3. _____
4. _____
5. _____
6. _____
7. _____
8. _____
9. _____
10. _____
11. _____
12. _____
13. _____
14. _____
15. _____
16. _____

Figure 10.5

1. _____
2. _____
3. _____
4. _____
5. _____
6. _____
7. _____
8. _____
9. _____

Figure 10.8

1. _____
2. _____
3. _____
4. _____
5. _____
6. _____
7. _____

B. Bones

Select the bones that are described or that have the structures noted in the statements below.

ethmoid   nasal   sphenoid
frontal   nasal conchae   temporal
lacrimal   occipital   vomer
mandible   palatine   zygomatic
maxilla   parietal

1. Upper bone of nasal septum.
2. Foramen magnum.
3. Lower jaw.
4. Cheekbone.
5. Mental foramen.
6. Scroll-like bones in nasal cavity.
7. Coronoid process.
8. Hypoglossal canal.
9. Internal acoustic meatus.
10. Carotid canal.
11. Mandibular fossa.
12. Mastoid process.
13. Ramus.
14. Sella turcica.
15. Foramen ovale.
16. Forehead bone.
17. Supraorbital foramen.
18. Upper jaw.
19. Optic canal.
20. Cribriform plates.

C. Sutures

Select the sutures described below.

coronal   median palatine   squamosal
lambdoidal   sagittal

1. Joins parietal bones.
2. Joins frontal and parietal bones.
3. Joins palatine processes of maxillae.
4. Joins parietal and occipital bones.
5. Joins parietal and temporal bones.

D. Completion

Write your response in the answer column.

1. Two bones forming the hard palate.
2. Two bones forming the zygomatic arch.
3. Allow compression of skull during birth.
4. Four bones that contain sinuses.
5. Point of attachment of skull with first vertebra.
6. Allow sound waves to enter skull.
7. Form external bridge of nose.
8. Contain groove for tear duct.

Bones

1. _____
2. _____
3. _____
4. _____
5. _____
6. _____
7. _____
8. _____
9. _____
10. _____
11. _____
12. _____
13. _____
14. _____
15. _____
16. _____
17. _____
18. _____
19. _____
20. _____

Sutures

1. _____
2. _____
3. _____
4. _____
5. _____

Completion

1. _____
   _____
2. _____
   _____
3. _____
4. _____
   _____
   _____
5. _____
6. _____
7. _____
8. _____

Laboratory Report **11**

Student ______________________

Lab Section ______________

The Vertebral Column and Thorax

A. *Figures*

List the labels for Figures 11.1 and 11.2.

Figure 11.1

1. ______________
2. ______________
3. ______________
4. ______________
5. ______________
6. ______________
7. ______________
8. ______________
9. ______________
10. ______________
11. ______________
12. ______________
13. ______________
14. ______________
15. ______________
16. ______________
17. ______________
18. ______________
19. ______________
20. ______________
21. ______________
22. ______________
23. ______________
24. ______________
25. ______________
26. ______________

Figure 11.2

1. ______________
2. ______________
3. ______________
4. ______________
5. ______________
6. ______________
7. ______________
8. ______________

B. Parts of the Vertebral Column

Select the response from the list at the right that matches the statement.

1. Facet on transverse processes.
2. Transverse foramina.
3. First cervical vertebra.
4. Floating rib.
5. Inferior tip of breastbone.
6. Mid-part of breastbone.
7. Second cervical vertebra.
8. Superior part of breastbone.
9. Vertebrae of the neck.
10. True rib.
11. False rib with costal cartilage.
12. Breastbone.
13. Tailbone.
14. Possesses odontoid process.
15. Has largest vertebral foramen.

atlas
axis
cervical vertebrae
coccyx
gladiolus
lumbar vertebrae
manubrium
sacrum
sternum
thoracic vertebrae
vertebral rib
vertebrochondral rib
vertebrosternal rib
xiphoid

Parts

1. _____
2. _____
3. _____
4. _____
5. _____
6. _____
7. _____
8. _____
9. _____
10. _____
11. _____
12. _____
13. _____
14. _____
15. _____

C. Numbers

Indicate the number of components of the following.

1. Cervical vertebrae.
2. Fused vertebrae composing coccyx.
3. Fused vertebrae composing sacrum.
4. Lumbar vertebrae.
5. Pairs of false ribs.
6. Pairs of floating ribs.
7. Pairs of true ribs.
8. Pairs of vertebrochondral ribs.
9. Spinal curvatures.
10. Thoracic vertebrae.

Numbers

1. _____
2. _____
3. _____
4. _____
5. _____
6. _____
7. _____
8. _____
9. _____
10. _____

D. Completion

Provide the terms or phrases that match the statements.

1. Cushion between vertebral bodies.
2. Articulates with inferior articulating facets.
3. Bony framework around vertebral foramen.
4. Part of vertebra bearing most load.
5. Vertebral type with largest bodies.
6. Forms posterior wall of pelvis.
7. Vertebrae with ribs.
8. Attach ribs to sternum.
9. Articulate with transverse facets.
10. Openings through which spinal nerves exit.

Completion

1. _____
2. _____
3. _____
4. _____
5. _____
6. _____
7. _____
8. _____
9. _____
10. _____

Student _____

Lab Section _____

The Appendicular Skeleton

A. Figures
List the labels for the figures.

Figure 12.1

1. _____
2. _____
3. _____
4. _____
5. _____
6. _____
7. _____
8. _____

Figure 12.3

1. _____
2. _____
3. _____
4. _____
5. _____
6. _____
7. _____
8. _____
9. _____
10. _____
11. _____
12. _____

Figure 12.2

1. _____
2. _____
3. _____
4. _____
5. _____
6. _____
7. _____
8. _____
9. _____
10. _____
11. _____
12. _____
13. _____
14. _____
15. _____
16. _____
17. _____
18. _____
19. _____
20. _____
21. _____
22. _____
23. _____
24. _____

Figure 12.4

1. _____
2. _____
3. _____
4. _____
5. _____
6. _____
7. _____
8. _____
9. _____
10. _____
11. _____
12. _____
13. _____
14. _____
15. _____
16. _____
17. _____
18. _____
19. _____
20. _____
21. _____
22. _____
23. _____

B. Completion

Provide the bones or structures described by the statements.

1. Fossa on scapula that articulates with humerus.
2. Condyle of humerus that articulates with the radius.
3. Scapular process that articulates with clavicle.
4. Large process lateral to head of humerus.
5. Process forming point of elbow.
6. Fossa of ulna that articulates with the humerus.
7. Wrist bones.
8. Bones of palm of the hand.
9. Scapular process projecting forward under clavicle.
10. Fossa on distal anterior surface of humerus.
11. Process on distal lateral margin of radius.
12. Bones of the fingers and toes.
13. Tuberosity just below neck of radius.
14. Fossa on distal posterior surface of humerus.
15. Part of radius that articulates with the humerus.
16. Upper bone of os coxa.
17. Process at angle of ischium.
18. Heelbone.
19. Large foramen in os coxa formed by pubis and ischium.
20. Joint between the pubic bones.
21. Large process lateral to neck of femur.
22. Fossa of os coxa that articulates with head of femur.
23. Joint between os coxa and sacrum.
24. Uppermost process on posterior edge of ischium.
25. Superior edge of ilium.
26. Carpal bone that articulates with tibia and fibula.
27. Process below neck on medial surface of femur.
28. Proximal anterior process on tibia below kneecap.
29. Process on medial surface at distal end of tibia.
30. Forms the instep of the foot.

C. Male and Female Pelves: True–False

Compare the male and female pelvic girdles side-by-side to determine if the following statements are true or false.

1. The opening of the female pelvis is larger and more oval than the male pelvis.
2. The acetabula of the female face more anteriorly.
3. The iliac crests of the female pelvis protrude laterally more than the male pelvis.
4. The distances between the ischial spines and ischial tuberosities are greater in the female pelvis.
5. The angle of the pubic arch below the symphysis pubis is wider in the female pelvis.
6. The sacrum curves more posteriorly in the female pelvis.

Completion

1. _____
2. _____
3. _____
4. _____
5. _____
6. _____
7. _____
8. _____
9. _____
10. _____
11. _____
12. _____
13. _____
14. _____
15. _____
16. _____
17. _____
18. _____
19. _____
20. _____
21. _____
22. _____
23. _____
24. _____
25. _____
26. _____
27. _____
28. _____
29. _____
30. _____

True–False

1. _____
2. _____
3. _____
4. _____
5. _____
6. _____

Laboratory Exercise 13

Student ______________________

Lab Section ______________

Articulations

A. Figures

List the labels for Figures 13.1 through 13.3.

Figure 13.1

1. ______
2. ______
3. ______
4. ______
5. ______
6. ______
7. ______
8. ______
9. ______
10. ______

Figure 13.2

1. ______
2. ______
3. ______
4. ______
5. ______
6. ______
7. ______
8. ______
9. ______

Figure 13.3

1. ______
2. ______
3. ______
4. ______
5. ______
6. ______
7. ______
8. ______
9. ______
10. ______
11. ______
12. ______
13. ______
14. ______
15. ______
16. ______
17. ______
18. ______
19. ______
20. ______

B. Types and Subtypes

Identify the basic articulation type to which the subtypes listed belong.

amphiarthrosis diarthrosis synarthrosis

1. Ball-and-socket.
2. Condyloid.
3. Gliding.
4. Hinge.
5. Pivot.
6. Saddle.
7. Suture.
8. Symphysis.
9. Synchondrosis.
10. Syndesmosis.

Types

1. ______
2. ______
3. ______
4. ______
5. ______
6. ______
7. ______
8. ______
9. ______
10. ______

C. Joint Characteristics

Select the joints that have the characteristics described by the statements.

ball-and-socket   saddle
condyloid   suture
gliding   symphysis
hinge   synchondrosis
pivot   syndesmosis

1. Bone ends are flat or slightly convex.
2. Slightly movable; fibrocartilaginous pad.
3. Slightly movable; bones joined by interosseous ligament.
4. Immovable; bones bonded by cartilaginous disk.
5. Rotational movement only.
6. Freely movable in one direction only.
7. Freely movable in two planes.
8. Angular movement in all directions.
9. Immovable; bonded by fibrous connective tissue.
10. Bone ends are convex in one direction and concave in the other.

D. Joint Classification

By examining and manipulating the bones of an articulated skeleton, classify the following joints according to the list in Section C.

1. Femur—tibia.
2. Humerus—ulna.
3. Vertebra—vertebra.
4. Phalanges—phalanges.
5. Phalanges—metacarpals.
6. Humerus—scapula.
7. Atlas—axis.
8. Carpals—carpals.
9. Tibia—fibula, distal ends.
10. Left pubis—right pubis.
11. Parietal—frontal.
12. Trapezium—metacarpal of thumb.
13. Diaphysis—epiphysis of femur in child.
14. Radius—ula, distal ends.
15. Femur—os coxa.

Characteristics

1. _____
2. _____
3. _____
4. _____
5. _____
6. _____
7. _____
8. _____
9. _____
10. _____

Classification

1. _____
2. _____
3. _____
4. _____
5. _____
6. _____
7. _____
8. _____
9. _____
10. _____
11. _____
12. _____
13. _____
14. _____
15. _____

Laboratory Report **14**

Student ____________________

Lab Section ____________

Muscle Structure and Body Movements

A. *Figures*

List the labels for Figure 14.1 and the types of body movements in Figure 14.2.

Figure 14.1	Figure 14.2
1. ______	A. ______
2. ______	B. ______
3. ______	C. ______
4. ______	D. ______
5. ______	E. ______
6. ______	F. ______
7. ______	G. ______
8. ______	H. ______
9. ______	I. ______
10. ______	J. ______
11. ______	K. ______
	L. ______
	M. ______

B. *Muscle Structure*

Match the terms below with the statements that follow.

aponeurosis	insertion
deep fascia	ligament
endomysium	origin
epimysium	perimysium
fasciculus	tendon

1. Bundle of muscle fibers.
2. Connective tissue around each fasciculus.
3. Movable end of a muscle.
4. Thin layer of connective tissue around each muscle fiber.
5. Broad sheet of connective tissue attaching a muscle to another muscle or bone.
6. Connective tissue that surrounds a group of fasciculi.
7. Immovable end of a muscle.
8. Outer layer of connective tissue that envelops entire muscle.
9. Narrow band of connective tissue attaching muscle to bone or other muscle.
10. Not part of a muscle.

Structure

1. ______
2. ______
3. ______
4. ______
5. ______
6. ______
7. ______
8. ______
9. ______
10. ______

C. Muscle Types

Match the terms in the list below to the statements that follow.

agonists   fixators
antagonists   synergists

1. Muscles that assist prime movers.
2. Prime movers.
3. Muscles opposing prime movers.
4. Muscles holding structures steady to allow prime movers to act smoothly.

D. Movements

Match the terms in the list below with the statements that follow.

abduction   hyperextension
adduction   inversion
circumduction   plantar flexion
dorsiflexion   pronation
eversion   rotation
extension   supination
flexion

1. Movement of hand from palm down to palm up.
2. Movement of leg away from midline of the body.
3. Movement of 200° around a central point.
4. Flexion of foot (specific term).
5. Decrease in the angle of bones at a joint.
6. Movement of hand from palm up to palm down.
7. Movement of limb subscribing a circle.
8. Turning the sole of the foot inward.
9. Increase in the angle of bones at a joint.
10. Backward movement of head.
11. Movement of arm toward midline of the body.
12. Extension of foot (specific term).
13. Turning the sole of the foot outward.

Types

1. _____
2. _____
3. _____
4. _____

Movements

1. _____
2. _____
3. _____
4. _____
5. _____
6. _____
7. _____
8. _____
9. _____
10. _____
11. _____
12. _____
13. _____

Student _____

Lab Section _____

Head and Trunk Muscles

A. *Figures*

List the labels for Figures 15.1 through 15.4.

Figure 15.1

1. _____
2. _____
3. _____
4. _____
5. _____
6. _____
7. _____
8. _____
9. _____
10. _____
11. _____

Figure 15.2

1. _____
2. _____
3. _____
4. _____
5. _____

Figure 15.3

1. _____
2. _____
3. _____
4. _____
5. _____
6. _____

Figure 15.4

1. _____
2. _____
3. _____
4. _____
5. _____
6. _____

B. *Head Muscles*

Select the muscle that matches the following statements.

buccinator
frontalis
masseter
orbicularis oculi
orbicularis oris
platysma
sternocleidomastoideus
temporalis
triangularis
zygomaticus

1. Primary muscle of mastication.
2. Draws corner of mouth downward only.
3. Wrinkles forehead and raises eyebrows.
4. Draws corner of mouth upward and backward.
5. Compresses cheeks.
6. Draws corner of mouth downward and backward.
7. Draws head downward toward chest.
8. Acts synergistically with the masseter.
9. Closes eyelids.
10. Closes and puckers lips.
11. Pulls scalp backward.
12. Assists in opening the mouth.
13. Holds food between teeth during chewing.
14. Contraction of muscle on right side turns face to the left.

Head Muscles

1. _____
2. _____
3. _____
4. _____
5. _____
6. _____
7. _____
8. _____
9. _____
10. _____
11. _____
12. _____
13. _____
14. _____

C. Trunk Muscles

Select the muscle described in the following statements.

deltoideus
external intercostals
external oblique
infraspinatus
internal intercostals
internal oblique
latissimus dorsi

pectoralis major
rectus abdominis
serratus anterior
supraspinatus
teres major
transversus abdominis
trapezius

1. Innermost abdominal muscle.
2. Rotates upper arm medially only.
3. Adducts and elevates scapula.
4. Rotates upper arm laterally only.
5. Extends, adducts, and rotates upper arm medially.
6. Elevates rib cage in breathing.
7. Draws scapula downward and forward.
8. Adducts and flexes upper arm medially.
9. Abducts and extends upper arm.
10. Draws head backward.
11. Primary muscle of the shoulder.
12. Large surface muscle of the upper chest.
13. Lowers rib cage during breathing.
14. Triangular surface muscle of upper back.
15. Assists deltoid in abducting arm.
16. Segmented muscle extending from ribs to pubis.
17. Hyperextends head.
18. Outermost abdominal muscle.
19. Superficial muscle covering lower back.
20. Lies just under the external oblique.

Trunk Muscles

1. _____
2. _____
3. _____
4. _____
5. _____
6. _____
7. _____
8. _____
9. _____
10. _____
11. _____
12. _____
13. _____
14. _____
15. _____
16. _____
17. _____
18. _____
19. _____
20. _____

Laboratory Report **16**

Student _____

Lab Section _____

Muscles of the Upper Extremity

A. Figures 16.1 and 16.2
List the labels for these figures.

Figure 16.1

A. _____

B. _____

C. _____

D. _____

E. _____

Figure 16.2

1. _____

2. _____

3. _____

4. _____

5. _____

6. _____

7. _____

8. _____

9. _____

10. _____

11. _____

12. _____

13. _____

14. _____

B. Arm Movements
Record the muscle described in the statements below.

biceps brachii	pronator quadratus
brachialis	pronator teres
brachioradialis	supinator
coracobrachialis	triceps brachii

1. Extends the forearm.
2. Flexes and adducts the upper arm.
3. Rotates forearm laterally.
4. Muscle at distal end of forearm that is a synergist of the pronator teres.
5. Three muscles that flex the forearm.
6. Antagonist of the brachialis.
7. Flexes and rotates the forearm laterally.
8. Superficial muscle on anterior surface of upper arm.
9. Covers posterior surface of humerus.
10. Superficial muscle on lateral surface of forearm.
11. Located just under biceps brachii.
12. Primary muscle rotating forearm medially.

Arm Movements

1. _____

2. _____

3. _____

4. _____

5. _____

6. _____

7. _____

8. _____

9. _____

10. _____

11. _____

12. _____

C. Hand Movements

Record the muscle described in the statements below.

abductor pollicis
extensor carpi radialis longus
extensor carpi ulnaris
extensor digitorum communis
extensor pollicis longus

flexor carpi radialis
flexor carpi ulnaris
flexor digitorum profundus
flexor digitorum superficialis
flexor pollicis longus

1. Flexes and adducts the hand.
2. Flexes and abducts the hand.
3. Abducts the thumb.
4. Extends distal phalanges of fingers 2–5.
5. Flexes distal phalanges of fingers 2–5.
6. Extends and abducts the hand.
7. Extends and adducts the hand.
8. Flexes middle phalanges of fingers 2–5.
9. Extends the thumb.
10. Flexes the thumb.

Hand Movements

1. _____
2. _____
3. _____
4. _____
5. _____
6. _____
7. _____
8. _____
9. _____
10. _____

Laboratory Report **17**

Student ____________________

Lab Section ____________

Muscles of the Lower Extremity

A. Figures

List the labels for Figures 17.1 through 17.4.

Figure 17.1

1. ____________
2. ____________
3. ____________
4. ____________
5. ____________
6. ____________
7. ____________
8. ____________

Figure 17.2

1. ____________
2. ____________
3. ____________
4. ____________
5. ____________
6. ____________
7. ____________
8. ____________
9. ____________
10. ____________
11. ____________

Figure 17.3

1. ____________
2. ____________
3. ____________
4. ____________
5. ____________

Figure 17.4

1. ____________
2. ____________
3. ____________
4. ____________
5. ____________
6. ____________
7. ____________

B. Thigh Movements

Select the muscle(s) from the list at the right that is described by the statements below.

adductor brevis	gluteus maximus
adductor longus	gluteus medius
adductor magnus	gluteus minimus
	piriformis

1. Extends and rotates thigh laterally.
2. Abducts and rotates thigh laterally.
3. Three muscles that adduct, flex, and rotate thigh laterally.
4. Two muscles that abduct and rotate thigh medially.
5. Strongest adductor muscle.
6. Superficial buttocks muscle.
7. Innermost gluteus muscle.
8. Longest adductor muscle.

Thigh Movements

1. ____________
2. ____________
3a. ____________
3b. ____________
3c. ____________
4a. ____________
4b. ____________
5. ____________
6. ____________
7. ____________
8. ____________

C. Thigh and Leg Movements

Select the muscle(s) that apply to the following statements.

biceps femoris   semitendinosus
gracilis   tensor fasciae latae
rectus femoris   vastus intermedius
sartorius   vastus lateralis
semimembranosus   vastus medialus

1. Four muscles extending the leg.
2. Adducts thigh and flexes leg.
3. Flexes the leg and thigh.
4. Flexes and abducts thigh.
5. Three hamstrings flexing the leg.
6. Deepest quadriceps muscle.
7. Anterior central quadriceps muscle.
8. Lateral hamstring muscle.
9. Medial hamstring muscle.

D. Leg and Foot Movements

Select the muscle(s) that apply to the following statements.

extensor digitorum longus   peroneus brevis
extensor hallucis longus   peroneus longus
flexor digitorum longus   peroneus tertius
flexor hallucis longus   soleus
gastrocnemius   tibialis anterior
     tibialis posterior

1. Two muscles with a common Achilles tendon.
2. Extends and inverts the foot.
3. Flexes the leg and extends the foot.
4. Two muscles that extend and evert the foot.
5. Flexes the great toe.
6. Extends the great toe.
7. Flexes toes 2–5.
8. Extends toes 2–5.
9. Dorsiflexes and everts the foot.
10. Dorsiflexes and inverts the foot.
11. Two calf muscles that extend the foot.
12. Supports the transverse arch of the foot.

Thigh and Leg Movements

1a. _____
1b. _____
1c. _____
1d. _____
2. _____
3. _____
4. _____
5a. _____
5b. _____
5c. _____
6. _____
7. _____
8. _____
9. _____

Leg and Foot Movements

1a. _____
1b. _____
2. _____
3. _____
4a. _____
4b. _____
5. _____
6. _____
7. _____
8. _____
9. _____
10. _____
11a. _____
11b. _____
12. _____

List the labels for this figure.

Anterior

1. _____
2. _____
3. _____
4. _____
5. _____
6. _____
7. _____
8. _____
9. _____
10. _____
11. _____
12. _____
13. _____
14. _____
15. _____
16. _____
17. _____
18. _____
19. _____

Posterior

1. _____
2. _____
3. _____
4. _____
5. _____
6. _____
7. _____
8. _____
9. _____
10. _____
11. _____
12. _____
13. _____
14. _____
15. _____
16. _____

Muscle and Nerve Tissues

A. Muscle Tissue Characteristics

Select the type of muscle tissue described by the statements below.

cardiac   skeletal   smooth

1. Attached to bones.
2. Rhythmic contractions.
3. Voluntary in function.
4. Cells tapered to a point at each end.
5. Cells without striations.
6. Cells with striations and many nuclei.
7. Occurs in walls of blood vessels.
8. Cells separated by intercalated disks.
9. Striated cells with single nucleus.
10. Spontaneous contractions without neural activation.
11. Fibers are grouped in fasciculi.
12. Smooth muscle tissue.
13. Cells often branched to form network.
14. Forms muscle of heart.

B. Drawings

Make drawings of the three types of muscle tissue from your microscopic studies. Label pertinent parts.

Muscle Tissue

1. _____
2. _____
3. _____
4. _____
5. _____
6. _____
7. _____
8. _____
9. _____
10. _____
11. _____
12. _____
13. _____
14. _____

C. Figure 18.4

List the labels for Figure 18.4.

1. _____
2. _____
3. _____
4. _____
5. _____
6. _____

7. _____
8. _____
9. _____
10. _____
11. _____
12. _____

13. _____
14. _____
15. _____
16. _____

D. Neuron Structure

Select the structure from the list that is described in the statements that follow.

axon
cell body
collateral
dendrite
myelin sheath

neurilemma
neurofibrils
Nissl bodies
nodes of Ranvier
Schwann cells

1. Contains neuron nucleus.
2. Side branch of an axon.
3. Myelinated process of a motor neuron.
4. Myelinated process of a sensory neuron.
5. Process receiving impulses.
6. Process transmitting impulses to an effector.
7. Thin filaments in the cell body and processes.
8. Clumps of RER in the cell body.
9. Gaps between Schwann cells.
10. Formed by inner wrappings of Schwann cells.
11. Outer layer of a Schwann cell.
12. Cells covering all peripheral neuron fibers.

E. Neuron Types

Select the type of neuron described in the statements that follow.

interneuron   motor   sensory

1. An afferent neuron.
2. An efferent neuron.
3. Totally within central nervous system.
4. Carries impulses to a muscle or gland.
5. Carries impulses to central nervous system.

F. Drawings

Make drawings of neuron cell bodies from your microscopic studies. Label pertinent parts.

Neuron Structure

1. _____
2. _____
3. _____
4. _____
5. _____
6. _____
7. _____
8. _____
9. _____
10. _____
11. _____
12. _____

Neuron Types

1. _____
2. _____
3. _____
4. _____
5. _____

The Nature of Muscle Contraction

A. Figure 19.1
List the labels for this figure.

1. _____   4. _____   7. _____

2. _____   5. _____   8. _____

3. _____   6. _____

B. Neuromuscular Junction
Select the terms that match the statements below.

acetylcholine   synaptic cleft
axon terminal branches   synaptic vesicle
sarcolemma

1. Space between axon tip and sarcolemma.
2. Embedded in invaginations of the sarcolemma.
3. Secreted into synaptic cleft when action potential reaches the axon tip.
4. Membrane covering muscle fiber.
5. Membranous container of acetylcholine.
6. Reacts with receptors on sarcolemma to initiate an action potential in the muscle fiber.

Neuromuscular Junction

1. _____

2. _____

3. _____

4. _____

5. _____

6. _____

C. Myofibril Ultrastructure
Select the terms that match the statements below.

A band   myosin
actin   sarcomere
I band   Z line

1. Part of myofibril between two Z lines.
2. Thick myofilament with cross-bridges.
3. Thin myofilament attached to Z lines.
4. Main myofilament within the A band.
5. Lies between two I bands.
6. Myofilament within I bands.

Myofibril Ultrastructure

1. _____

2. _____

3. _____

4. _____

5. _____

6. _____

D. Mechanics of Contraction
Select the terms that match the statements below.

actin   cross-bridges on myosin
ATP   sacroplasmic reticulum
calcium ions

1. Releases calcium ions when depolarized by action potential.
2. Expose the active sites on the actin myofilament.
3. Attach to active sites on actin myofilament.
4. Provides energy for contraction.
5. Pull actin myofilaments toward the center of the A band.

Mechanics of Contraction

1. _____

2. _____

3. _____

4. _____

5. _____

E. Experimental Muscle Contraction

1. Indicate the length of the muscle strands before and after contraction.

   Before contraction .... _____ mm

   After contraction .... _____ mm

2. Draw the pattern of the striations as they appear in relaxed and contracted muscle fibers.

   Relaxed   *Contracted*

Physiology of Muscle Contraction

A. Threshold Stimulus

1. Record the threshold stimulus (voltage) that produced a visible muscle twitch when using a Width (duration) of 2 msec.

   Direct stimulation of the muscle .... _____ volts

   Stimulation of the nerve .... _____ volts

2. Determine how much greater the threshold stimulus is for direct muscle stimulation than for nerve stimulation.

$$\frac{\text{Voltage for Direct Muscle Stimulation}}{\text{Voltage for Nerve Stimulation}} = \underline{\hspace{2cm}}$$

3. Record the minimal voltage required to extend the foot. _____ volts

B. Myogram

1. Attach the myogram in the space below.

2. Calculate from the myogram the duration of the following:

   Latent period .... _____ msec

   Contraction phase .... _____ msec

   Relaxation phase .... _____ msec

   Total duration .... _____ msec

C. Summation

1. Attach the record for motor unit summation below. Be sure that the voltage for each contraction is recorded on the chart.

2. Attach the record for wave summation below. Be sure that the stimulus frequency (Hz) is recorded on the chart for each set of contractions.

D. Definitions

Define the following terms.

1. Threshold stimulus _____

2. Maximal stimulus _____

3. Motor unit _____

4. Tetanus _____

5. Motor unit summation _____

_____

6. Wave summation _____

_____

Laboratory Report **21**

Student ______________________

Lab Section ______________

The Spinal Cord and Reflex Arcs

A. *Figures*

List the labels for Figures 21.1, 21.2, and 21.4.

Figure 21.1

1. ______
2. ______
3. ______
4. ______
5. ______
6. ______
7. ______
8. ______
9. ______
10. ______
11. ______
12. ______
13. ______
14. ______
15. ______
16. ______
17. ______
18. ______
19. ______
20. ______
21. ______

Figure 21.2

1. ______
2. ______
3. ______
4. ______
5. ______
6. ______
7. ______
8. ______
9. ______
10. ______
11. ______
12. ______
13. ______
14. ______
15. ______
16. ______

Figure 21.4

1. ______
2. ______
3. ______
4. ______
5. ______
6. ______

B. *Spinal Nerves and Plexuses*

Select the terms that match the statements below.

brachial plexus	1
cervical plexus	5
lumbar plexus	8
sacral plexus	12

1. The number of pairs of cervical nerves.
2. The number of pairs of thoracic nerves.
3. The number of pairs of lumbar nerves.
4. The number of pairs of sacral nerves.
5. The number of pairs of coccygeal nerves.
6. Formed of spinal nerves C_1 through C_4.
7. Formed of spinal nerves L_4 and L_5 and S_1 through S_3.
8. Formed of spinal nerves C_5 through C_8 plus T_1.
9. Formed of spinal nerves L_1 through L_4.

Spinal Nerves

1. ______
2. ______
3. ______
4. ______
5. ______
6. ______
7. ______
8. ______
9. ______

C. The Spinal Cord

Select the terms that match the statements below.

arachnoid mater
central canal
dura mater
epidural space
gray matter

pia mater
subarachnoid space
subdural space
vertebral canal
white matter

1. Meninx adhered to surface of spinal cord.
2. Outermost fibrous meninx.
3. Meninx between other meninges.
4. Channel continuous with ventricles of the brain.
5. Outer portion of spinal cord.
6. Inner portion of spinal cord.
7. Space between outer meninx and vertebrae.
8. Space around spinal cord containing cerebrospinal fluid.
9. Part of spinal cord containing unmyelinated neuron processes.
10. Part of dorsal cavity containing spinal cord.

Spinal Cord

1. _____
2. _____
3. _____
4. _____
5. _____
6. _____
7. _____
8. _____
9. _____
10. _____

D. Reflex Arcs

Select the terms that match the statements below.

effector
gray matter
interneuron
motor neuron
receptor

postganglionic efferent neuron
preganglionic efferent neuron
sensory neuron
spinal ganglion
white matter

1. A muscle or gland.
2. Carries impulses from receptor to spinal cord.
3. Carries impulses from spinal cord to an effector.
4. Receives stimuli and forms impulses.
5. Part of spinal cord with many myelinated neuron processes.
6. Contains cell bodies of sensory neurons.
7. Carries impulses from spinal cord to autonomic ganglia.
8. Neuron within the gray matter.
9. Carries impulses from autonomic ganglia to viscera.
10. Where synapses occur in spinal cord.

Reflex Arcs

1. _____
2. _____
3. _____
4. _____
5. _____
6. _____
7. _____
8. _____
9. _____
10. _____

E. Diagnostic Reflexes

Record the results of your responses.

Biceps reflex: left arm _____ right arm _____
Patellar reflex: left leg _____ right leg _____
Achilles reflex: left foot _____ right foot _____

Explain why reinforcement technique improves responses. _____

_____

Student _____

Lab Section _____

Brain Anatomy: External

A. Figures

List the labels for Figures 22.1, 22.2, and 22.4.

Figure 22.1

1. _____
2. _____
3. _____
4. _____
5. _____
6. _____
7. _____

Figure 22.2

1. _____
2. _____
3. _____
4. _____
5. _____
6. _____

7. _____
8. _____
9. _____
10. _____
11. _____
12. _____
13. _____

Figure 22.4

1. _____
2. _____
3. _____
4. _____
5. _____
6. _____

7. _____
8. _____
9. _____
10. _____
11. _____
12. _____
13. _____
14. _____
15. _____
16. _____
17. _____
18. _____
19. _____
20. _____

B. The Meninges

Select the structures described by the statements below.

arachnoid mater   epidural space   subarachnoid space
dura mater   pia mater   subdural space

1. The fibrous meninx attached to cranial bones.
2. Meninx attached to the brain surface.
3. The middle meninx.
4. Space containing cerebrospinal fluid.
5. Meninx containing sagittal sinus.

Meninges

1. _____
2. _____
3. _____
4. _____
5. _____

C. Fissures and Sulci

Select the structures described by the statements below.

central sulcus   longitudinal cerebral fissure
lateral cerebral fissure   parieto-occipital fissure

1. Between frontal and parietal lobes.
2. Separates cerebral hemispheres.
3. Between temporal and parietal lobes.
4. Between occipital and parietal lobes.

Fissures and Sulci

1. _____
2. _____
3. _____
4. _____

D. Brain Regions and Functions

Select the parts of the brain described below.

cerebellum   hypothalamus   midbrain
cerebrum   medulla oblongata   pons

1. Controls heart and breathing rates.
2. Functions in will, intelligence, and memory.
3. Controls body temperature and water balance.
4. Controls muscular coordination.
5. Pathway for fibers between the cerebrum and the pons, cerebellum, and medulla.

E. Functional Areas of the Cerebrum

Select the functional areas described below.

association areas   primary motor area
auditory area   primary sensory area
motor speech area   visual area

1. Occupies the precentral gyrus of the frontal lobe.
2. Occupies a portion of the superior temporal gyrus.
3. Controls muscular action.
4. Areas involved in reasoning and judgment.
5. Occupies the posterior part of the occipital lobe.
6. Controls speech production.
7. Sensations of the skin.
8. Hearing.
9. Sight.
10. Located in each lobe of the cerebrum.

F. Cranial Nerves

Select the cranial nerves described below.

I. Olfactory	VII. Facial
II. Optic	VIII. Statoacoustic
III. Oculomotor	IX. Glossopharyngeal
IV. Trochlear	X. Vagus
V. Trigeminal	XI. Accessory
VI. Abducens	XII. Hypoglossal

1. Innervates abdominal viscera.
2. Carries impulses from inner ear.
3. Controls swallowing reflexes.
4. Innervates teeth and gums.
5. Controls four muscles moving eyeball.
6. Two nerves each innervating one exterior eye muscle.
7. Innervates muscles of the tongue.
8. Carries impulses from light receptors in the eye.
9. Innervates facial muscles and salivary glands.
10. Innervates the trapezius and muscles of the pharynx and larynx.
11. Carries impulses interpreted as odors.

Brain Functions

1. _____
2. _____
3. _____
4. _____
5. _____

Functional Areas

1. _____
2. _____
3. _____
4. _____
5. _____
6. _____
7. _____
8. _____
9. _____
10. _____

Cranial Nerves

1. _____
2. _____
3. _____
4. _____
5. _____
6. _____
7. _____
8. _____
9. _____
10. _____
11. _____

Laboratory Report **23**

Student ______________________

Lab Section ____________

Brain Anatomy: Internal

A. *Figures*

List the labels for Figures 23.2 and 23.3.

Figure 23.2

1. ______________
2. ______________
3. ______________
4. ______________
5. ______________
6. ______________
7. ______________
8. ______________
9. ______________
10. ______________
11. ______________
12. ______________
13. ______________
14. ______________
15. ______________
16. ______________
17. ______________
18. ______________
19. ______________
20. ______________

Figure 23.3

1. ______________
2. ______________
3. ______________
4. ______________
5. ______________
6. ______________
7. ______________
8. ______________
9. ______________
10. ______________
11. ______________
12. ______________

B. *Location*

Select the part of the brain in which the following structures are located.

cerebellum
cerebrum
diencephalon
medulla oblongata
midbrain
pons

1. Corpus callosum
2. Aqueduct of Sylvius
3. Hypothalamus
4. Fornix
5. Cerebral peduncles
6. Third ventricle
7. Thalamus
8. Mammillary bodies
9. Lateral ventricles
10. Corpora quadrigemina

Location

1. ______________
2. ______________
3. ______________
4. ______________
5. ______________
6. ______________
7. ______________
8. ______________
9. ______________
10. ______________

C. General Questions

Record the name of the brain structure described below.

1. The second largest portion of the brain.
2. Composed of the midbrain, pons, and medulla.
3. Fluid within the subarachnoid space.
4. Secretes the hormone melatonin.
5. Provides an uncritical sensory awareness.
6. Generates impulses to keep the cerebrum alert.
7. Lowest portion of the brain stem.
8. The largest part of the brain.
9. Fissure separating the cerebral hemispheres.
10. Shallow furrows on the cerebrum.

General Questions

1. _____
2. _____
3. _____
4. _____
5. _____
6. _____
7. _____
8. _____
9. _____
10. _____

D. Structures

Select the structures that are described below.

aqueduct of Sylvius
arachnoid granulations
cerebral peduncles
choroid plexus
corpora quadrigemina
corpus callosum

foramen of Monroe
fornix
hypothalamus
infundibulum
optic chiasma
thalamus

1. Receives sensory impulses and relays them to the cerebrum.
2. Contains fiber tracts associated with the sense of smell.
3. Regulates body temperature.
4. Consists of fibers connecting cerebral hemispheres.
5. Stalk supporting the hypophysis.
6. Enables cerebrospinal fluid to diffuse into the blood.
7. Allows cerebrospinal fluid to pass from the lateral ventricles into the third ventricle.
8. Crossing of optic fibers.
9. Main pathway of motor fibers from the cerebrum.
10. Reflex center for movements in response to auditory and visual stimuli.
11. Connects third and fourth ventricles.
12. Secretes cerebrospinal fluid into the ventricles.

Structures

1. _____
2. _____
3. _____
4. _____
5. _____
6. _____
7. _____
8. _____
9. _____
10. _____
11. _____
12. _____

E. Cerebrospinal Fluid

Provide the words to correctly complete the paragraph.

Cerebrospinal fluid is released into the ventricles from the ____1____ in each ventricle. It passes from the lateral ventricles into the third ventricle via an opening, the foramen of ____2____ , and then into the fourth ventricle through the ____3____ . From the fourth ventricle, the fluid passes via the foramen of ____4____ into the ____5____ below the cerebellum. From here, some of the fluid moves upwards around the cerebellum and cerebrum, and diffuses into the sagittal sinus. Most of the fluid flows downward in the subarachnoid space along the ____6____ surface of the spinal cord, upward along the ____7____ surface of the spinal cord, around the cerebrum, and finally diffuses from the ____8____ into the blood in the ____9____ .

Cerebrospinal Fluid

1. _____
2. _____
3. _____
4. _____
5. _____
6. _____
7. _____
8. _____
9. _____

Laboratory Report **24**

Student ______________________

Lab Section ____________

The Eye

A. Figures

List the labels for Figures 24.1, 24.2, and 24.3.

Figure 24.1

1. ____________
2. ____________
3. ____________
4. ____________
5. ____________
6. ____________
7. ____________
8. ____________
9. ____________
10. ____________

Figure 24.2

1. ____________
2. ____________
3. ____________
4. ____________
5. ____________
6. ____________
7. ____________
8. ____________
9. ____________

Figure 24.3

1. ____________
2. ____________
3. ____________
4. ____________
5. ____________
6. ____________
7. ____________
8. ____________
9. ____________
10. ____________
11. ____________
12. ____________
13. ____________
14. ____________
15. ____________

B. Lacrimal Apparatus

Provide the words that correctly complete the paragraph.

The ___1___ lines the inner surface of the eyelids and covers the anterior eye surface. A fold near the medial corner of the eye is the ___2___, a remnant of a third eyelid. Tears, secreted by the ___3___, keep the ___4___ moist. After flowing across the eye surface, tears enter the ___5___ and the ___6___, are carried into the ___7___, and drain into the ___8___ via the ___9___.

Lacrimal Apparatus

1. ____________
2. ____________
3. ____________
4. ____________
5. ____________
6. ____________
7. ____________
8. ____________
9. ____________

C. Extrinsic Muscles

Select the muscles described by statements below.

inferior oblique

inferior rectus

lateral rectus

medial rectus

superior levator palpebrae

superior oblique

superior rectus

1. Rotates eye downward.
2. Raises the upper eyelid.
3. Inserted on the medial surface of the eye.
4. Passes through a cartilaginous loop above the eye.
5. Rotates eye outward.

D. Structures

Select the structures described by the statements below.

aqueous humor

choroid coat

ciliary body

cornea

fovea centralis

iris

lens

macula lutea

optic disc

pupil

retina

scleroid coat

suspensory ligaments

vitreous body

1. Contains light receptors: cones and rods.
2. Clear window at front of eye that bends light rays.
3. Fills cavity anterior to the lens.
4. Outer layer of the eye.
5. Opening in center of iris.
6. Attaches lens to ciliary body.
7. Fills cavity posterior to lens.
8. Precisely focuses light on retina.
9. Area of critical vision.
10. Controls shape of the lens.
11. Controls amount of light entering the eye.
12. Yellow spot containing only cones.
13. Blackish middle coat of the eyeball.
14. Provides strength to the eyeball wall.
15. Lacks rods and cones.
16. Layer of eyeball containing blood vessels.
17. Place where nerve fibers of retina leave eyeball.
18. Place where blood vessels enter eyeball.
19. Jelly-like substance within eyeball.
20. Colored ring anterior to the lens.

Structures

1. _____
2. _____
3. _____
4. _____
5. _____
6. _____
7. _____
8. _____
9. _____
10. _____
11. _____
12. _____
13. _____
14. _____
15. _____
16. _____
17. _____
18. _____
19. _____
20. _____

E. Beef Eye Dissection

1. Which layer of the eyeball is the toughest? _____

2. What is the shape of the pupil? _____

3. When you hold up the lens, look at a distant object. What is unusual about the image? _____

4. When placed on a printed page, does the lens make print appear larger or smaller? _____

5. Is the lens more dense near the center or periphery? _____

6. What is the function of the black pigment in the eye? _____

7. What is the function of the iridescent portion of the choroid coat in the beef eye? _____

The Ear

A. *Figure 25.1*

List the labels for this figure.

1. _____
2. _____
3. _____
4. _____
5. _____
6. _____
7. _____
8. _____

9. _____
10. _____
11. _____
12. _____
13. _____
14. _____
15. _____
16. _____

17. _____
18. _____
19. _____
20. _____
21. _____
22. _____
23. _____

B. *Hearing (Structures)*

Select the structures described in the statements below.

basilar membrane
cochlear duct
endolymph
hair cells
incus
malleus
middle ear
perilymph

pinna
round window
scala tympani
scala vestibuli
stapes
tectorial membrane
tympanum
vestibular membrane

1. Membrane receiving sound waves via air.
2. Ossicle attached to the tympanum.
3. Ossicle inserted into oval window.
4. Cochlear chamber into which oval window opens.
5. Cochlear chamber with the round window.
6. Fluid within the cochlear duct.
7. Fluid between the bony and membranous labyrinths.
8. Receptors for sound stimuli.
9. Contains 20,000 transverse fibers.
10. Cochlear chamber containing the organ of Corti.
11. Membrane in which hair tips are embedded.
12. Membrane allowing movement of perilymph.
13. Transfers vibrations from fluid to hair cells.
14. Cavity traversed by the ear ossicles.
15. Directs sound waves into external auditory meatus.
16. Activated by the movement of the perilymph.
17. The middle ear ossicle.
18. Communicates with the nasopharynx via the Eustachian tube.
19. Set in motion by the movement of the stapes in the oval window.
20. Generate impulses carried by the cochlear nerve.

Hearing (Structures)

1. _____
2. _____
3. _____
4. _____
5. _____
6. _____
7. _____
8. _____
9. _____
10. _____
11. _____
12. _____
13. _____
14. _____
15. _____
16. _____
17. _____
18. _____
19. _____
20. _____

C. Hearing (True–False)

1. Sound waves may reach the cochlea through the skull bones as well as via the tympanum.
2. The more hair cells sending impulses, the louder will be the sound sensation.
3. Pitch is determined by the part of the basilar membrane that is activated.
4. Bending of the hair cells causes the formation of impulses.
5. There is no cure for conduction deafness, but nerve deafness may often be corrected.

D. Equilibrium

Select the structures described by the statements below.

ampulla   otoconia
crista ampullaris   saccule
macula   utricle

1. Two portions of the membranous labyrinth containing sensory receptors for static equilibrium.
2. Sensory hair cells in ampullae of semicircular canals.
3. Hair cells stimulated by the movement of the otoconia.
4. Hair cells stimulated by the force of endolymph.
5. Sensory receptors for dynamic equilibrium.
6. Enlarged base of a semicircular canal.
7. Calcium carbonate crystals.

E. Hearing Tests

Record your results of the hearing tests.

Rinne Test

Left Ear _____

Right Ear _____

State a conclusion from this test. _____

Weber Test

Left Ear _____

Right Ear _____

State a conclusion from this test. _____

F. Static Equilibrium

Record the degree of swaying as small, moderate, or great.

Standing on both feet with eyes open. _____

Standing on both feet with eyes closed. _____

Standing on one foot with eyes closed. _____

Rate the importance of visual input in static balance. _____

_____

Hearing (True–False)

1. _____
2. _____
3. _____
4. _____
5. _____

Equilibrium

1a. _____
1b. _____
2. _____
3. _____
4. _____
5. _____
6. _____
7. _____

Blood Tests

A. Formed Elements

Select the blood cell type described below.

basophils   lymphocytes   neutrophils
eosinophils   monocytes   thrombocytes
erythrocytes

1. Agranulocytes with spherical nucleus and little cytoplasm.
2. Biconcave, enucleate cells.
3. Cells with 2–5 lobed nucleus and tiny lavender cytoplasmic granules.
4. Smallest of the formed elements.
5. Cells with red-orange cytoplasmic granules and bilobed or U-shaped nucleus.
6. Large agranulocytes with a lobed or U-shaped nucleus.
7. Cell with bilobed or U-shaped nucleus and bluish-purple cytoplasmic granules.
8. 4,000,000–6,000,000 per mm^3.
9. Initiate clotting process.
10. Produce antibodies.
11. Transport oxygen and carbon dioxide.
12. Counteract products of allergy reactions.
13. Release histamine in allergy reactions.
14. 50–70% of the leukocytes.
15. 20–30% of the leukocytes.
16. Two types of cells that phagocytize pathogens.
17. 250,000–400,000 per mm^3.
18. Contain hemoglobin.
19. 2–6% of the leukocytes.

Formed Elements

1. _____
2. _____
3. _____
4. _____
5. _____
6. _____
7. _____
8. _____
9. _____
10. _____
11. _____
12. _____
13. _____
14. _____
15. _____
16a. _____
16b. _____
17. _____
18. _____
19. _____

B. Differential White Cell Count

Tabulate your identification of the various types of leukocytes in the table below. Count a total of 100 white cells. Calculate the percentage of each type of leukocyte by dividing the total count for each type by 100.

Neutrophils	Lymphocytes	Monocytes	Eosinophils	Basophils
Totals				
Percent				

C. Blood Test Results

Record your test results in the table below.

Test	Normal Values	Test Results	Evaluation (over, under, normal)
Differential WBC Count	Neutrophils: 50%–70%		
	Lymphocytes: 20%–30%		
	Monocytes: 2%–6%		
	Eosinophils: 1%–5%		
	Basophils: 0.5%–1%		
Hematocrit (VPRC)	Males: 40%–54% (Av. 47%)		
	Females: 37%–47% (Av. 42%)		
Coagulation Time	2 to 6 minutes		

D. Blood Typing

1. Indicate your blood type.   ABO group _____   Rh _____

2. Complete the table below to indicate compatibility (o) and incompatibility (+) of possible transfusions. A plus (+) indicates agglutination of the erythrocytes.

Blood Type/Antigen of Donor	Blood Type and Antibodies of Recipient			
	O a, b	A b	B a	AB none
O				
A				
B				
AB				

Laboratory Report **27**

Student ______________________

Lab Section ______________

The Heart

A. Figures

List the labels for Figures 27.1 and 27.2.

Figure 27.1

1. ______________
2. ______________
3. ______________
4. ______________
5. ______________
6. ______________
7. ______________
8. ______________
9. ______________
10. ______________
11. ______________
12. ______________
13. ______________
14. ______________
15. ______________
16. ______________
17. ______________
18. ______________
19. ______________
20. ______________
21. ______________
22. ______________
23. ______________
24. ______________
25. ______________
26. ______________
27. ______________
28. ______________
29. ______________
30. ______________

Figure 27.2
Posterior Aspect

1. ______________
2. ______________
3. ______________
4. ______________
5. ______________
6. ______________
7. ______________
8. ______________
9. ______________
10. ______________

Figure 27.2
Anterior Aspect

1. ______________
2. ______________
3. ______________
4. ______________
5. ______________
6. ______________
7. ______________
8. ______________
9. ______________
10. ______________
11. ______________
12. ______________
13. ______________
14. ______________

B. Circulation

Provide the words to correctly complete the paragraph.

During diastole, deoxygenated blood returns to the ___1___ via the ___2___ and ___3___, and enters the right ventricle. Simultaneously, oxygenated blood returns to the ___4___ via the ___5___ and enters the left ventricle. During systole, the right ventricle pumps blood into the ___6___ that carries blood to the lungs. Simultaneously, the left ventricle pumps blood into the ___7___ that carries blood to all parts of the body, except the lungs.

Circulation

1. ______________
2. ______________
3. ______________
4. ______________
5. ______________
6. ______________
7. ______________

C. Structures

Select the structures described by the statements below.

aortic semilunar valve	myocardium
chordae tendineae	papillary muscles
endocardium	pulmonary semilunar valve
epicardium	right atrium
left atrium	right ventricle
left ventricle	tricuspid valve
mitral valve	ventricular septum

1. Partition between right and left ventricles.
2. Muscular portion of cardiac wall.
3. Lines interior of the heart.
4. Receives deoxygenated blood from the vena cavae.
5. Receives oxygenated blood from the pulmonary veins.
6. Prevents backflow of blood into right atrium.
7. Prevents backflow of blood into left atrium.
8. Prevents backflow of blood into right ventricle.
9. Prevents backflow of blood into left ventricle.
10. Pumps blood into the pulmonary artery.
11. Pumps blood into the aorta.
12. Chamber with the thickest walls.
13. Portion of ventricular wall to which chordae tendineae are attached.
14. Strands of fibrous tissue attaching valve cusps to the inner wall of the ventricles.
15. Receives blood from the coronary sinus.

Structures

1. _____
2. _____
3. _____
4. _____
5. _____
6. _____
7. _____
8. _____
9. _____
10. _____
11. _____
12. _____
13. _____
14. _____
15. _____

D. Sheep Heart Dissection

1. Indicate the number of cusps composing the heart valves.

   Mitral _____   Tricuspid _____   Aortic _____   Pulmonary _____

2. Compare the thickness of the valve cusps.

   Atrioventricular valves _____

   Semilunar valves _____

3. Which ventricle has the thicker wall? _____

4. Are the chordae tendineae fragile or strong? _____

   Describe their function. _____

   _____

5. Is the ventricular septum a thin sheet of connective tissue or a thick muscular wall? _____

   _____

Blood Vessels, Lymphatic System, and Fetal Circulation

A. Arteries

Select the arteries described by the statements below.

anterior tibial
aorta
axillary
brachial
brachiocephalic
celiac
common iliac
deep femoral

external iliac
femoral
inferior mesenteric
internal iliac
left common carotid
left subclavian
popliteal

posterior tibial
radial
renal
right common carotid
right subclavian
superior mesenteric
ulnar

1. Carries oxygenated blood directly from the heart.
2. Branches to form the radial and ulnar arteries.
3. Branches to form right subclavian and right common carotid.
4. Supplies the kidneys with blood.
5. Supplies most of small intestine and part of large intestine.
6. Arterial segment between the subclavian and the brachial arteries.
7. Gives rise to the femoral artery.
8. Branches from aorta to supply left side of head.
9. Supplies blood to the stomach.
10. Large arteries branching from inferior end of aorta.
11. Branches to form the anterior and posterior tibial arteries.
12. Supplies the large intestine and rectum.

B. Veins

Select the veins described by the statements below.

axillary
basilic
brachial
brachiocephalic
cephalic
common iliac
external iliac
external jugular

femoral
great saphenous
hepatic
hepatic portal
inferior mesenteric
inferior vena cava
internal iliac

internal jugular
median cubital
popliteal
posterior tibial
subclavian
superior mesenteric
superior vena cava

1. Vein in the armpit area.
2. Merge to form superior vena cava.
3. Veins of the neck emptying into the subclavians.
4. Lateral vein of the arm.
5. Medial vein of the arm.
6. Commonly used for venipuncture.
7. Empties into the inferior end of the inferior vena cava.
8. Large superficial vein of the leg.
9. Drains into the popliteal vein.
10. Carries blood from the intestines, stomach, and spleen to the liver.
11. Carries blood from liver to the inferior vena cava.
12. Deep vein of the thigh.

Arteries

1. _____
2. _____
3. _____
4. _____
5. _____
6. _____
7. _____
8. _____
9. _____
10. _____
11. _____
12. _____

Veins

1. _____
2. _____
3. _____
4. _____
5. _____
6. _____
7. _____
8. _____
9. _____
10. _____
11. _____
12. _____

265

C. Figures

List the labels for Figures 28.2, 28.4, and 28.5.

Figure 28.2: Arteries

1. _____
2. _____
3. _____
4. _____
5. _____
6. _____
7. _____
8. _____
9. _____
10. _____
11. _____
12. _____
13. _____
14. _____
15. _____
16. _____
17. _____
18. _____
19. _____
20. _____
21. _____
22. _____
23. _____

Figure 28.2: Veins

1. _____
2. _____
3. _____
4. _____
5. _____
6. _____
7. _____
8. _____
9. _____
10. _____
11. _____
12. _____
13. _____
14. _____
15. _____
16. _____
17. _____
18. _____
19. _____
20. _____
21. _____
22. _____

Figure 28.4

1. _____
2. _____
3. _____
4. _____
5. _____
6. _____
7. _____
8. _____

Figure 28.5

1. _____
2. _____
3. _____
4. _____
5. _____
6. _____
7. _____
8. _____
9. _____
10. _____
11. _____
12. _____
13. _____

D. Lymphatic System

Select the terms described by the statements below.

cisterna chyli   lymph capillaries   lymphatic vessels
lymph   lymph nodes   right lymphatic duct
                                           thoracic duct

1. Collects extracellular fluid from tissues.
2. Fluid in lymphatic vessels.
3. Empties lymph into right subclavian vein.
4. Sac-like structure at inferior end of thoracic duct.
5. Cleanses lymph as it passes through.
6. Carries lymph to left subclavian vein.

Lymphatic System

1. _____
2. _____
3. _____
4. _____
5. _____
6. _____

E. Fetal Circulation

Select the structures described by statements below.

ductus arteriosus   umbilical arteries
ductus venosus   umbilical vein
foramen ovale

1. Opening between the atria.
2. Extensions of the hypogastric arteries.
3. Brings oxygenated blood to the fetus.
4. Connects the pulmonary artery with the aorta.
5. Carry fetal blood to the placenta.
6. Becomes the ligamentum venosum after birth.
7. Becomes the round ligament after birth.
8. Becomes the ligamentum arteriosum after birth.

Fetal Circulation

1. _____
2. _____
3. _____
4. _____
5. _____
6. _____
7. _____
8. _____

Student _____

Lab Section _____

Cardiovascular Phenomena

A. Heart Sounds

Select the heart sound described by the statements below.

first sound   third sound
second sound   murmur

1. Vibration of ventricular walls and valve cusps.
2. Closure of the semilunar valves.
3. Closure of the atrioventricular valves.
4. Damaged heart valves.
5. Detected best in supine subject.

B. Auscultatory Areas

Select the auscultatory area described by the statements below.

aortic semilunar area   pulmonary semilunar area
mitral area   tricuspid area

1. About 2 inches left of sternum at fifth rib.
2. On left edge of sternum just below second rib.
3. On left edge of sternum at the fifth rib.
4. On right edge of sternum at the second rib.
5. Site used to detect the third heart sound.

C. Blood Pressure

1. Complete the following statement.

   The reading on the pressure gauge at the first Korotkoff sound is the _____ pressure; the

   reading when the Korotkoff sounds cease is the _____ pressure.

2. Indicate your blood pressure before and after exercise.

   Systolic: before exercise _____   after exercise _____

   Diastolic: before exercise _____   after exercise _____

3. Explain the cause of a higher blood pressure after exercise.

   _____

   _____

4. Explain how high blood pressure (hypertension) may cause the rupture of an artery in the brain. _____

   _____

   _____

Heart Sounds

1. _____
2. _____
3. _____
4. _____
5. _____

Auscultatory Areas

1. _____
2. _____
3. _____
4. _____
5. _____

D. Peripheral Blood Flow

1. What part of the brain controls peripheral circulation? _____

2. Describe the normal blood flow through a capillary. _____

_____

_____

3. Describe the observable changes in circulation in the frog's foot when histamine was applied. _____

_____

_____

4. Did histamine cause vasodilation or vasoconstriction? _____

5. Describe the observable effect of epinephrine on the circulation of the frog's foot. _____

_____

_____

6. Did epinephrine cause vasodilation or vasoconstriction? _____

7. What blood vessels did histamine and epinephrine *directly* affect?

_____

Laboratory Report 30

Student _____

Lab Section _____

The Respiratory Organs

A. Figures

List the labels for Figures 30.1 and 30.2.

Figure 30.1

1. _____
2. _____
3. _____
4. _____
5. _____
6. _____
7. _____
8. _____
9. _____
10. _____
11. _____

12. _____
13. _____
14. _____
15. _____
16. _____
17. _____
18. _____
19. _____
20. _____
21. _____

Figure 30.2

1. _____
2. _____
3. _____
4. _____
5. _____
6. _____
7. _____
8. _____
9. _____
10. _____
11. _____
12. _____

B. Respiratory Structures

Select the structures described by the statements below.

alveoli
bronchi
bronchioles
epiglottis
hard palate
larynx
lingual tonsils
nasal conchae
nasopharynx

oral cavity
oral vestibule
oropharynx
palatine tonsils
pharyngeal tonsils
pleural cavity
soft palate
trachea

1. Cavity between lips and teeth.
2. Tonsils in oropharynx.
3. Increase surface area of nasal cavity.
4. Sites of gas exchange.
5. Tubes branching from inferior end of trachea.
6. Prevents food from entering larynx.
7. Potential space between visceral and parietal pleurae.
8. Small tubes leading to alveoli.
9. Tonsils in nasopharynx.
10. Separated from oral cavity by hard palate.
11. Tiny air sacs in lung.
12. Portion of palate supported by bone.
13. Contains the vocal folds.
14. Where swallowing reflex is initiated.
15. Tonsils at base of the tongue.

Respiratory Structures

1. _____
2. _____
3. _____
4. _____
5. _____
6. _____
7. _____
8. _____
9. _____
10. _____
11. _____
12. _____
13. _____
14. _____
15. _____

C. Air Pathway

Complete the following paragraph.

During inspiration, air passes through the ____1____ into the nasal cavity where it is warmed. The surface area of the nasal cavity is increased by the three ____2____ . From the nasal cavity, air passes through the ____3____ and ____4____ , and into the larynx, which contains the ____5____ folds. From the larynx, air passes down the ____6____ , which divides into the primary ____7____ that enter each lung. The tubes continue to branch and ultimately form small tubules called ____8____ which lead to the alveoli. Gas exchange occurs between ____9____ in the alveoli and ____10____ in the capillaries surrounding the alveoli. During ____11____ , air flow is reversed through this same pathway.

Air Pathway

1. _____
2. _____
3. _____
4. _____
5. _____
6. _____
7. _____
8. _____
9. _____
10. _____
11. _____

D. Sheep Pluck Dissection

1. Describe the shape of the cartilaginous rings of the trachea.

   _____

2. What contributes to the smooth and slimy nature of the inside of the trachea? _____

   _____

3. Indicate the number of lobes in the:

   left lung _____   right lung _____

4. Lung tissue seems to be full of small spaces. What pulmonary structures contribute most to this appearance?

   _____

E. Microscopic Study

1. Diagram a section of tracheal wall at 100× magnification and label the ciliated epithelial lining and hyaline cartilage.
2. Diagram the epithelial lining at 400×.
3. Diagram 2–3 alveoli in lung tissue at 400× magnification.

   Tracheal Wall   Epithelial Lining   Alveoli

Respiratory Physiology

A. Control of Breathing

Complete the following paragraph.

Breathing is controlled by the ____1____ center located in the ____2____ of the brain. The most important stimulus for breathing is the direct stimulation of this center by an ____3____ concentration of ____4____ in the blood and cerebrospinal fluid. ____5____ also acts directly on the control center. In contrast, ____6____ acts indirectly on the control center. Its concentration is detected by ____7____ in the carotid arteries and aorta, which send impulses to the control center. Usually, ____8____ concentration is of little importance in stimulating inspiration.

Control of Breathing

1. _____

2. _____

3. _____

4. _____

5. _____

6. _____

7. _____

8. _____

B. Breathing Experiments

Hyperventilation

1. Describe how you felt after hyperventilating. _____

_____

2. Explain why it was easier to breathe into the paper bag. _____

_____

_____

3. Indicate how long you were able to hold your breath.

after normal breathing _____ sec.   after hyperventilating _____ sec.

Explain your results. _____

_____

B. Breathing Experiments (continued)

Deglutition Apnea

1. Indicate whether the urge to breathe decreased or increased when you sipped the water through the straw.

_____

2. Is this phenomenon a reflex or a voluntary response? _____

C. Spirometry and Lung Volumes

Record your lung volumes determined by spirometry in the table below. Circle those values that are 22% or more below normal.

Lung Capacities	Normal (ml.)	Your Capacities
Tidal Volume (TV)	500	
Minute Respiratory Volume (MRV) (MRV × TV × Resp. Rate)	6,000	
Expiratory Reserve Volume (ERV)	1,100	
Vital Capacity (VC) (VC = TV + ERV + IRV)	See Appendix	
Inspiratory Capacity (IC) (IC = VC − ERV)	3,000	
Inspiratory Reserve Volume (IRV) (IRV = IC − TV)	2,500	

The Digestive Organs

A. *Figures*

List the labels for Figures 32.1, 32.2, 32.3, 32.4, and 32.5.

Figure 32.1

1. _____
2. _____
3. _____
4. _____
5. _____
6. _____
7. _____

Figure 32.4

1. _____
2. _____
3. _____
4. _____
5. _____
6. _____

Figure 32.2

1. _____
2. _____
3. _____
4. _____
5. _____
6. _____
7. _____
8. _____
9. _____
10. _____
11. _____
12. _____

Figure 32.3

1. _____
2. _____
3. _____
4. _____
5. _____
6. _____
7. _____
8. _____
9. _____

Figure 32.5

1. _____
2. _____
3. _____
4. _____
5. _____
6. _____
7. _____
8. _____
9. _____
10. _____

11. _____
12. _____
13. _____
14. _____
15. _____
16. _____
17. _____
18. _____
19. _____
20. _____

21. _____
22. _____
23. _____
24. _____
25. _____
26. _____
27. _____
28. _____
29. _____

B. Oral Cavity

Select the structures described by the statements below.

cementum
crown of tooth
dentin
enamel
gingiva
oropharynx
papillae, filiform
papillae, fungiform

papillae, vallate
parotid gland
pulp cavity
root of tooth
sublingual gland
submandibular gland
tonsils, lingual
tonsils, palatine

1. Portion of a tooth covered with enamel.
2. Papillae giving rough texture to the tongue.
3. Mucous membrane covering the neck of each tooth.
4. Salivary gland located anterior to the ear.
5. Substance forming most of a tooth.
6. Bone-like substance between root and peridontal membrane.
7. 8–12 large papillae located at back of tongue.
8. Salivary gland located in anterior floor of oral cavity.
9. Contains nerves and blood vessels of a tooth.
10. Where swallowing reflex is initiated.
11. Papillae with taste buds scattered over surface of tongue.
12. Tonsils located in oropharynx.
13. Salivary gland secreting salivary amylase.
14. Salivary glands secreting mucin.

C. Esophagus to Anus

Select the structures described by the statements below.

cardiac valve
cecum
colon
common bile duct
cystic duct
duodenum
esophagus
hepatic duct

ileocecal valve
ileum
jejunum
pancreatic duct
pyloric valve
rectum
stomach
villi

1. Valve that opens to let food into the stomach.
2. Valve that opens to let chyme out of stomach.
3. Valve between the small and large intestines.
4. Pouch from which appendix extends.
5. Tube that carries food to the stomach.
6. The first 10–12 inches of the small intestine.
7. Duct that carries bile from the liver.
8. Duct that carries bile to and from gall bladder.
9. The terminal 5–6 inches of the large intestine.
10. Duct that carries bile and pancreatic juice to the duodenum.
11. Microscopic projections through which absorption of nutrients occurs.
12. Site of bacterial decomposition of food residue.
13. Middle portion of the small intestine.
14. Where food is mixed with gastric juice.

Oral Cavity

1. _____
2. _____
3. _____
4. _____
5. _____
6. _____
7. _____
8. _____
9. _____
10. _____
11. _____
12. _____
13. _____
14a. _____
14b. _____

Esophagus to Anus

1. _____
2. _____
3. _____
4. _____
5. _____
6. _____
7. _____
8. _____
9. _____
10. _____
11. _____
12. _____
13. _____
14. _____

Laboratory Report **33**

Digestion

A. Completion

1. Explain the meaning of enzymatic hydrolysis. _____
   _____
   _____
   _____

2. What determines the three-dimensional shape of an enzyme? _____
   _____

3. Explain the importance of the shape of an enzyme. _____
   _____

4. Explain how temperature and pH changes may alter the shape of an enzyme. _____
   _____
   _____

B. Digestive Enzymes

Select the enzyme described below.

amylase	lipase
carboxypeptidase	pepsin
chymotrypsin	peptidase
lactase	sucrase
maltase	trypsin

1. Enzyme present in saliva and pancreatic juice.
2. Converts lipids to fatty acids and glycerol.
3. Intestinal enzyme converting peptides to amino acids.
4. Converts starch to maltose.
5. Converts maltose to glucose.
6. Converts sucrose to glucose and fructose.
7. Converts lactose to glucose and galactose.
8. Gastric enzyme converting proteins to peptides.
9. Two pancreatic enzymes converting proteins to peptides only.
10. Pancreatic enzyme converting peptides to peptide residues and amino acids.
11. Enzyme found in gastric, pancreatic, and intestinal juices.

Enzymes

1. _____
2. _____
3. _____
4. _____
5. _____
6. _____
7. _____
8. _____
9a. _____
9b. _____
10. _____
11. _____

C. Completion
Indicate the end products of digestion for:

carbohydrates _____

lipids _____

proteins _____

D. Starch Digestion Experiments
Record your data from the experiments in the tables below. In the starch digestion row, use a "0" to indicate no digestion, a "+" to indicate some digestion, and a "C" to indicate complete digestion.

Table I. Amylase Activity at 20° C.

Tube No.	1	2	3	4	5	6	7	8	9	10
Color										
Starch Digestion										

Table II. Amylase Activity at 37° C.

Tube No.	1	2	3	4	5	6	7	8	9	10
Color										
Starch Digestion										

Table III. Amylase Activity When Boiled.

Tube No.	1	2	3
Color			
Starch Digestion			

Table IV. pH and Amylase Activity.

Time	2 min.			4 min.			4 min.		
Tube No.	1	2	3	4	5	6	7	8	9
pH	5	7	9	5	7	9	5	7	9
Color									
Starch Digestion									

E. Conclusions

1. How long did it take for starch digestion to be:

   evident?   20° C _____ min.   37° C _____ min.

   complete?   20° C _____ min.   37° C _____ min.

2. Did you expect this pattern of results? _____ Why? _____

   _____

3. What effect did boiling have on amylase? _____

   _____

4. At which pH was digestion most rapid? _____ How does this compare with the normal pH of saliva?_____

   _____

5. Gastric juice has a pH of 2 or less. What will its effect be on salivary amylase? _____

   _____

Do you expect any starch digestion in the stomach? _____

The Urinary Organs

A. *Figures*

List the labels for Figures 34.1, 34.2, and 34.3.

Figure 34.1

1. _____
2. _____
3. _____
4. _____
5. _____
6. _____
7. _____
8. _____
9. _____
10. _____

Figure 34.2

1. _____
2. _____
3. _____
4. _____

5. _____
6. _____
7. _____
8. _____
9. _____
10. _____
11. _____
12. _____
13. _____
14. _____
15. _____
16. _____
17. _____
18. _____
19. _____
20. _____
21. _____

Figure 34.3

1. _____
2. _____
3. _____
4. _____
5. _____
6. _____
7. _____
8. _____
9. _____
10. _____
11. _____

B. *Anatomy*

Select the structures described by the statements below.

Bowman's capsule	medulla	renal pelvis
calyces	nephron	renal pyramids
collecting duct	renal capsule	ureters
cortex	renal column	urethra
glomerulus	renal papilla	

1. Tubes that drain the kidneys.
2. Tube that drains the urinary bladder.
3. Portion of kidney that contains renal corpuscles.
4. Cone-shaped portions of the medulla.
5. Part of kidney containing collecting ducts.
6. Functional unit of the kidney.
7. Capillary tuft within renal corpuscle.
8. Tip of renal pyramid.
9. Tube receiving urine from several nephrons.
10. Thin, fibrous covering of the kidney.
11. Cortical tissue between renal pyramids.
12. Short tubes receiving urine from renal pyramids.
13. Part of kidney composed of renal pyramids and renal columns.
14. Funnel-like structure receiving urine from calyces.
15. Cup-like structure enveloping a glomerulus.

Anatomy

1. _____
2. _____
3. _____
4. _____
5. _____
6. _____
7. _____
8. _____
9. _____
10. _____
11. _____
12. _____
13. _____
14. _____
15. _____

C. Table 34.1

Calculate the concentrations (grams/liter) of the following substances in Table 34.1 and record them below.

Substance	Plasma	Filtrate	Urine
Proteins	_____	_____	_____
Chloride ions	_____	_____	_____
Potassium ions	_____	_____	_____
Sodium ions	_____	_____	_____
Glucose	_____	_____	_____
Creatinine	_____	_____	_____
Urea	_____	_____	_____
Uric acid	_____	_____	_____

1. Why do so few proteins enter the filtrate? _____

_____

2. What substances have different concentrations in the plasma and in the filtrate? _____

_____

3. What substances are completely reabsorbed? _____

_____

4. What substances are more concentrated in urine than in the filtrate? _____

_____

5. Does urine formation remove all nitrogenous wastes or only reduce their concentrations in the plasma? _____

_____

6. What percentage of the filtrate is reabsorbed? _____

The reabsorption of what substance accounts for most of the volume reduction and the concentration of urine

solutes? _____

7. List the three processes of urine formation.

_____   _____   _____

D. Microscopic Study

Diagram a glomerulus and Bowman's capsule as observed on the prepared slide of kidney tissue.

Laboratory Report **35**

Urine and Urinalysis

A. Terminology

Write the term defined below in the answer column.

1. Inflammation of the kidney (general).
2. Albumin in the urine.
3. A measure of the concentration of solutes in urine.
4. Erythrocytes in the urine.
5. Inflammation of the urinary bladder.
6. Most abundant inorganic compound in urine.
7. Leukocytes or pus components in the urine.
8. pH range of normal urine.
9. Hemoglobin in the urine.
10. More than a trace of glucose in the urine.
11. Ketones in the urine.
12. Inflammation of the kidney involving glomeruli.
13. Accumulations of materials hardened in tubules.
14. Most abundant nitrogenous waste.
15. Excessive urine production.
16. Bile pigment in the urine.
17. Inflammation of the urethra.
18. Kidney stones.
19. Little or no urine production.
20. Most abundant inorganic solute in urine.

B. Clinical Significance

Select the name of the possible clinical condition from the list below that is indicated by the urinalysis results. Write your answer in the answer column.

Clinical Conditions

calculi	hemolytic anemia
diabetes insipidus	liver pathology
diabetes mellitus	renal failure
glomerulonephritis	urinary infection

Urinalysis Results

1. Anuria
2. Pyuria, hematuria, kidney/ureter pain
3. Hemoglobinuria
4. Polyuria, very low specific gravity
5. Urobilinogenuria and/or bilirubinuria
6. Polyuria, glycosuria, ketonuria, high specific gravity
7. Pyuria, bacteriuria
8. Low pH, albuminuria, pyuria, casts

Terminology

1. _____
2. _____
3. _____
4. _____
5. _____
6. _____
7. _____
8. _____
9. _____
10. _____
11. _____
12. _____
13. _____
14. _____
15. _____
16. _____
17. _____
18. _____
19. _____
20. _____

Clinical Significance

1. _____
2. _____
3. _____
4. _____
5. _____
6. _____
7. _____
8. _____

C. Urinalysis

Record on the chart below the results of the analysis of each urine specimen. All of the following tests except color and turbidity are easily done with a 10SG-Multistix.

TEST	NORMAL VALUES	UNKNOWNS No. _____	UNKNOWNS No. _____	STUDENT'S URINE
Leukocytes	None			
Nitrite	None			
Urobilinogen	<2 mg/dl			
Protein	None			
pH	4.8 to 7.5			
Blood	None			
Sp. Gravity	1.002–1.035			
Ketone	None			
Bilirubin	None			
Glucose	None or Trace			
Color	Pale yellow to amber			
Turbidity	Clear			

Indicate the possible clinical conditions associated with the unknowns.

No. _____ _____

No. _____ _____

Laboratory Report **36**

Student ____________________

Lab Section ____________

The Endocrine Glands

A. *Figure 36.1*

List the labels for Figure 36.1.

Figure 36.1

1. ____________
2. ____________
3. ____________
4. ____________
5. ____________
6. ____________
7. ____________
8. ____________
9. ____________
10. ____________
11. ____________
12. ____________
13. ____________
14. ____________
15. ____________
16. ____________
17. ____________
18. ____________
19. ____________
20. ____________
21. ____________

B. *Hormone Sources*

Select the glandular tissue that produces the hormones listed.

adrenal cortex
adrenal medulla
hypothalamus
ovary: corpus luteum
ovary: follicles
pancreas
parathyroid gland
pineal gland
pituitary: ant. lobe
pituitary: post. lobe
testis
thymus gland
thyroid gland

1. ACTH
2. ADH
3. Aldosterone
4. Calcitonin
5. Cortisol
6. Epinephrine
7. Estrogens
8. FSH
9. Glucagon
10. Growth hormone
11. Insulin
12. LH
13. Melatonin
14. Oxytocin
15. Parathyroid hormone
16. Prolactin
17. Progesterone
18. Testosterone
19. Thymosin
20. Thyrotropin
21. Thyroxine

Hormone Sources

1. ____________
2. ____________
3. ____________
4. ____________
5. ____________
6. ____________
7. ____________
8. ____________
9. ____________
10. ____________
11. ____________
12. ____________
13. ____________
14. ____________
15. ____________
16. ____________
17. ____________
18. ____________
19. ____________
20. ____________
21. ____________

C. Hormone Functions

Select the hormones that perform the functions listed below.

Group 1

aldosterone   epinephrine
cortisol   melatonin

1. Increases resistance to stress.
2. Promotes excretion of K^+.
3. Increases reabsorption of Na^+.
4. Increases blood glucose level (2).
5. Prepares body for response to emergencies.
6. Increases heart and respiration rate.

Group 2

calcitonin   parathyroid hormone
glucagon   somatostatin
insulin   thyroid hormone

1. Promotes bone deposition.
2. Promotes bone reabsorption.
3. Increases blood glucose levels.
4. Decreases blood glucose levels.
5. Stimulates metabolism.
6. Raises blood calcium level.
7. Lowers blood calcium level.
8. Aids entrance of glucose into cells.

Group 3

ACTH   LH
ADH   oxytocin
estrogens   progesterone
FSH   testosterone
growth hormone   thyrotropin

1. Stimulates descent of testes.
2. Promotes breast development.
3. Promotes release of milk.
4. Stimulates thyroxin production.
5. Stimulates follicle development.
6. Stimulates glucocorticoid production.
7. Stimulates growth of body cells.
8. Prepares uterus for implantation of embryo.
9. Rapid increase causes ovulation.
10. Stimulates spermatogenesis.
11. Promotes water reabsorption by kidneys.
12. Strengthens uterine contractions.

Hormone Functions
Group 1

1. _____
2. _____
3. _____
4a. _____
4b. _____
5. _____
6. _____

Group 2

1. _____
2. _____
3. _____
4. _____
5. _____
6. _____
7. _____
8. _____

Group 3

1. _____
2. _____
3. _____
4. _____
5. _____
6. _____
7. _____
8. _____
9. _____
10. _____
11. _____
12. _____

D. Hormonal Imbalance

Select the hormones involved in the maladies listed and indicate if the condition is due to an excess (+) or deficiency (−).

glucagon
glucocorticoids
growth hormone
insulin

mineralocorticoids
parathyroid hormone
thyroid hormone

1. Acromegaly
2. Addison's disease
3. Cretinism
4. Cushing's syndrome
5. Diabetes mellitus
6. Dwarfism
7. Gigantism
8. Hypoglycemia
9. Myxedema
10. Tetany

E. Figure 36.9

List the labels for Figure 36.9.

Hormone Imbalance

1. _____
2. _____
3. _____
4. _____
5. _____
6. _____
7. _____
8. _____
9. _____
10. _____

Figure 36.9

1. _____
2. _____
3. _____
4. _____
5. _____
6. _____
7. _____
8. _____
9. _____
10. _____
11. _____
12. _____
13. _____
14. _____

The Reproductive Organs

A. *Figures*

List the labels for Figures 37.1, 37.3, 37.4, and 37.5.

Figure 37.1

1. _____
2. _____
3. _____
4. _____
5. _____
6. _____
7. _____
8. _____
9. _____
10. _____
11. _____
12. _____
13. _____
14. _____
15. _____
16. _____
17. _____
18. _____
19. _____
20. _____

Figure 37.3

1. _____
2. _____
3. _____
4. _____
5. _____
6. _____
7. _____
8. _____
9. _____

Figure 37.4

1. _____
2. _____
3. _____
4. _____
5. _____
6. _____
7. _____
8. _____
9. _____

10. _____
11. _____
12. _____
13. _____
14. _____

Figure 37.5

1. _____
2. _____
3. _____
4. _____
5. _____
6. _____
7. _____
8. _____
9. _____
10. _____
11. _____
12. _____
13. _____
14. _____
15. _____

B. Gametogenesis

Select the cells described by the statements below.

Spermatogenesis

primary spermatocytes
secondary spermatocytes
spermatids
spermatogonia
spermatozoa

Oogenesis

oogonia
ovum
polar bodies
primary oocyte
secondary oocyte

1. Haploid male gametes.
2. Haploid female gamete.
3. Divide mitotically to form primary spermatocytes.
4. Cells composing germinal epithelium in ovary.
5. Large haploid cell released in ovulation.
6. Male cells formed by 1st meiotic division.
7. Large female cell formed by 1st meiotic division.
8. Male cells in which 1st meiotic division occurs.
9. Female cells in which 1st meiotic division occurs.
10. Large female cell formed by 2nd meiotic division.

C. Male Organs

Select the structures described by the statements below.

bulbourethral gland
corpora cavernosa
corpus spongiosum
ejaculatory duct
epididymis
penis
prepuce

prostate gland
prostatic urethra
seminal vesicle
seminiferous tubule
testes
vas deferens

1. Male copulatory organ.
2. Male gonads.
3. Sheath of skin over the glans penis.
4. Carries sperm from epididymis to ejaculatory duct.
5. Tube from urinary bladder into prostate gland.
6. Secretions emptied into base of penile urethra.
7. Secretions emptied into prostatic urethra.
8. Secretions emptied into ejaculatory duct.
9. Erectile tissue forming glans penis.
10. Two dorsally located cylinders of erectile tissue.
11. Slender tube containing sperm-forming cells.
12. Site of sperm storage and maturation.

Gametogenesis

1. _____
2. _____
3. _____
4. _____
5. _____
6. _____
7. _____
8. _____
9. _____
10. _____

Male Organs

1. _____
2. _____
3. _____
4. _____
5. _____
6. _____
7. _____
8. _____
9. _____
10. _____
11. _____
12. _____

D. Female Organs

Select the structures described by the statements below.

broad ligament
cervix
clitoris
fimbriae
hymen
infundibulum
labia majora
labia minora
mons pubis

myometrium
ovarian ligament
ovaries
suspensory ligament
uterosacral ligament
uterine tubes
vagina
vestibular glands

1. Female copulatory organ.
2. Muscular wall of the uterus.
3. Narrow neck of the uterus.
4. Female gonads.
5. Finger-like projections around infundibulum.
6. Source of ovum.
7. Provides vaginal lubrication during intercourse.
8. Duct carrying oocyte toward uterus.
9. Inner folds of vulva surrounding vestibule.
10. Nodule of erectile tissue.
11. Outer fatty folds of vulva merging with mons pubis.
12. Funnel-like opening of uterine tube.
13. Narrow ligament attaching ovary to uterus.
14. Sheet-like ligament supporting ovary, uterine tubes, and uterus.
15. Attaches uterus to posterior pelvic wall.
16. Ligament from ovary to lateral pelvic wall.

E. Completion

1. The inner lining of the uterus is the _____ .
2. The mixture of sperm and secretions of accessory glands is called _____ .
3. Testes are located in a sac called the _____ .
4. Fertilization usually occurs in the upper third of the _____ .

Female Organs

1. _____
2. _____
3. _____
4. _____
5. _____
6. _____
7. _____
8. _____
9. _____
10. _____
11. _____
12. _____
13. _____
14. _____
15. _____
16. _____

Completion

1. _____
2. _____
3. _____
4. _____

Calibrating the Physiograph
Exercise 20

Balancing the Myograph to Physiograph Transducer Coupler

Balancing (matching) of the transducer signal with the amplifier is necessary to get the recording pen to stay on a preset baseline when the RECORD button is in either the OFF or ON position. Proceed as follows:

1. Before starting, be sure that the RECORD switch is OFF and that the myograph jack is inserted into the transducer coupler.
2. Set the outer knob of the SENSITIVITY control to its lowest numbered setting, i.e. its highest sensitivity.
3. Set the paper speed at 0.5 cm/sec and lower the pen lifter.
4. Adjust the PEN POSITION control so that it is *exactly* on the center line.
5. Place the RECORD switch in the ON position. The pen will probably be moving a large distance either up or down.
6. With the BALANCE control, adjust the pen so that it is again writing *exactly* on the center line.
7. Check the balance by placing the RECORD button in the OFF position. *The system is balanced if the pen stays on the center line.*
8. If the pen moves from the center line, readjust the pen to the center line with the PEN POSITION control. Place the RECORD button in the ON position and repeat the balancing procedure. The system is balanced only when the pen remains at the center line when the RECORD button is in both the ON or OFF positions.

Calibration of the Myograph Transducer

Calibration of the transducer is necessary so that you know exactly how much tension, in grams, is being exerted by the muscle during contraction. The channel amplifier must be balanced first. This calibration will be to obtain a 5 cm pen deflection with a tension of 100 gm. *The following procedures apply to couplers that lack built-in calibration buttons.* Figure A-1 illustrates the steps.

Materials

100 gram weight
50 gram weight

1. Start the paper drive at 0.1 cm/sec and lower the pen lifter to record on the desired channel.
2. Check to see that the channel amplifier is balanced.
3. With the PEN POSITION control, set the baseline exactly 2.5 cm (5 blocks) below the channel center line.
4. Rotate the outer knob of the SENSITIVITY control fully counterclockwise to the lowest sensitivity setting (the 1000 setting). Be sure that the inner knob of the SENSITIVITY control is in its fully clockwise "clicked" position.
5. Place the RECORD button in the ON position.
6. Suspend a 100 gm weight to the actuator of the myograph. Note that the addition of the weight has caused the baseline to move upward approximately 0.5 cm (one block). See Step C in Figure A-1.
7. Rotate the *outer* knob of the SENSITIVITY control clockwise until the pen exceeds 5 cm (10 blocks) of deflection from the original baseline. See Step F in Figure A-1.
8. Rotate the *inner* knob of the SENSITIVITY control counterclockwise to bring the pen downward until it is *exactly* 5 cm (10 blocks) of deflection from the original baseline.
9. Remove the weight. The pen should return to the original baseline. If it does not, reset the baseline with the PEN POSITION control, reapply the weight, and rotate the *inner* knob of the SENSITIVITY control until the pen is exactly 5 cm from the original baseline.
10. Attach a 50 gm weight to the actuator of the myograph to see if you get 2.5 cm (5 blocks) of deflection as shown in Steps I and J in Figure A-1. If so, the unit is properly calibrated.

Calibration of Myograph Transducer with a Calibration Button

When the coupler has a calibration button, it is not necessary to use a 100 gm weight. To calibrate, just press the 100 gm button and follow the steps to produce a 5 cm pen deflection.

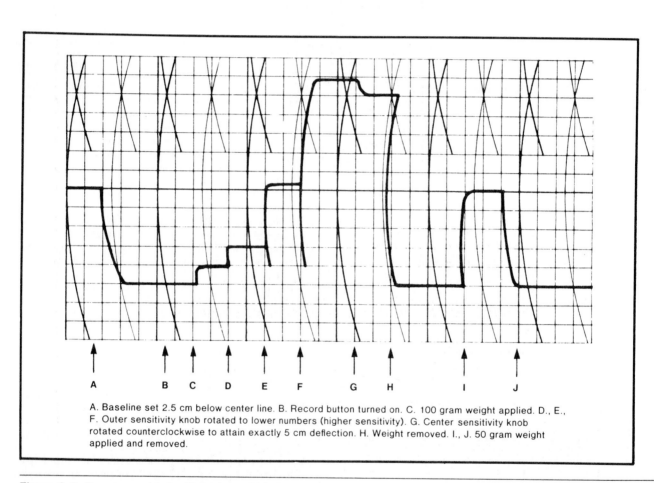

A. Baseline set 2.5 cm below center line. B. Record button turned on. C. 100 gram weight applied. D., E., F. Outer sensitivity knob rotated to lower numbers (higher sensitivity). G. Center sensitivity knob rotated counterclockwise to attain exactly 5 cm deflection. H. Weight removed. I., J. 50 gram weight applied and removed.

Figure A-1 Sample record of calibration at 100 grams per 5 centimeters.

Sample Physiograph Records

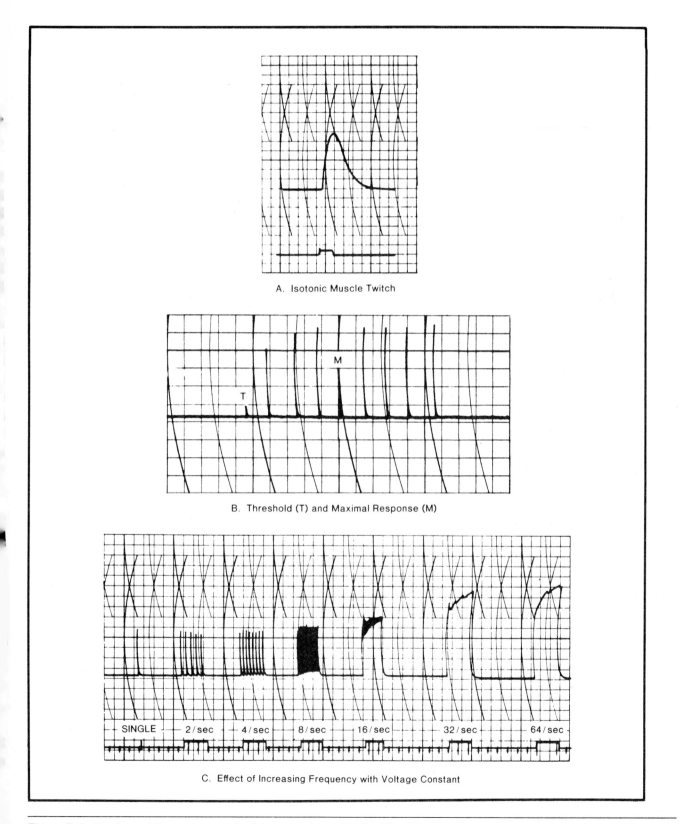

A. Isotonic Muscle Twitch

B. Threshold (T) and Maximal Response (M)

| SINGLE | 2/sec | 4/sec | 8/sec | 16/sec | 32/sec | 64/sec |

C. Effect of Increasing Frequency with Voltage Constant

Figure B-1 Physiograph® sample records of frog muscle contraction (Ex. 20).

Vital Capacities

Table 1 Predicted vital capacities for females.

HEIGHT IN CENTIMETERS AND INCHES

AGE	CM. 152 / IN. 59.8	154 60.6	156 61.4	158 62.2	160 63.0	162 63.7	164 64.6	166 65.4	168 66.1	170 66.9	172 67.7	174 68.5	176 69.3	178 70.1	180 70.9	182 71.7	184 72.4	186 73.2	188 74.0
16	3,070	3,110	3,150	3,190	3,230	3,270	3,310	3,350	3,390	3,430	3,470	3,510	3,550	3,590	3,630	3,670	3,715	3,755	3,800
17	3,055	3,095	3,135	3,175	3,215	3,255	3,295	3,335	3,375	3,415	3,455	3,495	3,535	3,575	3,615	3,655	3,695	3,740	3,780
18	3,040	3,080	3,120	3,160	3,200	3,240	3,280	3,320	3,360	3,400	3,440	3,480	3,520	3,560	3,600	3,640	3,680	3,720	3,760
20	3,010	3,050	3,090	3,130	3,170	3,210	3,250	3,290	3,330	3,370	3,410	3,450	3,490	3,525	3,565	3,605	3,645	3,695	3,720
22	2,980	3,020	3,060	3,095	3,135	3,175	3,215	3,255	3,290	3,330	3,370	3,410	3,450	3,490	3,530	3,570	3,610	3,650	3,685
24	2,950	2,985	3,025	3,065	3,100	3,140	3,180	3,220	3,260	3,300	3,335	3,375	3,415	3,455	3,490	3,530	3,570	3,610	3,650
26	2,920	2,960	3,000	3,035	3,070	3,110	3,150	3,190	3,230	3,265	3,300	3,340	3,380	3,420	3,455	3,495	3,530	3,570	3,610
28	2,890	2,930	2,965	3,000	3,040	3,070	3,115	3,155	3,190	3,230	3,270	3,305	3,345	3,380	3,420	3,460	3,495	3,535	3,570
30	2,860	2,895	2,935	2,970	3,010	3,045	3,085	3,120	3,160	3,195	3,235	3,270	3,310	3,345	3,385	3,420	3,460	3,495	3,535
32	2,825	2,865	2,900	2,940	2,975	3,015	3,050	3,090	3,125	3,160	3,200	3,235	3,275	3,310	3,350	3,385	3,425	3,460	3,495
34	2,795	2,835	2,870	2,910	2,945	2,980	3,020	3,055	3,090	3,130	3,165	3,200	3,240	3,275	3,310	3,350	3,385	3,425	3,460
36	2,765	2,805	2,840	2,875	2,910	2,950	2,985	3,020	3,060	3,095	3,130	3,165	3,205	3,240	3,275	3,310	3,350	3,385	3,420
38	2,735	2,770	2,810	2,845	2,880	2,915	2,950	2,990	3,025	3,060	3,095	3,130	3,170	3,205	3,240	3,275	3,310	3,350	3,385
40	2,705	2,740	2,775	2,810	2,850	2,885	2,920	2,955	2,990	3,025	3,060	3,095	3,135	3,170	3,205	3,240	3,275	3,310	3,345
42	2,675	2,710	2,745	2,780	2,815	2,850	2,885	2,920	2,955	2,990	3,025	3,060	3,100	3,135	3,170	3,205	3,240	3,275	3,310
44	2,645	2,680	2,715	2,750	2,785	2,820	2,855	2,890	2,925	2,960	2,995	3,030	3,060	3,095	3,130	3,165	3,200	3,235	3,270
46	2,615	2,650	2,685	2,715	2,750	2,785	2,820	2,855	2,890	2,925	2,960	2,995	3,030	3,060	3,095	3,130	3,165	3,200	3,235
48	2,585	2,620	2,650	2,685	2,715	2,750	2,785	2,820	2,855	2,890	2,925	2,960	2,995	3,030	3,060	3,095	3,130	3,160	3,195
50	2,555	2,590	2,625	2,655	2,690	2,720	2,755	2,785	2,820	2,855	2,890	2,925	2,955	2,990	3,025	3,060	3,090	3,125	3,155
52	2,525	2,555	2,590	2,625	2,655	2,690	2,720	2,755	2,790	2,820	2,855	2,890	2,925	2,955	2,990	3,020	3,055	3,090	3,125
54	2,495	2,530	2,560	2,590	2,625	2,655	2,690	2,720	2,755	2,790	2,820	2,855	2,885	2,920	2,950	2,985	3,020	3,050	3,085
56	2,460	2,495	2,525	2,560	2,590	2,625	2,655	2,690	2,720	2,755	2,790	2,820	2,855	2,885	2,920	2,950	2,980	3,015	3,045
58	2,430	2,460	2,495	2,525	2,560	2,590	2,625	2,655	2,690	2,720	2,750	2,785	2,815	2,850	2,880	2,920	2,945	2,975	3,010
60	2,400	2,430	2,460	2,495	2,525	2,560	2,590	2,625	2,655	2,685	2,720	2,750	2,780	2,810	2,845	2,875	2,915	2,940	2,970
62	2,370	2,405	2,435	2,465	2,495	2,525	2,560	2,590	2,620	2,655	2,685	2,715	2,745	2,775	2,810	2,840	2,870	2,900	2,935
64	2,340	2,370	2,400	2,430	2,465	2,495	2,525	3,555	2,585	2,620	2,650	2,680	2,710	2,740	2,770	2,805	2,835	2,865	2,895
66	2,310	2,340	2,370	2,400	2,430	2,460	2,495	2,525	2,555	2,585	2,615	2,645	2,675	2,705	2,735	2,765	2,800	2,825	2,860
68	2,280	2,310	2,340	2,370	2,400	2,430	2,460	2,490	2,520	2,550	2,580	2,610	2,640	2,670	2,700	2,730	2,760	2,795	2,820
70	2,250	2,280	2,310	2,340	2,370	2,400	2,425	2,455	2,485	2,515	2,545	2,575	2,605	2,635	2,665	2,695	2,725	2,755	2,780
72	2,220	2,250	2,280	2,310	2,335	2,365	2,395	2,425	2,455	2,480	2,510	2,540	2,570	2,600	2,630	2,660	2,685	2,715	2,745
74	2,190	2,220	2,245	2,275	2,305	2,335	2,360	2,390	2,420	2,450	2,475	2,505	2,535	2,565	2,590	2,620	2,650	2,680	2,710

From: Archives of Environmental Health
February 1966, Vol. 12, pp. 146–189
E. A. Gaensler, MD and G. W. Wright, MD

Table 2 Predicted vital capacities for males.

HEIGHT IN CENTIMETERS AND INCHES

AGE	CM. 152	154	156	158	160	162	164	166	168	170	172	174	176	178	180	182	184	186	188
	IN. 59.8	60.6	61.4	62.2	63.0	63.7	64.6	65.4	66.1	66.9	67.7	68.5	69.3	70.1	70.9	71.7	72.4	73.2	74.0
16	3,920	3,975	4,025	4,075	4,130	4,180	4,230	4,285	4,335	4,385	4,440	4,490	4,540	4,590	4,645	4,695	4,745	4,800	4,850
18	3,890	3,940	3,995	4,045	4,095	4,145	4,200	4,250	4,300	4,350	4,405	4,455	4,505	4,555	4,610	4,660	4,710	4,760	4,815
20	3,860	3,910	3,960	4,015	4,065	4,115	4,165	4,215	4,265	4,320	4,370	4,420	4,470	4,520	4,570	4,625	4,675	4,725	4,775
22	3,830	3,880	3,930	3,980	4,030	4,080	4,135	4,185	4,235	4,285	4,335	4,385	4,435	4,485	4,535	4,585	4,635	4,685	4,735
24	3,785	3,835	3,885	3,935	3,985	4,035	4,085	4,135	4,185	4,235	4,285	4,330	4,380	4,430	4,480	4,530	4,580	4,630	4,680
26	3,755	3,805	3,855	3,905	3,955	4,000	4,050	4,100	4,150	4,200	4,250	4,300	4,350	4,395	4,445	4,495	4,545	4,595	4,645
28	3,725	3,775	3,820	3,870	3,920	3,970	4,020	4,070	4,115	4,165	4,215	4,265	4,310	4,360	4,410	4,460	4,510	4,555	4,605
30	3,695	3,740	3,790	3,840	3,890	3,935	3,985	4,035	4,080	4,130	4,180	4,230	4,275	4,325	4,375	4,425	4,470	4,520	4,570
32	3,665	3,710	3,760	3,810	3,855	3,905	3,950	4,000	4,050	4,095	4,145	4,195	4,240	4,290	4,340	4,385	4,435	4,485	4,530
34	3,620	3,665	3,715	3,760	3,810	3,855	3,905	3,950	4,000	4,045	4,095	4,140	4,190	4,225	4,285	4,330	4,380	4,425	4,475
36	3,585	3,635	3,680	3,730	3,775	3,825	3,870	3,920	3,965	4,010	4,060	4,105	4,155	4,200	4,250	4,295	4,340	4,390	4,435
38	3,555	3,605	3,650	3,695	3,745	3,790	3,840	3,885	3,930	3,980	4,025	4,070	4,120	4,165	4,210	4,260	4,305	4,350	4,400
40	3,525	3,575	3,620	3,665	3,710	3,760	3,805	3,850	3,900	3,945	3,990	4,035	4,085	4,130	4,175	4,220	4,270	4,315	4,360
42	3,495	3,540	3,590	3,635	3,680	3,725	3,770	3,820	3,865	3,910	3,955	4,000	4,050	4,095	4,140	4,185	4,230	4,280	4,325
44	3,450	3,495	3,540	3,585	3,630	3,675	3,725	3,770	3,815	3,860	3,905	3,950	3,995	4,040	4,085	4,130	4,175	4,220	4,270
46	3,420	3,465	3,510	3,555	3,600	3,645	3,690	3,735	3,780	3,825	3,870	3,915	3,960	4,005	4,050	4,095	4,140	4,185	4,230
48	3,390	3,435	3,480	3,525	3,570	3,615	3,655	3,700	3,745	3,790	3,835	3,880	3,925	3,970	4,015	4,060	4,105	4,150	4,190
50	3,345	3,390	3,430	3,475	3,520	3,565	3,610	3,650	3,695	3,740	3,785	3,830	3,870	3,915	3,960	4,005	4,050	4,090	4,135
52	3,315	3,355	3,400	3,445	3,490	3,530	3,575	3,620	3,660	3,705	3,750	3,795	3,835	3,880	3,925	3,970	4,010	4,055	4,100
54	3,285	3,325	3,370	3,415	3,455	3,500	3,540	3,585	3,630	3,670	3,715	3,760	3,800	3,845	3,890	3,930	3,975	4,020	4,060
56	3,255	3,295	3,340	3,380	3,425	3,465	3,510	3,550	3,595	3,640	3,680	3,725	3,765	3,810	3,850	3,895	3,940	3,980	4,025
58	3,210	3,250	3,290	3,335	3,375	3,420	3,460	3,500	3,545	3,585	3,630	3,670	3,715	3,755	3,800	3,840	3,880	3,925	3,965
60	3,175	3,220	3,260	3,300	3,345	3,385	3,430	3,470	3,500	3,555	3,595	3,635	3,680	3,720	3,760	3,805	3,845	3,885	3,930
62	3,150	3,190	3,230	3,270	3,310	3,350	3,390	3,440	3,480	3,520	3,560	3,600	3,640	3,680	3,730	3,770	3,810	3,850	3,890
64	3,120	3,160	3,200	3,240	3,280	3,320	3,360	3,400	3,440	3,490	3,530	3,570	3,610	3,650	3,690	3,730	3,770	3,810	3,850
66	3,070	3,110	3,150	3,190	3,230	3,270	3,310	3,350	3,390	3,430	3,470	3,510	3,550	3,600	3,640	3,680	3,720	3,760	3,800
68	3,040	3,080	3,120	3,160	3,200	3,240	3,280	3,320	3,360	3,400	3,440	3,480	3,520	3,560	3,600	3,640	3,680	3,720	3,760
70	3,010	3,050	3,090	3,130	3,170	3,210	3,250	3,290	3,330	3,370	3,410	3,450	3,480	3,520	3,560	3,600	3,640	3,680	3,720
72	2,980	3,020	3,060	3,100	3,140	3,180	3,210	3,250	3,290	3,330	3,370	3,410	3,450	3,490	3,530	3,570	3,610	3,650	3,680
74	2,930	2,970	3,010	3,050	3,090	3,130	3,170	3,200	3,240	3,280	3,320	3,360	3,400	3,440	3,470	3,510	3,550	3,590	3,630

From: Archives of Environmental Health
February 1966, Vol. 12, pp. 146–189
E. A. Gaensler, MD and G. W. Wright, MD

Index

deltoid tuberosity, 60
dendrite, 84
dermis, 42
diabetes insipidis, 171
diabetes mellitus, 171, 176
diaphragm, 6, 151
diaphysis, 45
diarthrotic joints, 64
diencephalon, 111
differential WBC count, 126
diffusion, 30
digestion, 162
digestive system, 10, 156
dilation, 68
directional terms, 3
DNA, 23
dorsal cavity, 6
dorsiflexion, 68
ductus arteriosus, 144
ductus venosus, 144
dura mater, 99, 103
dural sac, 97

ear anatomy, 119
effector, 86
elastic cartilage, 39, 195
electrodes, 92, 94
endocardium, 132
endocrine glands, 174
endocrine system, 10
endolymph, 121
endomysium, 66
endoplasmic reticulum, 25
endosteum, 45
epicardium, 132
epidermis, 42
epidural space, 99
epimysium, 66
epinephrine, 178
epiphyseal disk, 45
epiphysis, 45
epithelial tissues, 34
equilibrium, 122
ER, 25
esophagus, 159
estrogen, 180
Eustachian tubes, 119, 149
eversion, 38
extension, 38
external auditory meatus, 119
extremities, 4, 60, 62
extrinsic eye muscles, 114
eye anatomy, 113

fascia, 38, 66
fasciculus, 66
fat tissue, 38
fetal circulation, 144
fetal skull, 53
fibroblasts, 37
fibrocartilage, 39
fibrous connective tissue, 38, 194
flagellum, 25
flexion, 68
fontanels, 53
foramen
  incisive, 52
  infraorbital, 52

intervertebral, 55
jugular, 50
lacerum, 50
Magendie, 112
magnum, 49
mandibular, 53
mental, 53
Monroe, 111
obturator, 62
olfactory, 51
optic, 51
ovale, 51, 144
palatine, 52
rotundum, 51
supraorbital, 49
transverse, 55
vertebral, 55
formed elements, 125
fornix, 111
fovea centralis, 114

gallstones, 161
ganglia, 86, 100
glands, skin, 44
glenoid cavity, 58
glomerulus, 168, 205
glucagon, 176
goblet cells, 35, 161, 201, 202
Golgi apparatus, 25
gonadocorticoids, 177
Graafian follicles, 180, 190
gray matter, 99, 190
growth hormone, 181
gyri, 105

H Zone, 89
hair, 44
Haversian system, 39
hearing physiology, 122
hearing tests, 123
heart anatomy, 132
heart dissection, 136
heart sounds, 145
hematocrit, 127
hemoglobin, 127
hemolysis, 23
histology atlas, 191
hormones, 174
hyaline cartilage, 39, 194
hydrolysis, 162
hyperglycemia, 176
hypertonic solution, 31
hypodermis, 44
hypoglossal canals, 49
hypoglycemia, 176
hypophysis, 107, 111, 181
hypothalamus, 107, 111
hypotonic solution, 31

I Band, 89
iliotibial ligament, 78
incontinence, 167
infundibulum, 107, 111
inguinal ligament, 73
insertion, 66
insulin, 176
integument, 9, 42, 196
intercalated disks, 84

internal auditory meatus, 50
interneurons, 86
interphase, 27
intervertebral disks, 55
intestinal wall, 203
intestines, 161
inversion, 68
iris, 115
Ishihara color-blind test, 118
isotonic solution, 31

joint types, 63

keratin, 42
kidney anatomy, 168
kidney dissection, sheep, 169
kidney histology, 204
knee joint, 65

laboratory reports, 207
lacrimal apparatus, 113
lacunae, 38
lamellae, 39
lamina, basilar, 34
larynx, 150
latent period, 94
lens of the eye, 115
ligaments, 64, 65
linea alba, 73
liver, 141, 161, 203
lumbar plexus, 99
lung capacities, 154
lung tissue, 201
lungs, 151
lymph, 141
lymph nodes, 141
lymphatic system, 10, 141
lysosomes, 25

magnification, 20
malleolus, 62
mammillary bodies, 111
marrow, 45
matrix, 37, 39
mediastinum, 6
medulla oblongata, 107, 111, 153
medullary cavity, 45
Meissner's corpuscle, 44
melanin, 42
melanoma, 42
melatonin, 179
membranes, cavity, 7
membranous labyrinth, 121
meninges, 99, 103
mesentery, 8
metaphase, 29
microfilaments, 25
microscope, 19
microscope use, 22
microtubules, 25
microvilli, 25
micturition, 167
midbrain, 107, 111
mineralcorticoids, 177
mitochondria, 25, 87
mitosis, 27
motor neurons, 86, 87
motor unit, 95